Open Questions in Quantum Physics

Fundamental Theories of Physics

A New International Book Series on The Fundamental Theories of Physics: Their Clarification, Development and Application

Open Questions in Quantum Physics

Invited Papers on the Foundations of Microphysics

edited by

Gino Tarozzi
Departments of Philosophy and Mathematics, University of Bologna, Italy

and

Alwyn van der Merwe
Department of Physics, University of Denver, U.S.A.

D. Reidel Publishing Company

A MEMBER OF THE KLUWER ACADEMIC PUBLISHERS GROUP

Dordrecht / Boston / Lancaster

Library of Congress Cataloging in Publication Data

Main entry under title:

CIP

Open questions in quantum physics.

(Fundamental theories of physics)

Proceedings from a workshop, held in Bari, Italy, in the Dept. of Physics of the University, during May 1983.

Includes index.

1. Quantum theory–Congresses. 2. Physics–Philosophy–Congresses. 3. Stochastic processes–Congresses. 4. Wave-particle duality–Congresses. I. Tarozzi, G. II. Van der Merwe, Alwyn. III. Title: Microphysics. IV. Series.

QC173.96.064 1984 530.1'2 84–18009

ISBN 90–277–1853–9

Published by D. Reidel Publishing Company
P.O. Box 17, 3300 AA Dordrecht, Holland

Sold and distributed in the U.S.A. and Canada
by Kluwer Academic Publishers,
190 Old Derby Street, Hingham, MA 02043, U.S.A.

In all other countries, sold and distributed
by Kluwer Academic Publishers Group,
P.O. Box 322, 3300 AH Dordrecht, Holland

Printed in The Netherlands

CONTENTS

PREFACE

Due to its extraordinary predictive power and the great generality of its mathematical structure, quantum theory is able, at least in principle, to describe all the microscopic and macroscopic properties of the physical world, from the subatomic to the cosmological level. Nevertheless, ever since the Copenhagen and Göttingen schools in 1927 gave it the definitive formulation, now commonly known as the orthodox interpretation, the theory has suffered from very serious logical and epistemological problems. These shortcomings were immediately pointed out by some of the principal founders themselves of quantum theory, to wit, Planck, Einstein, Ehrenfest, Schrödinger, and de Broglie, and by the philosopher Karl Popper, who assumed a position of radical criticism with regard to the standard formulation of the theory.

The aim of the participants in the workshop on *Open Questions in Quantum Physics*, which was held in Bari (Italy), in the Department of Physics of the University, during May 1983 and whose Proceedings are collected in the present volume, accordingly was to discuss the formal, the physical and the epistemological difficulties of quantum theory in the light of recent crucial developments and to propose some possible resolutions of three basic conceptual dilemmas, which are posed respectively by:

(a) the physical developments of the Einstein-Podolsky-Rosen argument and Bell's theorem, i.e., the problem of the logical and empirical contrast between local realism and quantum mechanics, with special reference to a simplified new experiment presented by Popper in his opening lecture;
(b) the stochastic interpretation of quantum processes, both according to the de Broglie school and to the approaches of Bohm and Vigier and its implications for the theory of measurement;
(c) the wave-particle duality and the testability of the physical properties of quantum waves on the basis of the experimental proposals advanced by Selleri and others.

The Universita' degli Studi di Bari, the Istituto Nazionale di Fisica Nucleare (I.N.F.N.), the Regione Puglia, and the Comune and the Provincia di Bari made possible the organization of the workshop through their financial support, while expenses incurred in the editing of the present Proceedings were partially covered by a faculty research grant from the University of Denver.

Our thanks are expressed here to all these institutions and in particular to Professor Luigi Ambrosi, Rector of the Bari University and to Professor Antonino Zichichi, President of I.N.F.N.

Gino Tarozzi
Alwyn van der Merwe

Part 1

Quantum Mechanics, Reality and Separability: Physical Developments of the Einstein-Podolsky-Rosen Argument

REALISM IN QUANTUM MECHANICS AND A NEW VERSION OF THE EPR EXPERIMENT*

Karl Popper

London School of Economics & Political Science
Houghton Street
London WC2 2AE

ABSTRACT

Since 1927, physicists with few exceptions (Einstein, Podolsky and Rosen; de Broglie; Landé) accepted the Copenhagen interpretation of the quantum formalism, due to Heisenberg, Bohr and Pauli. Since 1934 I opposed this. I contended that the quantum formalism in its statistical interpretation (Einstein, Born) had no special epistemological consequences; that it could, and should, be interpreted realistically (and, of course, locally); and that the so-called indeterminacy relations were just scatter relations; and that the so-called 'collapse of the wave packet' was something that occurred in any probabilistic theory and had nothing to do with Planck's quantum h and even less with an action at a distance. I now submit a new variant of the EPR experiment that can decide between a realistic interpretation and Copenhagen.

I have been asked to open this meeting because I am a realist. Indeed, I am not only a realist but a metaphysical realist, as I want to admit at once. That is, my realism is not based on physics; but physics, I think, is based on realism.

* 1 I am a realist in a very simple sense: I conjecture that I shall soon die, as we all shall sooner or later. (I do not wish to say here anything against the possibility of solving the biological problem of ageing one day.) And I expect the

G. Tarozzi and A. van der Merwe (eds.), Open Questions in Quantum Physics, 3–25.

world to go on after my death, and also after your deaths. This is a simple way of saying that I am a *metaphysical* realist. For it is, at least for us, impossible empirically to test our expectation that the world will go on after we are dead.

I am a realist, and I believe in the reality of matter, of energy, of particles, of fields of forces, of wavelike disturbances of these fields, and of propensity fields (de Broglie fields). (These remarks are conjectural, of course.) And I suggest that quantum mechanics is misinterpreted when it is not interpreted realistically. I also suggest that quantum mechanics says nothing whatever about epistemology, about our knowledge and its limits, no more than Newtonian dynamics. And I hold that Heisenberg's famous indeterminacy or uncertainty relations inform us neither about indeterminacy nor about uncertainty: properly interpreted, they simply inform us about the *scatter* of physical particles, such as photons or electrons or neutrons, after they have, for example, passed through a narrow slit. Thus the Heisenberg relations are simply scatter relations, and they have no special significance for the theory of knowledge. This is part of my realism.

* 2 I suppose that most human beings, and certainly most experimental physicists and biologists, share the realistic attitude I have described. And most physicists believe that their own interpretation of quantum mechanics, which is fundamentally realistic, is identical with the "official" interpretation, the so-called Copenhagen interpretation due to Bohr and Heisenberg.

But this is an historical blunder which would not matter much were it not for the fact that at least three important doctrines are still lingering on from the non-realistic ("idealistic" or "positivistic") Copenhagen interpretation. What is - or rather was - characteristic for it is the thesis that quantum physics is not so much a theory of micro particles than a theory of *our knowledge* (of micro particles). This leads, among others, to the following three doctrines:

(1) The doctrine that the Heisenberg formulae

$$\Delta p_x \; \Delta q_x \geqslant \frac{h}{2\pi}$$

etc. are about limits to human knowledge or to the precision of possible measurements on particles. This Copenhagen thesis I deny. As a realist I assert that the formula is about the lower limits of the scatter of particles: if we arrange an experiment so that all particles selected or prepared have their position within q_x and $q_x + \Delta q_x$, then their momenta will show a statistical scatter $\Delta p_x = h/2\pi \Delta q_x$. The particles themselves possess sharp positions and, at the same time, sharp momenta.

(2) The Copenhagen interpretation asserts the doctrine that particles and waves are "complementary" views of *the same* microphysical entities: these entities reveal themselves to us, or appear to our knowledge, or in some of our measurements, as particles and in other measurements as waves.

As opposed to this, I as a realist assert, with Louis de Broglie, that particles, which are carriers of energy, are always accompanied by waves, while waves are perhaps not always accompanied by particles: *there may be empty waves.* The Copenhagen interpretation has prevented many physicists from investigating the tremendously interesting consequences of this possibility.

(3) The Copenhagen interpretation asserts that if in an experiment a particle is reflected by a semi-transparent mirror instead of passing through it, then the part of the wave packet which has passed through the mirror is being destroyed by our knowledge that it cannot accompany a particle (so that its amplitude is zero). This is the famous "collapse of the wave packet". I, as a realist, believe that this view derives from a trivial misinterpretation of probability theory, and I have, I believe, cleared up this issue more than 50 years ago.

I suggest that most of the doctrines of the Copenhagen interpretation have died a natural death; but these three theses, (1) to (3), are still pretty much alive.

The famous paper of Einstein, Podolsky and Rosen (EPR) was designed, in my opinion (confirmed by Einstein himself in 1950), to establish that a particle may possess at the same time position and momentum, as against the Copenhagen interpretation. Now, although I am a *metaphysical* realist, I shall try to present to you a *physical* experiment. The version of the EPR experiment which I am going to propose here has the same aim: to establish that a particle has both position and momentum at the same time.

* 3 My EPR experiment in this particular form dates, I think from 1978 or 1977. However, I have been trying to design an experiment to show the existence of particle paths (position *and* momenta) since at least 1934 (1); at any rate, before the publication of the experiment of Einstein, Podolsky and Rosen (EPR).

The form in which I shall propose it here has, I think, gained something from the long discussion of EPR. It can be regarded as a simplification of this experiment.

We have a source S in the centre of the experimental arrangement (see Fig. 1 below). There is a screen on the right and another on the left of the source. The x-axis runs from the centre to the right (positive) and from the centre to the left (negative); and the y-axis runs up (positive) and down (negative). On the screen there are slits, parallel to the y-axis, and these can be opened and closed by varying their width Δy. The screen on the left can be removed. Behind each of the two screens we have a battery of counters. It is an Aspect experiment with slitted screens instead of polarizers.

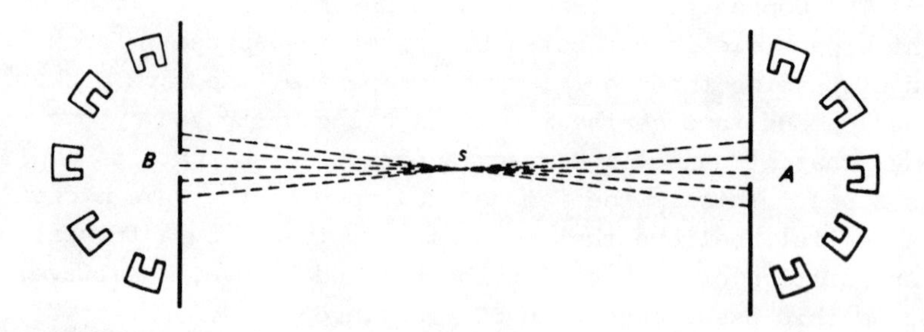

Fig. 1

(See my (1983) *Quantum Theory and the Schism in Physics,* Hutchinson, London, p. 28, Fig. 2.)

The diagram must not be interpreted to mean that there is a need to keep the source S in the centre very small: even if the source is somewhat extended, each pair of correlated particles emanates from an extremely small region, and this is sufficient for the argument. There is no need for us to identify this small region within the perhaps somewhat larger region occupied by the source.

* 4 In his book on *The Physical Principles of the Quantum Theory* (2), Heisenberg gives a picture of such a source and one screen; and he writes that if we make the slit small we measure (perhaps we would say "select") the q_y (position) coordinate.

The width of the slit determines the accuracy of Δq_y.

So the narrower we make this slit, the smaller Δq_y becomes; and Heisenberg writes that, the smaller Δq_y becomes, the greater will be the "uncertainty" of Δp_y. (I would, of course, say "the statistical scatter Δp_y", which makes the assertion a realistic and testable prediction.)

He suggests that his "uncertainty" (my scatter) in momentum is due to his indeterminacy relations. We can forget about the indeterminacy relations and simply assume that there is a scatter: this is the objectivist's and the realist's point of view. The projection of the scatter upon the y-axis is our Δp_y, the range of the scatter of the momentum in the y-direction. So there will be a Δp_y, related to the Δq_y according to the known Heisenberg relations (our scatter relations as I first proposed to call them in 1934, in the German original of my book *The Logic of Scientific Discovery,* then called *Logic der Forschung* (3)).

To the right and the left of the slits are Geiger counters wired as coincidence counters. By the coincidence on the Geiger counters, we will be able to get an idea of the width of the scatter. That is to say, if one of the two slits is wide open, then only the counters near the centre will be "talking"; but if we almost close the slit, then some counters at far larger angles will also begin to talk.

Now since we have a coincidence arrangement, only those pairs of particles are counted which arrive at the same time on both the left and right sides; and we may play around by opening or closing our slits. What we ought to find is that certain results about the scatter will correspond on both sides.

I shall say a little more about this correlation later, a matter which I discussed for several days with Professor Jean-Pierre Vigier.

Now in the EPR paper of 1935 an argument is advanced which has been accepted by Bohr, elaborated by Schrödinger, and, to my knowledge, has never been contradicted. It is the argument that, if we measure the position of the particle A on the right along the y-coordinate, we obtain information about the y-coordinate of the correlated particle B on the left, the particle which, having interacted with A has gone in the opposite direction. (The source is a source of particles from which pairs of particles that have interacted are emitted in opposite directions.)

So, if we measure the position of the interacting particle going to the right, we obtain information about the position of the correlated particle going to the left; and this prediction we can check when we play around with the arrangement.

Now let us remove the screen on the left. I assert that this means that we obtain an EPR situation in the following sense: when we remove the screen on the left and measure a particle on the right with the help of the screen on the right, then we shall expect (in the majority of cases) another particle to have gone to the left; and our "measurement" of y on the right informs us of of the y-position of the correlated particle that has gone to the left.

I propose that the thesis which EPR tried to establish was this: a particle possesses sharp position and momentum, and thus a trajectory; and our *knowledge* of a particle's position cannot, *qua knowledge,* disturb its momentum: the particle's momentum remains undisturbed. It remains a particle, having position *and* momentum and a trajectory, a path. But the Copenhagen interpretation wishes us to accept that our *knowledge* of the position of B (obtained by measuring A) must, *merely as knowledge,* make the momentum of B "indeterminate", since no particle can have both. But if this is so, then *our knowledge would make the momenta on the left scatter,* upon repetition (even with the screen on the left removed).

This was the EPR argument as I see it, and I wish to apply it to my experiment. On the other hand, according to the Copenhagen interpretation (and also according to Bohr, if I understand him), we are presented with the following paradox: *If we obtain knowledge of the position on the right, then this knowledge should disturb, or make indeterminate, the momentum of the particle on the left. And this would make the momenta on the left scatter upon repetition.* This is an inescapable consequence of the Copenhagen interpretation; and it is implicit in Bohr's three replies to EPR (4). But it has not, I think, been stated before explicitly (except that it was alluded to by myself (5)).

I believe that Einstein wanted to say, essentially, that our *knowledge* has no physical consequences - it cannot disturb the momentum of a distant particle (of course, a screen would disturb the momentum).

* 5 So far I have described the experiment, and stated the problem. Now, what will happen to the particle on the left after we have removed the left screen?

There are several possibilities:

(1) Perhaps the particle will go on and hit one of the central counters, while its twin-brother on the right, after passing the slit, shows a more radical scatter. This would indicate that nothing has happened to the momentum of the B-particle. So the first possibility is that our *knowledge* has no physical effect. *Nothing happens:* the B-particle goes on undisturbed, and will be detected by one of the central counters, because it is a particle that would have gone through the central slit if the screen had not been removed.

This possibility (1) is the one which I believe will be the outcome of the experiment, and I think that it is what the quantum formalism predicts.

(2) The second possibility is that, upon repetition, the particles going to the left will scatter (like those on the right) as a result of our knowledge of their position. This, I assert, is what Bohr and Heisenberg were committed to assert: the particles on the left will scatter, as if they had passed through a slit on the (removed) screen, equal in width to the slit of the screen on the right. That is to say, upon repetition of the experiment in which the A-particle scatters, the B-particle will also scatter. Now, we may identify various possibilities concerning the two scatters. According to Heisenberg who links his indeterminacy relations to a wave packet, and to Born's statistical interpretation accepted by Heisenberg, and in particular according to his book in which the scatter experiments are mentioned (6), the second particle will have to scatter in directions that do *not* depend on the direction of the scatter of the first particle. In other words,

(3) the directions of the two scattering coincident particles are uncorrelated. That is, the direction of the scatter of the B-particle on the left will not be correlated with that of the A-particle on the right.

(4) The scatters of the particles are correlated. This was the position taken up by Jean-Pierre Vigier when we discussed my experiment. That is to say, when the A-particle on the left scatters upwards, the B-particle on the right scatters downwards. Vigier argued that this would correspond to the situation in the Bohm version (measurement of polarization rather than position) of the EPR experiment. Perhaps; but I do not think that the Copenhagen interpretation is committed to Vigier's prediction.

I suppose that the quantum formalism yields possibility (1);

and I predict that the experiment will confirm this; the Copenhagen interpretation certainly yields possibility (2).

So we have these four possibilities to discuss; I think they are sufficiently interesting for recommending that this experiment be carried out.

* 6 According to Heisenberg, we have two position measurements (the slit and the counter) of the particle on the right. The first position measurement which can be interpreted either as a state preparation of the A-particle that has gone through the slit, or a reduction of its wave packet, would destroy, or be the end of, the correlation between the "right" and the "left" particles.

According to Vigier (who bases his argument on his interpretation of the Aspect experiment (7)), nothing would be destroyed, and we would have the following situation: if the A-particle is, say, deflected upwards, then even after the screen on the left has been removed, the second particle going to the left will be deflected downwards, and in such a way that the sum of the two momenta is conserved (Vigier's reason is that the conservation of linear momentum in my experiment corresponds to that of angular momentum in the Aspect experiment). I do not think this is convincing as the screen on the right absorbs part of the momentum of the A-particle which it deflects, thereby conserving the momenta.

Thus both Heisenberg and Vigier would say that our measurement of the particle on the right has an influence on the particle on the left, even with the left screen removed.

* 7 My own conjecture is that *nothing will happen;* that is, I vote for possibility (1). But of course I have an open mind; I am aware that quantum theory harbours surprises for common sense physics. I plead here only that my experiment should be conducted by somebody. My reasons for voting for (1) are these.

1. If we make the slit A very narrow so that we get a wide scatter of the momentum, the law of conservation of momentum is preserved by the absorption of the screen at A of the component of the momentum in the y-direction (as mentioned above). If the screen at B is removed, nothing is there to absorb the momentum vector; so we cannot get any increased scatter of the momentum at B merely by narrowing the slit at A, even though our *knowledge* of the position of the particle at B is quite considerable.

2. This consideration can be generalized. As Bohr always stressed, it is the total experimental arrangement that must be

considered. But this altogether different with the screen B in place or with the screen B removed. It seems clear that the removal of the screen will affect the result of the measurement of the momentum which is achieved by the battery of counters beyond B.

* 8 Should, quite against my expectations, the result of the experiment support the subjectivist Copenhagen interpretation, then this could be interpreted as indicative of action at a distance. This would indeed be similar to the published results of the Aspect experiment. I am very sceptical about this experiment; not so much because of action at a distance (Newton's theory was a marvelous theory) but because of an action at a distance that does not diminish with increasing distance. Before we admit such a theory we must have overwhelming evidence. I do not think that there is sufficient evidence available for saying that quantum mechanics is not a local theory. (For 50 years I have argued against the "collapse" of the wave packet as an example of action at a distance.)

* 9 Another argument in favour of the view that quantum mechanics involves action at a distance is Bell's inequality. This is often interpreted to imply the following assertion, called by T. Angelidis (8) "the Universality Claim" (U.C.): U.C.= *"All possible local theories (of the emission and propagation of light or of particles in opposite directions) lead to statistical predictions that differ from the predictions of the quantum formalism".*

"Local theories" are theories (or models) that do not imply action at a distance. So all theories are by definition either local theories or action at a distance theories (and never both); and it is clear that, if U.C. is right, then quantum mechanics *must* be an action at a distance theory (as asserted by Heisenberg and the Copenhagen interpretation on other grounds). If, on the other hand, U.C. is mistaken, then there is no reason whatsoever to believe that quantum mechanics is not a local theory (provided my view of the so-called "collapse of the wave packet" is accepted).

According to this analysis of the situation, everything depends on a *theoretical* issue. It is the problem whether or not U.C. is true. And nothing depends on the experiments: they have, as could be expected, supported quantum mechanics, which, if U.C. is false, may be just one of the possible local theories.

I do not know of any physicist with the exception of

Angelidis, involved in the discussion of the Einstein Podolsky Rosen argument or of the Bell inequality (or of the Clauser Horne inequality) who is prepared to look at the issue in the way advocated here. Everybody, even those who criticize the Bell inequality, seem to be convinced that quantum mechanics is non-local; and they are, perhaps for this reason, not prepared to attribute to U.C. the crucial importance given to it here. I can only appeal to them to admit that the thesis that quantum mechanics is non-local is a conjecture, and to review the very weak (in my opinion non-existing) reasons that speak in favour of this shattering conjecture.

* 10 It is sometimes said that, as long as we cannot exploit instantaneous action at a distance for the transmission of signals, special relativity (Einstein's interpretation of the Lorentz transformations) is not affected. I think that this is mistaken: we would have to introduce an inertial frame relative to which two events occur simultaneously in an absolute sense: with the idea of infinite velocity, a Newtonian - Lorentzian absolute space with an absolute coordinate system becomes , I think, almost unavoidable. (Of course, there are other reasons in favour of an absolute space; they existed long before Penzias' and Wilson's discovery of 1965).

But if, against my expectation, we *have* to interpret Aspect's experiment (or my experiment) as indicative of instantaneous action at a distance, the most reasonable way would be to accept Lorentz's own interpretation of his formalism of special relativity and to give up Einstein's interpretation (this would not, I think, necessarily affect general relativity). Thus Aspect's experiment (or my experiment) could perhaps become the first experiments that are crucially decisive between Lorentz's and Einstein's interpretations of the formalism of special relativity.

All the well known experiments supporting special relativity could be retained. They are not crucial between Lorentz's and Einstein's interpretations.

* 11 I have tried to give you the arguments for all the four possibilities (1) to (4). I may perhaps add just a few words.

Heisenberg's view would be based on his thesis that "objective reality has evaporated" (9). That is a literal quotation from Heisenberg, and he goes on to say that quantum theory is not about particles but only *about our knowledge of particles*.

This leads to one of the (in my opinion) absurd consequences of possibilities (2) to (4) discussed above. For example, if it is our knowledge that creates the scatter, then the particle on the left side (where the screen has been removed) is supposed to scatter *because of our knowledge*. But our "knowledge" (it is after all, only conjectural knowledge) may come long after the event; at least, I should say, half a minute later, for even in the best of circumstances we have to check the evidence. Accordingly, Heisenberg ought to say that the second particle does not scatter where the screen has been; rather, it should scatter, say, 30 light seconds away from that place.

These are just some of the difficulties due to the non-realistic position taken up by Heisenberg. As to the problem of realism, I do not like to bring in highly emotional arguments, but I think that this particular question was decided in Hiroshima. Hiroshima was destroyed not when physicists who knew what may happen heard that the bomb had been exploded; Hiroshima was destroyed before they had any news, before they had any knowledge.

So, I think that the events are connected in a more or less classically causal way (understanding propensities as generalizations of causality), in an ordinary physical way, and not dependent on our knowledge of them. It should hardly be necessary to stress this point.

*12 However, the idealistic and non-realistic way of speaking has bitten deep into physics. I have here, for example, an excellent book by Professor Franco Selleri entitled *Die Debatte um die Quantentheorie* (10). The book is very outspoken and strongly realistic. I greatly admire his whole approach.

Nevertheless, at certain places, the book does adopt the idealistic terminology, and even a little more than just the terminology. For example, Selleri speaks of the "paradox of de Broglie" (11) but I think that it ought to be called an argument of de Broglie showing the incompleteness of all probabilistic theories.

De Broglie's argument is this; we have a box B containing an electron. The walls of the box are completely reflecting. After inserting in the middle of the box a double sided dividing wall with appropriate mirrors we call the two parts B_1 and B_2. B_1 is taken to Tokyo and B_2 is taken to Paris. De Broglie gives a detailed and elaborate argument amounting to the assessment of the probabilities $p(a_1,b) = 1/2 = (a_2,b)$, where a_1 means the particle is in Tokyo and a_2 that it is in Paris and where b

states the conditions of the experiment.

If we open the box in Paris, say, then we will either find an electron or an empty box. If we find an electron in the box in Paris then the probability of finding the electron in Tokyo immediately becomes zero. In other words the wave packet collapses due to our observation, or due to our knowledge: a kind of action at a distance. This is the result of the rather complicated quantum mechanical argument of de Broglie from which he rightly deduces the incompleteness of quantum mechanics.

However, the probabilities depend upon the experimental conditions which we choose to establish and which constitute our information b, for example. If we choose to open the box in Tokyo, say, then we have new information, c, which we may prefer to b as being more up to date; and $p(a_1,b) = 1/2$ does not contradict $p(a_1,c) = 0$. If our information c contains the result $c = a_1$ then the probability of a_1 relative to the information $c = a_1$ becomes $p(a_1,a_1) = 1$ and $p(a_2,a_1) = 0$. It is, therefore, not the original wave packet $p(a_1,b)$ which collapses, due to "our knowledge"; but there are two wave packets to be considered, the wave packet $p(a_1,b)$ and the wave packet $p(a_1,c)$ which may both be real propensities provided b and c were both true.

This holds not only for quantum theory but for all probabilities as one sees if we substitute a pea in place of our electron: precisely the same considerations hold. This shows that the "effect" (if we may so describe it) is not a quantum theoretical effect but a general probabilistic effect.

(If it is objected that we could have found out by shaking the box that the pea is in B_1, then we can reply that instead of having mirrors in the box we can sound proof the box; and if it is objected that we may find by weighing the two halves of the box wheter the pea is in B_1 or in B_2 then we can reply that de Broglie never assumed that the two halves of the box will have precisely the same weight: such an assumption is, in fact, quite unnecessary for his argument and for our argument.)

But although de Broglie's argument is not specific for quantum theory, it certainly shows the incompleteness of quantum theory: quantum theory is incomplete in exactly the same way in which all probabilistic theories are incomplete. But I do not regard this as important. In my opinion it is trivial (at least since Gödel) that all physical descriptions and theories are incomplete: the issue of completeness is, in my opinion, a red herring and as weighty as a green pea.

Clearly the decisive event happened when we split the box into two parts: there is no action at a distance in these experiments. (All this was anticipated in my *Logic der Forschung* fifty years ago.)

Selleri also speaks in his book of "the paradox of EPR" (12); and so does the programme of this Conference. But, again, I think it ought to be called "the argument of EPR"; for I regard it as a simple argument for the existence of particles and their trajectories; and it creates difficulties only for Copenhagen. (Note that Einstein in a letter that describes EPR (3) drops the definition of "physical reality".)

So there is no EPR paradox; *what is paradoxical is only the Copenhagen interpretation*. But many physicists now identify it with their own interpretation of quantum mechanics.

* 13 The EPR argument is an attempt to *expose the Copenhagen paradox*. But the current language of quantum theory and, unfortunately, the current thinking of many brilliant physicists is deeply affected by what I can only call the Copenhagen doublethink. It begins with talk of the so-called "observables". Of course, we are all used to this term to such an extent that we no longer notice that the term "observable" introduces a *subjectivist ideology* into physics - a really completely inadmissible kind of ideology: it is the ideology of von Neumann, of the positivistic philosophy (against which I have been fighting at least since 1925). It is an ideology that suggests to physicists that one can say in advance what is observable, and what is a "hidden variable"; that one can give a list of the (GOOD) observables, encluding everything which is not in that list as (BAD) non-observables - as if physics was not capable of progressing, and of introducing *new* kinds of variables.

Now physics has indeed introduced many new variables since this ideology was introduced into physics by von Neumann, in the days when there were only electrons and protons, together with their positions, momenta, spins, and energies. As a consequence, no experimentalist takes the ideology of the GOOD "observables" any longer seriously. The corresponding term is the BAD "hidden variable", a term which was, I think, also introduced by von Neumann.

Of course, *all variables are hidden;* and nothing could be more hidden, from von Neumann's point of view than, say, the colour of the charm of a quark. (In its early days, Landé's spin was equally well hidden.) What we do in physics is, quite genererally, to invent hidden variables by way of our theories. They cease to be "hidden" when our theories turn out to be successful: this is, precisely, how theory and experiment cooperate.

So, the talk about hidden variables is ideological and misleading: I do not think that anybody ever really knew which new variables were "hidden" and which were not.

The whole of quantum physics is beset by this kind of doubletalk, and the talk about the "paradox" of EPR shows how far this kind of ideology has worked its way even into the thinking of genuine realists.

Now, I have mentioned before, Vigier and I have two different expectations of the outcome of my experiment, but we are both realists and we both welcome new theories with new variables, whether or not they are called "observable" or "hidden",or GOOD, BAD, or FORBIDDEN, and we both think that particles possess position *and* momentum, whatever happens.

How did it come about that this view was abandoned? It was abandoned only in the year 1927, some two years after the birth of quantum mechanics and after Pauli's unfortunate idea that electrons in an atom had no trajectories (while what is true, I conjecture, is that their trajectories are not fixed or regular orbits, uniquely prescribed by a quasi Newtonian theory: the trajectories will scatter upon repetition of the experiment). This led to the idea that the particle itself has neither position nor momentum, but that by "measuring" the particle (and thus interfering with it), we impose upon it the duty to take up a position and forget about its momentum, or alternatively the duty to take up a momentum and forget about its position. If we measure one parameter, then the other becomes smeared. At the same time, we were told to attribute all this not to the particle but to our measurement, to our interference with the particle or, rather, the wave packet (which was thought to be another incarnation of the particle and, in some sense, identical with it). For the so-called wave-particle dualism greatly enforced the idea of the trackless particle.

* 14 To show that this was indeed the official view, I want to quote a passage from a book by Pascual Jordan about the wave-particle dualism. Most of you will know that Jordan was one of the three people (the others were Heisenberg and Born) who together published in 1925 the first paper on the new quantum mechanics (matrix mechanics). (Heisenberg's earlier paper only provided the stimulus.) So Jordan is one of the originators of quantum

mechanics. He is a very great physicist, a creative genius; as are also Born and Heisenberg whom Jordan follows closely.

Now, Jordan formulates at a later date Bohr's principle of complementarity in the following way: "Any one experiment which would bring forth *at the same time* both the wave properties and the particle properties of light would not only contradict the classical theories (we have got used to contradictions of this kind), but it would, over and above this, be absurd in a logical and mathematical sense." (13) (14)

So Jordan asserts that according to Bohr's principle of complementarity it is logically and mathematically impossible to devise an experiment which brings forth at the same time wave and particle properties. This is what he means. He does not, as one may think, mean "all the wave and all the particle properties at the same time": it would, of course, be impossible to bring forth either all wave properties at the same time or all particle properties at the same time.

As is well known, Bohr's favourite experiment in all his discussions about complementariry was the two-slit experiment (a variant of Young's two pinhole experiments). We have two slits, and waves or particles coming from a source pass through the slits, to impinge on a detecting screen where we obtain interference fringes.

Now the two-slit experiment ought to be, according to Jordan, *logically and mathematically impossible.*

For, what does the two-slit experiment tell us?

It tells us that there are particles: we can see their impacts. And it tells us that the particles build up interference fringes by way of the different densities with which they arrive at different parts of the screen. We can see the localized impact of particles, and at the same time we can see that the density of the impacts of the particles is determined by the waves, according to Born's statistical interpretation. That is exactly what the two-slit experiment shows us. So it is *one* experiment; it is undertaken to show us the waves, but the waves are density waves of particles. So it is an experiment that is impossible, according to Jordan; that is, according to the Copenhagen interpretation.

So I wish to ask you to speak no longer of the EPR paradox, but rather to call it an experiment or argument or something similar, and I should like to ask you to regard electrons as particles and their densities or probabilities as determined by waves, until such time as better arguments have refuted this view. Waves, whether in water or in a field of corn, are as a rule changing densities of many particles: this kind of duality needs no complementarity. But, if the waves are waves of a propensity to behave in this way or in that way (as I have proposed (15)) then they are real, and might possibly be observed even if they are empty of particles.

* 15 *Added 1984:*

I have repeated this lecture with some minor alterations at the International Congress for Logic, Method and Philosophy of Science in Saltzburg in June 1983, in the Trieste Institute for Theoretical Physics on September 22nd 1983 and in the Imperial College of Science in London on December 7th 1983. On the latter two occasions I distributed a summary of 19 points before the lecture which, with slight alterations, follow here.

1. Quantum theory is a probabilistic or statistical theory. It is as objective and realistic as Newtonian mechanics or as Boltzmann's Gas Theory.

2. The opposite view, and almost all the existing difficulties, arise from a misunderstanding of probability theory; especially from (1) the old tendency of interpreting probability subjectively and (2) the neglect of the calculus of relative or conditional probabilities.

3. Argument: take any quantum mechanical experiment you like. Make experimental arrangements very carefully, multiply these arrangements ten times or ten million times, as required by the statistical precision of the statistical prediction. Make the experiment completely self-contained by arranging for a source of electrical current within each of these ten or ten million arrangements. Put each of them in a wooden box and nail it down; and then, let them go. There is no observer, the experiments are self-contained and are self-registering, as such experiments are these days. When the experiments have run off, let each box signal the result by a big macro signal. Simply count the signals and make your statistical calculation.

4. The relations between particles and waves are insufficiently explored. They are not of the character of complementarity. (Complementarity is not of the character of a theory; at the most it is an ideology. I think it is an empty word that could be abandoned.) There are many reasons to support de Broglie's view that, though there are no particles without waves,

there may be waves without particles - empty waves (propensity waves).

5. Quantum mechanics has no more and no less epistemological import than any classical physical theory. More precisely, the so-called indeterminacy relations have no special epistemological import. They do not indicate anything like limits to our knowledge. They are merely *scatter relations.*

6. The opposite view due to Heisenberg arises from a misunderstanding of probability theory.

7. Von Neumann's so-called proof of the non-existence of hidden variables is invalid (as is admitted now by almost everybody). The concept of hidden variables is highly ambiguous and can be abandoned without loss. Incidentally, all reality is hidden: it is the task of science to discover the hidden reality. The opposite view, which is characteristic of positivism, is sheer ideology.

8. There are two types of measuring experiments. Those like polarization or measuring the spin of a particle, which may change the measured state of the object unless an identically oriented polarizer precedes the experiment; they can be called non-classical measurements. They should be distinguished from classical measurements such as the measurement of position which does not, in general, change the position of the object measured and therefore may be called classical.

9. Accordingly, there are two types of Einstein Podolsky and Rosen experiments. The non-classical experiments, as for example Aspect's experiment, may be called EPRB where B stands for Bohm and the classical is just the EPR in its original form. The latter type of experiment is obviously simpler, and easier to interpret. Also, the two types of experiment *may* possibly lead to different results, although this does not seem likely to me.

10. I have proposed myself an EPR experiment of the original type. Question: why should it not be possible to carry it out as some people assert? It should be far simpler than Aspect's experiment.

11. Its outcome should be predictable by quantum mechanical calculations. Some people have calculated it and if their calculations are right, and if my prediction is right, then this should refute the Copenhagen interpretation while leaving the quantum formalism intact.

12. My prediction, if correct, should show that at least

one interpretation - Heisenberg subjectivist interpretation - of Heisenberg's so-called uncertainty relations is false: the interpretation that they are limitations to our (predictive) knowledge.

13. Franco Selleri has suggested (continuing the work of de Broglie)that waves without particles may exist. I have proposed a somewhat similar theory, the theory of the existence of propensity fields and propensity waves. There are in this context very interesting experiments by U. Bonse and H. Rauch of Vienna on the reality (probabilistic and even causal effectiveness) of neutron fields without neutrons.(16)

14. The consequences of 13 would seem to be revolutionary. They would establish in place of the "complementary" character of particles and waves (wavicles) the interaction of two kinds of real objects: waves and particles.

15. Another point I may mention here is the problem of the (rational) *understanding of quantum theory.*

From Bohr's Como address at the International Congress of Physics in 1927 (a few weeks before the Solvay Conference in Brussels) it is well known that, in atomic theory, he argued that we have to *renounce the hope of understanding our theories* since the conditions permitting our understanding of macro physics were absent in micro physics. During a lecture (unpublished) which I gave in Cambridge, in approximately 1949 or 1950 (upon the invitation of N.H. Hanson who was then a lecturer in Cambridge), on my views of quantum mechanics, I explained, and have been explaining ever since, that "understanding" was not a matter of picturesque imagination or intuition or a matter of being aware of the logical functions of a theory: especially of the open *problems* which a theory was expected to solve and of the problems which were newly created by it; and comparing and evaluating the various competing theories from the point of view of their power to solve old problems and to create promising new ones.

16. Quantum mechanics is not an action at a distance theory. Action at a distance is far from absurd as we can see from Newtonian mechanics; but field theories that avoid action at a distance are preferable, for many reasons.

17. The so-called "collapse of the wave packet" is one of the two bases of the assertion that quantum theory is an action at a distance theory. But this so-called "collapse of the wave-packet" is in reality something that can occur in every probabilistic theory and has nothing whatever to do with action at a distance.

18. Bell's (17) and Clauser-Horne's (18) inequalities constitute the second of the two bases of the assertion that quantum theory is non-local. But I do not think that these inequalities prove the non-local character of the quantum formalism.

19. Whatever Aspect's (7) experiments may show, I am quite unconvinced that they can establish action at a distance.

* 16 Two points made at the two later meetings (Trieste and London) follow here. These two points oppose the view that quantum mechanics is a non-local theory.

(1) I have shown in 1934, that the so-called reduction of the wave packet is a myth; and I therefore believe that quantum mechanics should be regarded as a local theory unless we have very strong new arguments against this view. Heisenberg discusses the famous "collapse of the wave-packet" in his book *The Physical Principles of the Quantum Theory* (2). One of the obvious questions is: did the wave packet collapse when I gained knowledge of what had happened, or did the wave packet collapse when the experiment happened which later led to my knowledge?

These are two actually different situations. Although Heisenberg discusses some points very close to this formulation in considerable detail, he never really comes to my very simple and decisive question in his discussion: he implies that it is our knowledge that is crucial, but he does not discuss the decisive question connected with the fact that our knowledge is not simultaneous with the experiment that creates the knowledge. Moreover, I should say that NO EXPERIMENTAL PHYSICIST would agree that it is "knowledge" that influences things and not the local experimental conditions.

According to Heisenberg (19), it was Einstein who (almost certainly in an attempt to refute Heisenberg's subjectivist interpretation of quantum mechanics) suggested the idea that quantum mechanics involves action at a distance. Heisenberg reports it as follows:

> 'By reflection [of a wave packet] at a semi-transparent mirror, it is possible to recompose it into ... a reflected and a transmitted packet. After a sufficient time the two parts will be separated by any distance desired; now, if an experiment yields the result that the photon is, say, in the reflected part ..., then the probability of finding the photon in the other part ... immediately becomes zero. The experiment ... thus ex-

erts a kind of action (reduction of the wave packet) at the distant point ... with a velocity greater than that of light." (19)

I first quoted and discussed this passage in 1934 (20) pointing out that it created no problem for a probabilistic interpretation of quantum mechanics, and Einstein agreed. I have discussed it many times since. There are two possibilities: the first - it is the more exciting case - is that the wave packet is physically real (a propensity field that can be tested by the statistical distributions upon frequent repetitions of the experiment). In this case it does not collapse, but remains unaffected, because the probability (propensity) that, upon repetition, the photon will be found not to be reflected remains unaffected. (Remember that $p(a,b) = 1/2$ does not conflict with $p(a,c) = 0$.) The second possibility is that the wave packet is an unreal, and merely mathematical representation of our probabilistic estimate. In this case (which is not, I think, the true case, for various logical and physical reasons) there can be no physical action upon the wave packet. The problem arises only if we regard the wave packet somehow as complementary and identical with the particle; which is a view that ought to have been eliminated with the adoption of Born's statistical interpretation.

So this argument has been dead since 1934. However, it seems that some physicists who in 1935 or 1936 agreed with my counter argument forgot all about it afterwards; perhaps because I myself was anxious to stress that I had made some mistakes elsewhere in my book. (In fact, I stressed these mistakes more than my positive results, such as the objectivist interpretation of Heisenberg's indeterminacy relations as "scatter relations", or the distinction between state preparations or predictive measurements, and retrodictive measurements that test the predictions.)

(2) A non-local theory, like Newton's theory, may be inherently acceptable (even though a field theory has great advantages). But what is demanded from us (in my opinion, without any reason whatsoever) is the acceptance of actions at a distance whose intensity does not decrease with the distance; moreover actions which, we are told, cannot be used (and will never be usable) for the transmission of signals.

Among the reasons against what Angelidis calls the Universality Claim (and therefore among the reasons against the nonlocality of quantum mechanics) are the following:

(a) Clauser and Horne (18) found a counter example to the Universality Claim (U.C.), or more precisely, a local theory that, according to themselves, leads to the same predictions as the quantum formalism. But their reaction to this discovery was remarkable. Instead of describing it as a counter example, they used it as a proof of the *necessity* of an admittedly intuitively quite reasonable *ad hoc* assumption (called the no-enhancement assumption) in order to exclude their counter example. They offered no proof, in fact no argument, supporting the view that other counter examples do not exist (perhaps satisfying the no-enhancement assumption). This should make it clear that they can not even claim to have established what Angelidis calls the Universality Claim (U.C.).

(b) F. Selleri and G. Tarozzi (21) found a model that satisfies Bell's definition of locality but not the Clauser-Horne definition of locality (also known as the "factorizability condition"); this seems to show again that Clauser and Horne have not established the Universality Claim (U.C.).

(c) Although not strong enough to be independently testable, the no-enhancement assumption is strong enough to exclude several other local theories or models. For example, T.W. Marshall, E. Santos and F. Selleri (22) recently found a very convincing model that violates the Clauser-Horne inequality. It does not, however, satisfy the no-enhancement assumption.

(d) T. Angelidis (8) has now published his argument against the Universality Claim (U.C. see above), and he has proposed, in a later paper (not yet published at the time of writing), an infinite set of counterexamples that do satisfy the no-enhancement assumption of Clauser and Horne. If Angelidis is right, then any reason against a local interpretation of the quantum formalism is removed, and with it the theoretical basis for those world shattering interpretations of experiments like those of Aspect with which we have been threatened in recent years.

REFERENCES

1. K.R. Popper, 'Zur Kritik der Ungenauigkeitsrelationen', *Die Naturwissenshaften, 22,* 807 (1934); see also C.F. von Weizsäcker's reply in the same place.
2. W. Heisenberg, *The Physical Principles of the Quantum Theory,*

Dover Publications (1930), p. 30.

3. K.R. Popper, *The Logic of Scientific Discovery,* Hutchinson, London (1959), (translation of *Logic der Forschung,* Julius Springer, Vienna (1934)). For the "Scatter Relations" see especially the introduction to Chapter 9, section 75 and section 76. For the "collapse of the wave packet" see the end of section 76 (from footnote 8 on).
4. N. Bohr, *Phys. Rev., 48,* 696 (1935); *Dialectica, II,* 312 (1948); in P.A. Schilpp (ed.) *Albert Einstein: Philosopher Scientist, The Library of Living Philosophers,* (1949), p.201. Concerning the "Copenhagen Interpretation", see especially also W. Heisenberg in ref.(2) above and (9) below.
5. K.R. Popper, *op. cit.* (3).
6. W. Heisenberg, *op. cit.* (2), p. 33.
7. A. Aspect, P. Grangier, and G. Roger, *Phys. Rev. Lett. 47,* 460 (1981), and *Phys. Rev. Lett. 49,* 91 (1982); A. Aspect, J. Dalibard, and G. Roger, *Phys. Rev. Lett. 49,* 1804 (1982).
8. T. Angelidis, *Phys. Rev. Lett. 51,* 1819 (1983).
9. W. Heisenberg, *Daedalus, 87*, 95 (1958).
10. F. Selleri, *Die Debatte um die Quantentheorie,* Vieweg, Braunschweig (1983).
11. *Ibid.,* p. 39.
12. *Ibid.,* p. 83.
13. P. Jordan, *Anschauliche Quantentheorie,* Springer, Berlin (1936), p. 282.
14. See the English editions of *The Logic of Scientific Discovery* and the German editions from the second edition onwards. The New Appendices 11, section 11 (English edition p. 454; German edition p. 409).
15. K.R. Popper, *Quantum Theory and the Schism in Physics,* Hutchinson, London (1983), (Volume II of the *Postscript* to *The Logic of Scientific Discovery,* edited by W.W. Bartley III). See also Popper, 'Quantum Mechanics without "The Observer"', in M. Bunge (ed.) *Quantum Theory and Reality,* Springer Verlag, Berlin, Heidelberg (1967), p. 7; Popper 'The propensity interpretation of the calculus of probability and the quantum theory', in S. Körner (ed.) *Observation and Interpretation,* Butterworth Scientific Publications, London (1957), p. 65; and Popper, 'Proposal for a Simplified New Variant of the Experiment of Einstein, Podolsky and Rosen', in K.M. Meyer-Abich (ed.) *Physik, Philosophie und Politik,* Carl Hanser Verlag, München (1982), (Festschrift für Carl

Friedrich von Weizsäcker.), p. 310.

16. In U. Bonse, and H. Rauch (eds.), *Proceedings of the International Workshop* (5th to 7th June 1978) at the Institute Max von Laue, Paul Langevin, Grenoble, Oxford University Press, Oxford (1979).
17. J.S. Bell, *Physics, 1,* 195 (1965).
18. J.F. Clauser, and M.A. Horne, *Phys. Rev. D10,* 526 (1974).
19. W. Heisenberg, *op. cit.* (2), p. 39.
20. K.R. Popper, *Logic der Forschung,* Julius Springer, Vienna (1934), p. 171f.; J.C.B. Mohr (Paul Siebeck), Tübingen (1966), p. 184f.; also pp. 400, 403, 411 of *The Logic of Scientific Discovery,* 11th edn., Hutchinson, London (1983).
21. F. Selleri, and G. Tarozzi, *Lett. Nuovo Cimento, 29,* 533 (1980).
22. T.W. Marshall, E. Santos, and F. Selleri, *Physics Lett. 98A,* 5 (1983).

DISCUSSION

M. Cini, F. De Martini, K. Kraus, T.W. Marshall,
K.R. Popper, H. Rauch, M.C. Robinson, F. Selleri,
J. Six, G. Tarozzi, J.-P. Vigier

VIGIER: You did me great honor by mentioning my name in your lecture, but I am going to say that the result of the experiment is the same as the quantum-mechanical prediction: I mean that there are going to be *correlations* between the two measurements. I want to ask you this: What do you think is measured in your form of the EPR experiment? The quantum-mechanical people would say: One has waves *or* particles, never the two at the same time; with your counters, you measure the particle aspect of your two photons. Now the question is: Are these measurements correlated or not? According to quantum-mechanics, they are correlated. You measure the deflection due to the wave going through the slit. Therefore, what you measure is p_y (vertical component of the photon momentum in your figure), but this is the same as measuring spin because, in a particle moving at the speed of light, spin and momentum are either parallel or antiparallel. In the Aspect experiment spin correlations were found which are something like action at a distance. Therefore, also in your experiment there will be nonlocal correlations between the two photons. If one slit twists the spin of one photon, there will be an immediate action on the other photon going in the opposite direction. This implies that we have nonlocality. Now, is nonlocality compatible with realism or not? My point is that it is compatible, provided certain restrictions are imposed on the nonlocal potential. In some papers in collaboration with Cufaro-Petroni, we have shown that the quantum potential gives rise to an action

G. Tarozzi and A. van der Merwe (eds.), Open Questions in Quantum Physics, 26–32.

at a distance between the two photons. What happens, however, is that the exchanged momentum is orthogonal to the four-momentum of the center of mass of the two photons. This means that there can not be a retroaction in time: In particular one cannot kill one's grandfather!

When Newton constructed classical mechanics, he introduced action at a distance, but this concept was perfectly causal, deterministic, and realistic essentially because its use allowed the Cauchy problem to be solved.

Since, in your case, the Hamiltonians of the two photons have a zero Poisson bracket, we are also able to solve the Cauchy problem. Therefore, only positive energy propagates in time. This is an essential result because one had otherwise to accept a fantastic picture of the world where even past events could be changed by our acts of observation: As a consequence, realism would not be possible because one could not claim that an observer-independent reality existed.

POPPER: Of course a measurement is made by the counter, but at the same time we indirectly measure something else as well, as we always do in every measurement. So we have direct-indirect measurement. In a sense, of course, the final measurement is direct, but it is always just a position measurement of something arriving at some place. The question is: How do we interpret that measurement and by which interpretation do we get the physical result? A Copenhagen idealist would not care about this problem: He would, how shall I say, grudgingly admit that somehow the thing came from the slit. He would prefer to say that it had traversed the whole of space anyway. But, of course, in reality, the two measurements are being made.

I agree that my experiment is in some ways similar to Aspect's experiment, but I came to Bari to claim that my experiment should be carried out. I do think that it is simpler than Aspect's and, in any case, since it is fundamentally the same thing, but with a completely different and very much simpler arrangement, I think that it is a *necessary* experiment. I must say that I admire Vigier's own theory of action at a distance, which rescues a great deal of Einstein's view, although it also makes some corrections to it. And I think I would be happy if the experiment goes in favour of Vigier's prediction.

CINI: In the EPR experiment, there were two variables: One was the total momentum of the two particles and the other was their relative distance. Now, total momentum and relative distance commute, and this is the reason why, in the Einstein experiment, if one measured the momentum of one particle, one could predict the momentum of the other one; or if one chose to measure the position of one particle, one could predict the position of the other one. In your experiment, one measures the y position of the first particle and expects that the y position of the second particle should come out to be known. The problem is that there is actually no conservation law for the relative distance, but only for total momentum. Conservation of relative distance was, of course, forced in the wave function of the two particles chosen by EPR, which they chose in the form of a δ function of the relative position. Therefore, I do not understand why these two particles emanating from the source, i.e., in a concrete physical situation, should have strictly correlated positions.

POPPER: The two particles travel with the velocity of light in opposite directions, so that relative position will be conserved all the time, up to a scale factor.

SIX: In my opinion, this sort of experiment is not feasible. If we consider the source to consist of positronium annihilating into two γ-rays, each of them receives an energy of 0.5 Mev, to which corresponds a wave length λ of about 0.02 Å. Even if we admit perfect collinearity of the two photons (and I do not really see how collinear emissions could be compatible with Heisenberg's relations), the problem that nobody knows the solution of is how to make slits so narrow that photons with such short wave length will undergo diffraction by crossing them. If one chooses a slit of 1 μ, a simple calculation shows that a diffraction pattern of about 1 μ should occur at a distance of 1 meter. This is not observable in practice.

POPPER: I could take the same sourcc as in Aspect's experiment.

SIX: But in this case there are no correlations between the directions of the two photons.

SELLERI: Yes, in the atomic cascade experiments there are three-body decays, so that the two photons do not have strict correlations in momentum. Well, there are of course technical difficulties, and we are here also for investigating such difficulties, but I do not see anything, *in principle,* forbidding an experiment of the type Professor Popper is suggesting.

DE MARTINI: Professor Popper has proposed the following experiment: We have a source, let us say a positronium source, and two diffraction slits, as well as two sets of counters. We look for a correlation between directions (that is, between momenta) of the photons emerging from the two slits. But photons obey the Maxwell equations, which are time-reversible. Therefore, the foregoing experiment is equivalent to the following one: Two lasers send light through the slits from behind. The light in between the two screens will give rise to interference, and the interference maximum will, for symmetry reasons, lie along the line joining the two slits. This is essentially the reverse of Vigier's interpretation of Popper's experiment, and I agree with Vigier's interpretation because I see it is correct in this reversed experiment. This is also, in my opinion, the best way to carry out Popper's experiment.

ROBINSON: But in this way you are changing the experiment in an essential way. In Popper's experiment there should be a deviation of the second particle even if the second screen is not there, but this is not so with your approach.

MARSHALL: Professor Popper, as I understand, thinks that his experiment can discriminate between a realistic and an idealistic interpretation of Heisenberg's relations. Vigier does not think this is the case and believes that the experimental predictions are the same in the two philosophies.

As regards to the problem of the connection between causality and locality (a problem which I do intend to explore further during this workshop), I believe that if one is committed to causality and locality, then one must be led to a certain type of stochastic theory where, for instance, no negative probabilities appear. This is the main reason for my disagreement with Professor Vigier.

KRAUS: Well, I think that quantum mechanics is well established and has very simple calculation rules: If one knows the wave functions of the pairs that are emerging from the source, one can calculate everything. The screens impose definite boundary conditions on the wave function, and so it is a matter of a little exercise to calculate what quantum mechanics predicts. If it is a wave function of definite total momentum zero, then the localization is very bad, and one can expect in such a case to have a good but not a perfect correlation since momentum conservation is violated by the screen. Momentum conservation comes from translational invariance, and in the direction of the screen there cannot be conservation of momentum: There is, in fact, an exchange of momentum between the particle going through the slit and the screen. One cannot expect perfect correlations since total momentum conservation is violated. The position correlation is obtained in the original EPR paper by a particular δ-like configuration space wave function. I see no analogy with the situation here and therefore do not see any reason why the positions of the screens of these two photons should be correlated.

SELLERI: I wish to ask a technical question of the experimentalists here. I think that if a certain technical device were possible, then Popper's experiment would be feasible.

Neglecting for a moment the problems connected with the source, I suppose that correlated photons are coming out with total momentum equal to zero. Then I propose to add behind the second slit a cone with rounded edges and made of a perfectly absorbing substance, in such a way that large angle diffraction becomes *totally* impossible if Popper's realistic projection of the outcome of his experiment is correct (as I personally do believe). With Heisenberg's prediction for Popper's experiment, one should by contrast have some large-angle diffraction due to the photons that pass through the central part of the second slit. In this way one gets a sharp difference between the two interpretations of quantum mechanics.

RAUCH: I do not believe that such a tube can be built.

(Professor Mandel later suggested the use of an apodization device which has exactly the same effect as Selleri's cone, that is, it cuts large-angle diffraction down to zero.)

TAROZZI: I remain strongly impressed by the great relevance, both physical and epistemological, of Professor Popper's experimental proposal for testing Einstein locality versus Heisenberg's indeterminacy relations.

Physical relevance, since one is faced for the first time with a direct empirical confrontation between the two fundamental principles of contemporary physics, that is, between Einstein's locality condition, which is one of the two basic postulates of relativity theory, and Heisenberg's indeterminacy relations, the principle on which the whole logical structure of the orthodox quantum mechanics of the Copenhagen and Göttingen schools has been founded. Now, such a direct confrontation did not take place either in the original version of the EPR argument, where one has to introduce the additional assumption of the completeness of the quantum formalism, or in its subsequent developments, like the modern formulation of the argument due to Bohm, Aharonov, Bell, Selleri, d'Espagnat, and others, in which completeness no longer represents a necessary condition, but the incompatibility between locality and quantum mechanichal predictions arises from the use of (nonfactorable) state vectors of the second type. According to several notable, even if unsuccessful, attempts -- the so-called Bohm-Aharonov or Furry hypothesis whose research program can be condensed in Jauch's statement that "mixtures of the second kind do not exist" -- these vectors should be eliminated from the mathematical structure of quantum mechanics,they being the cause of the main quantal paradoxes, whereas the status of fundamental postulate of the theory attributed to the indeterminacy relations had never been questioned.

Epistemological relevance, since Professor Popper has stressed very clearly from the beginning of his lecture that the EPR principle of reality he applies in his reasoning, by attributing to the second unmeasured particle the predictable value of its position, is not in any way questioned in his proposed experiment. Now, such a conception is in open contrast with the, in my opinion, wrong viewpoint expressed by authoritative experts on the foundations of quantum mechanics, according to whom the (supposed) conclusive experimental falsification of the Bell theorem -- which seems in this perspective to replace von Neumann's logical "proof" against hidden variables with an empirical one -- must be considered if one wants to maintain the validity of the locality condition as an experimental refutation of the reality principle and, moreover, since this principle represents only a sufficient condition for reality as a refutation of realism *tout court.*

I discussed this question in Brussels with d'Espagnat during the 1981 conference of the Académie Internationale de Philosophie des Sciences, and he maintained that, if one agrees to express the realist point of view in the form of the EPR deterministic criterion, or even in the form of an analogous probabilistic principle no longer requiring the idealized notion of predictability with certainty of EPR, such as the one I was proposing in my lecture, then one must admit the possibility of an experimental falsification of the realist philosophy. In a similar way,

Shimony has spoken of "a quite decisive experimental test of a philosophical theory."

Now, I think -- and it seems to me that such a conclusion is implied also by Professor Popper's lecture -- that such a viewpoint is completely misleading, for it is not possible for a *single* experiment to falsify a philosophical principle on which depends the very interpretation of *every* experiment. As a matter of fact, if the experimental refutation of the Bell theorem were really conclusive -- that is, if Bell's observable $B = 2\sqrt{2}$ were a well-established empirical result -- and if, moreover, this result were interpreted as a refutation of the reality principle, then this very result would appear as completely irrelevant since, if predictability were no longer a guarantee of (physical) reality, then even $B = 2\sqrt{2}$, though predictable, would not be a property of quantal systems.

THE PHYSICAL ORIGIN OF THE EINSTEIN-PODOLSKY-ROSEN PARADOX

Dirk Aerts

Theoretische Natuurkunde
Vrije Universiteit Brussel
B-1050 Brussels, Belgium

ABSTRACT

We show that quantum mechanics cannot describe separated physical systems and that this incapacity is at the origin of the Einstein-Podolsky-Rosen (EPR) paradox. We refer to a more general theory where separated systems can be described. We analyse again the EPR problem and the EPR experiments in the light of this result. We give an example of a macroscopic physical system that violates Bell inequalities and show that this system is a good analogy to the two particles in the singlet spin state used for the EPR experiments.

1. INTRODUCTION

We intend to show that quantum mechanics is incapable of describing separated systems and that this incapacity is at the origin of the Einstein-Podolsky-Rosen (EPR) paradox.

The shortcoming of quantum mechanics is due to its mathematical structure and not to its interpretation. Hence, as a consequence, also the EPR paradox is due to the mathematical structure of quantum mechanics and cannot be solved by changing the interpretation. It is the mathematical structure of quantum mechanics that has to be changed, and this change does not seem possible in the existing quantum theory. We can affirm this because we made a study of the description of separated systems in the field of quantum logic. Quantum logic is a more general theory than quantum mechanics, in the sense that one can define axioms such that, when these axioms are satisfied, quantum logic reduces to quantum mechanics. (1) What one could show is that two of the foregoing axioms are not satisfied in the description of separated systems. On dropping them, one can describe separated systems, but of course a theory different from quantum mechanics is then being used.

One of the axioms involved is equivalent to the fact that the set of states

G. Tarozzi and A. van der Merwe (eds.), Open Questions in Quantum Physics, 33–50.

of a physical system forms a vector space ; hence it is also equivalent to the superposition principle. The other axiom is more or less equivalent to the statement that observables are represented by operators (e.g. von Neumann algebra's satisfy this axiom). It is clear that if one drop these two axioms almost nothing of the mathematical structure of quantum mechanics remains. A detailed study of this situation can be found in Ref. (2) and (3). In this paper it will be shown in a more direct way that quantum mechanics cannot describe separated physical systems without the use of quantum logic. We shall then analyse the influence of these findings on the EPR paradox reasoning. In their reasoning, Einstein, Podolsky and Rosen use the physical situation of two separated physical systems and they apply quantum mechanics to describe these two physical systems. (4)In doing so they advance the hypothesis that quantum mechanics describes correctly two separated systems, and then they construct, using this description, elements of reality of the subsystems that are not contained in the quantum mechanical description of these subsystems. Thus they can conclude that quantum mechanics is incomplete or does not describe correctly separated systems. Since in their paper EPR suppose quantum mechanics to be correct and hence also quantum mechanics to give a correct description of separated systems, they can conclude that quantum mechanics is not complete EPR have touched here upon a serious deficiency of quantum mechanics. However, as we shall demonstrate, the deficiency of quantum mechanics lies not in the description of the subsystems as indicated by the EPR reasoning, but in the description of the joint system of two separated systems.

2. THE EPR REASONING AND ELEMENTS OF REALITY

From the start I would like to make a distinction between two aspects of the EPR paradox. The first aspect is the reasoning of EPR that leeds to their conclusion that quantum mechanics is not a complete theory (4) ; I shall call this the EPR reasoning. The second aspect is a strange effect predicted by quantum mechanics and stated explicitely in Ref.(4). This effect is the following : If we have a physical system S consisting of two physical systems S_1 and S_2 and described by quantum mechanics, then the system S can be put in such a state, that an experiment, performed on one of the two subsystems determines the possible states of the other subsystem. Schrödinger studied this effect in detail Ref. (5) ; it may therefore be called the *Schrödinger effect*. In the EPR reasoning, a physical system S consisting of two separated physical systems S_1 and S_2 is considered. Also in this case quantum mechanics predicts the existence of the Schrödinger effect. Because our intuition says that in this case the Schrödinger effect should not be present, the term *EPR paradox* was introduced by Schrödinger. But EPR go further in their reasoning than just remarking on the strangeness of this Schrödinger effect in the case of separated systems. They use this Schrödinger effect and the situation of separated systems to show that quantum mechanics does not represent all elements of reality of one of the systems. Central to this reasoning is the concept of *element of reality*. An element of reality is defined by EPR as follows :

> If without in any way disturbing a system, we can predict with certainty the value of a physcial quantity, then there exists an element of reality corresponding to this physical quantity.

Many of the misunderstandings concerning the EPR paradox are due to a misunderstanding of this reality concept. Let us therefore very carefully analyse this concept. If we consider an experiment that can be performed on a physical system, then there are two different aspects of such an experiment that are often confused: the collection of the outcome resulting from the experiment and the prediction of the outcome of the experiment before it is performed. The preformance of the experiment consists in checking whether the prediction was correct.

As an example, let us consider the measurement of the volume of the water contained in a vessel. The experiment here consists of emptying the vessel by means of a siphon and collecting the water in a calibrated cylinder or other reference vessel. When shall we say the vessel contains 5 liters of water? This is not after we have performed the experiment and found 5 liter in the reference vessel. Indeed, the performance of the experiment has changed the state of the system vessel + water in such a way that it does not even contain any water anymore. No, we shall say that the vessel contains 5 liter of water when we can predict the outcome of the experiment to be 5 liter. So, the property or element of reality identified as "the vessel contains 5 liter of water" is something the system has, before we perform the experiment, and even independently of whether we perform the experiment or not. It is, for example, very possible for a system to have an element of reality corresponding to a certain quantity even when this quantity cannot be measured without disturbing the system. It must only be possible to predict the value without disturbing the system.

Whether the measurement disturbs the system or not is of no consequence. In a classical theory, these two aspects are confused, as we have shown in Refs. (6) and (7), and it is quantum mechanics that distinguishes for the first time between these two aspects. In a classical theory every quantity always corresponds to an element of reality of the system, while in quantum mechanics a system can be in a state such that the value of a certain quantity cannot be predicted. When the system is in this state, the quantity in question does not correspond to an element of reality of the system. There is nothing mysterious about this state of affairs, and situations of this kind are encountered not only in the microworld, as we have shown in Ref. (7). Let us now briefly recall the EPR reasoning.

Suppose we have a physical system S composed of two physical systems S_1 and S_2. The state of S is represented by the wave function $\psi(x_1, x_2)$. If A_2 is an observable, corresponding to an experiment on S_2, with eigenvalues a_i and eigenstates $u_i(x_2)$, then we can write

$$\psi(x_1, x_2) = \Sigma_i \psi_i(x_1) u_i(x_2).$$

If we perform an experiment A_2 on the system and find an outcome a_k, then the state of S_1 after the experiment is $\psi_k(x_1)$. This effect on the system S_1, caused by the performance of an experiment A_2 on the system S_2, is what we called the Schrödinger effect. The Schrödinger effect is not mysterious when S_1 and S_2 are not separated systems. Indeed, in this case it is natural that an experiment performed on one of the two systems changes the state of the other system. EPR suppose however that S_1 and S_2 are separated. In this case the experiment A_2 on S_2 does not change the state of the system S_1. So, in this case we can use the experiment A_2 on S_2 to prepare a prediction for the system S_1, and this prediction can then be made without disturbing the system S_1. Thus we can then predict that the system S_1 is in one of the states $\psi_i(x_1)$. Since the experiment A_2 did by hypothesis not change the state of the system S_1, the system S_1 was in one of the states $\psi_i(x_1)$ before we made the experiment A_2. Hence an observable A_1 corresponding to an experiment on S_1 with eigenfunctions $\psi_i(x_1)$, corresponds then to an element of reality of S_1. After this EPR consider another experiment on the system S_2 involving an observable B_2. By the same reasoning as above we can predict before the measurement of B_2 the value of an observable B_1 of S_1. Again we use the fact that S_1 and S_2 are separated. EPR then show that it is very easy to construct in this way two noncompatible observables A_1 and B_1. So, there exists an element of reality of the system S_1 corresponding to both observables A_1 and B_1. This means that there exists a state of the system S_1 such that when S_1 is in this state the values of both noncompatible observables A_1 and B_1 can be predicted. As we know, such a state is not described by the wave function in quantum mechanics. Therefore, EPR conclude, quantum mechanics is not a complete theory, since the wave function does not describe all elements of reality.

We want to remark again that this EPR reasoning is only possible if we suppose the systems S_1 and S_2 to be separated. One can wonder what EPR mean by separated systems. Probably they mean systems that are localized in separated regions of space. And it is indeed very plausible to think that, for systems localized in separated regions of space, an experiment on one of the systems does not change the state of the other system. What we are interested in is however to identify the relation between the two systems S_1 and S_2 that is needed to be able to carry out the EPR reasoning. The EPR reasoning is possible if and only if the two systems S_1 and S_2 are related in the following sense: An experiment performed on one of the systems does not change the state of the other system. This is so because measurements of the observables A_2 and B_2 on S_2 are used to prepare the predictions for the corresponding observables A_1 and B_1 on S_1. So, only in the case where this preparation of the predictions does not change the state of S_1 is it possible to attribute elements of reality corresponding to A_1 and B_1. We shall use the concept of separated systems for systems that are related in this sense. Whether two systems that are localized in separated regions of space are also separated in our sense is an interesting question, but it does not affect the EPR reasoning.

3. SEPARATED PHYSICAL SYSTEMS AND SEPARATED EXPERIMENTS

As we already remarked, we employ the concept of separated physical systems in the following sense: Two systems S_1 and S_2 are separated if an experiment performed on one of the systems does not change the state of the other system.

To be able to use this concept we first of all introduce the concept of separated experiments. Suppose we have two experiments e and f and let us denote by E and F the sets of their outcomes. In general it is not possible to perform e and f together. This is so because sometimes the performance of one of the experiments changes the state of the system in such a way, that it becomes impossible to perform the other experiment. Sometimes it is however possible to perform e and f together. This means that there exists an experiment,

which we shall denote by exf, with a set of outcomes ExF. If we perform the experiment exf and find an outcome $(x,y) \in$ ExF, then we interpret this as an outcome x for the experiment e and an outcome y for the experiment f.
Example 1: Consider a system S of two spin 1/2 particles in the singlet state. We perform a measurement e of the spin of one of the particles in a certain direction and in one region of space and a measurement f of the spin of the other particle in the same direction but in different region of space. The outcome sets of experiments e and f are {0, 1}, where "0" means that the particle is absorbed and "1" that the particle has passed the Stern Gerlach (Fig.1). What we have in mind is the well-known experiment that was proposed by Bohm (8) and since carried out several times to test the Bell inequalities.

Fig. 1

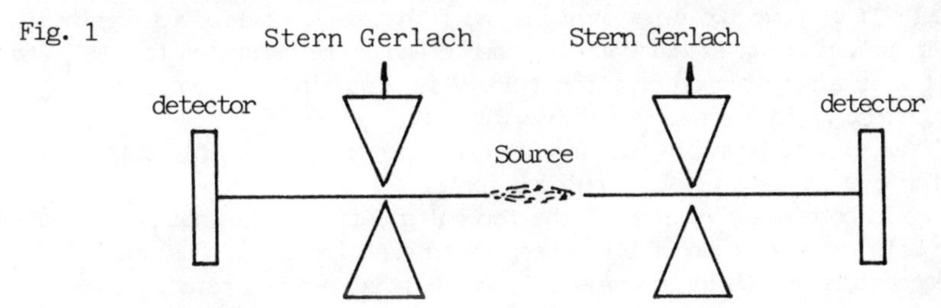

The experiment exf consists of performing e and f simultaneously, and its outcome set is {(0,0), (0,1), (1,0), (1,1)}. As is shown by experiments (9) and is also predicted by quantum mechanics, for exf we always find one of the outcomes (0,1) or (1,0). For e, however, one can find the outcome 0 and 1, both with the probability 1/2; and the situation is similar for f.
Example 2 : Consider a system S consisting of two vessels containing each 10 liter of water and connected by a tube, as shown in Fig. 2. The experiment e consists in testing whether the volume of the water contained in the first vessel is more than 10 liter. We perform this experiment by emptying the vessel, using a siphon, and collecting the water in a reference vessel. We call the outcome "1" if the water stops flowing after its volume exceeds 10 liter in the reference vessel, and we call the outcome "0" is the water stops flowing before it reaches the 10 liter mark. The experiment f consists of determining whether the volume contained in the second vessel exceeds or equals 10 liter.

Fig. 2

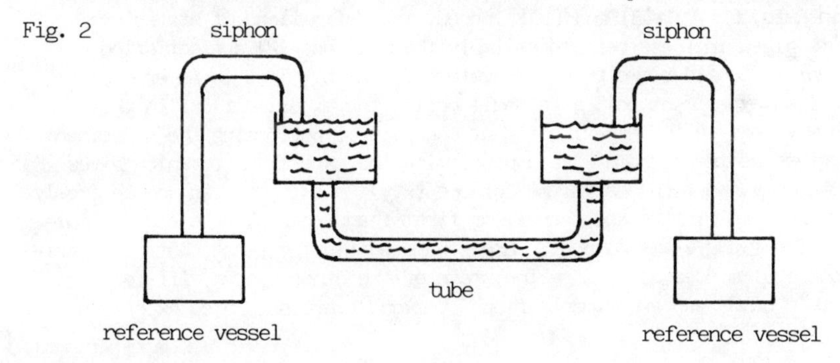

The experiment exf consists of performing e and f at the same time, and its outcome set is $\{(0,0),(0,1),(1,0),(1,1)\}$. Again it is seen that for exf the outcomes always are (0,1) or (1,0). However, for e one always find the outcome 1, and likewise for f.

In both examples 1 and 2 we see that some of the conbinations of outcomes of e and f are not possible for the experiment exf. Indeed, in both cases 1 is a possible outcome for e and 1 is a possible outcome for f, but (1,1) is not possible for exf. This indicates that the experiments e and f influence each other in a certain sense. When this kind of influence is present we will say that e and f are not separated experiments. Hence we can then give the following intuitively clear definition of separated experiments.

Definition 1: If we have two experiments e and f that can be performed together and hence there exists an experiment exf, then e and f are separated for exf iff:

(i) If x is a possible outcome for e and y is a possible outcome for f, then (x,y) is a possible outcome for exf.

(ii) If (x,y) is a possible outcome for exf, then x is a possible outcome for e and y is a possible outcome for f.

Clearly in both examples 1 and 2 the foregoing definition is not satisfied, so that in both examples e and f are nonseparated experiments. Let us now show that if we consider two physical systems S_1 and S_2 that are separated, in the sense that an experiment on one of the systems does not change the state of the other system, then two experiments e, performed on S_1, and f, performed on S_2, are separated. To be able to show this we have to point out the relation between the state of a physical system and the possible outcomes of an experiment on this physical system. There are two remarks that we would like to make:

(1) The state of a physical system S determines the possible outcomes of an experiment performed on S.

(2) If p and q are two different states of the physical system S, then there is at least one experiment e performable on S such that the possible outcomes of e when S is in state p are different from the possible outcomes of e when S is in state q.

In a classical theory every experiment has only one possible outcome for a given state of the system. Hence assertion (1) is clearly satisfied. In assertion (2) we suppose that there are enough experiments to distinguish between all the states. In quantum mechanics the possible outcomes of an experiment are those outcomes that correspond to eigenstates that are not orthogonal to the given state of the system. Clearly, then (1) and (2) are satisfied in quantum mechanics. That (1) and (2) are satisfied in much more general approaches can be seen in Refs. (1),(2),(3) and (10). Indeed, the definitions of states and experiments given in these references imply that (1) and (2) are satisfied.

Theorem 1: Consider two physical systems S_1 and S_2. If e is an experiment on S_1 and f an experiment on S_2, we shall define the experiment exf in the following way. exf is the experiment that consists of performing the experiment e, which gives us the outcome x, and performing the experiment f, which gives us the outcome y. We then assign the outcome (x,y) to exf. One can choose freely whether to perform first e and then f, or first f and then e, or e and f simultaneously. The two systems S_1 and S_2 are separated (in the sense that an experiment on one system does not change the state of the other system) iff for experiments e on S_1 and experiments f on S_2, e and f are separated experiments for exf.

Proof. Suppose that S_1 and S_2 are separated systems. If x is a possible outcome of e and y is a possible outcome of f, then (x,y) is a possible outcome of exf, because the performance of one of the experiments does not change the possible outcomes of the other one. For the same reason, if (x,y) is a possible outcome of exf, then x is a possible outcome of e and y is a possible outcome of f. This shows that e and f are separated experiments. Suppose that S_1 and S_2 are not separated systems. This means that there exists an experiment e on one of the systems and a state p of the other system such that, after the performance of e, the state p has been changed into a different state q. Suppose e is an experiment on S_1 and p and q are states of S_2. Then there exists an experiment f on S_2 such that the set of possible outcomes when S_2 is in state p is different from the set of possible outcomes when S_2 is in state q. Two situations can occur. The first situation obtains when there is an outcome y of f that is possible when S_2 is in state p and not possible when S_2 is in state q. If x is an outcome of e and exf denotes an experiment where we first perform e and then f, then x and y are possible outcomes of e and f, while (x,y) is not a possible outcome of exf. The second situation obtains when there is an outcome y of f that is possible when S_2 is in state q and not possible when S_2 is in state p. In this case, (x,y) is a possible outcome of exf, but y is not a possible outcome of f. Hence e and f are not separated experiments.

This theorem 1 shows that we can characterize separated physical systems by means of separated experiments and also shows that the definition of separated physical systems given in Refs. (2) and (3) is equivalent with the definition adopted in this paper. It moreover shows that the physical systems of example 1 and of example 2 are not separated.

3. BELL INEQUALITIES AND SEPARATED PHYSICAL SYSTEMS

We shall next show that Bell's inequalities test whether two systems are separated or not. But let us first explain briefly how these inequalities are defined.

Suppose we have a physical system described by a hidden variable formalism and that Σ denotes the set of states of the system. We consider experiments e, f, g and h on the system with two possible outcomes, "yes" and "no".

Associate with every yes-no experiment e two random variables

$$X_e : \Sigma \rightarrow \{-1, +1\} \qquad Y_e : \Sigma \rightarrow \{0, 1\}$$

such that $X_e(p) = Y_e(p) = +1$ if e gives the outcome "yes" and $X_e(p) = -1$ and $Y_e(p) = 0$ if e gives the outcome "no". If we have two yes-no experiments e and f that can be performed together, we define for the experiment exf the following random variables:

$X_{ef}(p) = +1$ if exf gives one of the outcomes (yes,yes) or (no,no);
$X_{ef}(p) = -1$ if exf gives one of the outcomes (yes,no) or (no,yes);

$Y_{ef}(p) = +1$ if exf gives the outcome (yes,yes);
$Y_{ef}(p) = 0$ if exf gives one of the outcomes (yes,no), (no,yes) or (no,no).

Suppose now that we have four yes-no experiments e, f, g, h such that exf, exg, hxf, hxg exist. The Bell inequalities (11) are then

$$|X_{ef}(p) - X_{eg}(p)| + |X_{hf}(p) + X_{hg}(p)| \leqslant 2 ,$$

while those of Clauser and Horne (12) read

$$\frac{Y_{eg}(p) - Y_{ef}(p) + Y_{hf}(p) + Y_{hg}(p)}{Y_h(p) + Y_g(p)} \leqslant 1.$$

We first of all should like to reformulate these inequalities for the case of two yes-no experiments e and f. One can immediately find the corresponding inequalities by taking for g and h the trivial experiments that give always "yes". Then

$$X_{eg}(p) = X_e(p),\ Y_{hf}(p) = Y_f(p),$$

$$X_{hg}(p) = 1,\ Y_{eg}(p) = Y_e(p),$$

$$X_{hf}(p) = X_f(p),\ Y_{hg}(p) = 1,$$

and Bell's inequalities become

$$| X_{ef}(p) - X_e(p) | + | X_f(p) + 1| \leqslant 2$$

The Clauser-Horne inequalities become

$$Y_e(p) + Y_f(p) - Y_{ef}(p) \leqslant 1.$$

If e is a yes-no experiment, one can consider the yes-no experiment that consists of performing the same experiment but interchanging the outcomes "yes" and "no". Let us denote this experiment by $\tilde{e}$. We can then of course also write the Bell and Clauser-Horne inequalities for $\tilde{e}$, f, $\tilde{e}$xf, for $\tilde{e}$, $\tilde{f}$, $\tilde{e}$x$\tilde{f}$, and for e, $\tilde{f}$, ex$\tilde{f}$. Hence with the two yes-no experiments e and f that can be performed together are connected four Bell inequalities and four Clauser-Horne inequalities.

Theorem 2: If e and f are two yes-no experiments that can be performed together on a hidden variable formalism, then e and f are separated iff e and f satisfy all simplified Clauser-Horne inequalities.

Proof: Suppose that e and f are separated, then clearly $Y_{ef}(p) = Y_e(p).Y_f(p)$ and the Clauser-Horne inequalities are satisfied.
Suppose that e and f are not separated. Then two situations are possible. In the first situation there exists a state p such that a certain outcome for the coincidence experiment, say the outcome (yes,no), never occurs, while e gives the outcome "yes" and f the outcome "no". Then

$$Y_{e\tilde{f}}(p) = 0,\ Y_e(p) = 1,\ Y_{\tilde{f}}(p) = 1.$$

Hence

$$Y_e(p) + Y_{\tilde{f}}(p) - Y_{e\tilde{f}}(p) = 2,$$

which is a violation of the Clauser-Horne inequalities.

In the second situation there exists a state p such that a certain outcome for one of the experiments, say the outcome "no" for the experiment f, never occurs, while exf gives one of the outcomes (yes,no) or (no,no). In this case, $Y_{ef} = Y_{\tilde{e}f} = 0$. If e gives "yes", we have $Y_e + Y_f - Y_{ef} = 2$, and if e gives "no", we have $Y_{\tilde{e}} + Y_f - Y_{\tilde{e}f} = 2$.

A consequence of theorem 2 is that for classical physical systems (which are always described by a hidden-variable formalism) Clauser-Horne inequalities test exactly whether the systems are separated. For non classical systems, Clauser-Horne inequalities are satisfied if the systems are separated, but they can still be satisfied if the systems are nonseparated. In any case, a violation of the inequalities shows that the systems are not separated. Let us show that the inequalities are indeed violated for the classical macroscopic example 2.

4. CLASSICAL MACROSCOPICAL EXAMPLE THAT VIOLATES BELL INEQUALITIES (13)

We consider example 2 and the state p of the two vessels described in this example. Then:

$$X_{ef}(p) = -1, \; Y_{ef}(p) = 0,$$

$$X_e(p) = +1, \; Y_e(p) = +1,$$

$$X_f(p) = +1, \; Y_f(p) = +1.$$

Hence $|X_{ef}(p) - X_e(p)| + |X_f(p) + 1| = 4 > 2$,

$$Y_e(p) + Y_f(p) - Y_{ef}(p) = 2 > 1.$$

This example shows that the violation of Bell's inequalities does not indicate some mysterious property of the microworld, but just indicates that the system under study consists of two nonseparated systems.

5. SEPARATED EXPERIMENTS CANNOT BE DESCRIBED BY QUANTUM MECHANICS OR EXPERIMENTS DESCRIBED BY QUANTUM MECHANICS ARE NEVER SEPARATED

We shall now proof that an experiment of the type exf, where e and f are separated experiments cannot be described by quantum mechanics.

Theorem 3: If e and f are experiments on a physical system S described by quantum mechanics, then e and f are never separated experiments.

Proof: If we describe the situation by quantum mechanics, then there exists a Hilbert space H and self-adjoint operators R and S corresponding to the experiments e and f. Suppose that e and f are separated experiments, then there also exists a self-adjoint operator O corresponding to the experiment exf. If E is the outcome set of e and F the outcome set of f, then ExF is the outcome set of exf. We will first of all show that if e and f are separated, then R and S commute. To do this we will use the spectral projections of O, R, and S. Consider two arbitrary spectral projections P_A and P_B of R and S. Hence $A \subset E$ and $B \subset F$. To AxB corresponds a spectral projection P_{AxB} of O. Consider a state w such that $w \perp P_A(H)$. If the system is in state w and $x \in A$, then x is not a possible outcome for e. If y is an arbitrary outcome for f, then (x,y) is not a possible outcome of exf. This shows that $w \perp P_{AxF}(H)$. So $1 - P_A \subset 1 - P_{AxF}$. From this follows that $P_{AxF} \subset P_A$ and $P_{(E\setminus A)xF} \subset P_{E\setminus A}$. Now clearly

$$P_{(E\setminus A)xF} = P_{ExF\setminus AxF} = 1 - P_{AxF}$$

This shows that $P_{AxF} = P_A$

In an analoguous way we show that $P_{ExB} = P_B$. From this follows that $[P_A, P_B] = 0$ since P_{AxF} and P_{ExB} are spectral projections of the same selfadjoint operator O. As a consequence $[R,S] = 0$. We also have:

$$P_{AxB} = P_{AxF}.P_{ExB} = P_A.P_B$$

Consider now the closed subspace $P_A.(1-P_B)(H)$. If $P_A.(1-P_B)(H) = 0$ then $P_A = P_A.P_B \neq 0$. Hence $P_A.P_B(H) \neq 0$. In an analoguous way we show that $(1-P_A).P_B(H) \neq 0$ in this case. Hence there are two possibilities:

$$P_A.(1-P_B)(H) \neq 0 \text{ and } (1-P_A).P_B(H) \neq 0 \text{ or}$$

$$P_A.P_B(H) \neq 0 \text{ and } (1-P_A).(1-P_B)(H) \neq 0.$$

We shall show that in both cases the superposition principle allows us to construct a state such that when the system is in this state the experiments e and f are not separated. Suppose $P_A.(1-P_B)(H) \neq 0$ and $(1-P_A).P_B(H) \neq 0$. Take $u \in P_A.(1-P_B)(H)$ and $v \in (1-P_A).P_B(H)$. Consider the state $w = u + v$. Then:

$$P_A(w) = u, \; (1-P_A)(w) = v, \; P_B(w) = v, \; (1-P_B)(w) = u$$

On the other hand:

$$P_{AxB}(w) = P_A.P_B(w) = P_A(v) = 0$$

$$P_{(E\setminus A)x(F\setminus B)}(w) = (1-P_A).(1-P_B)(w) = (1-P_A)(w) = 0$$

$$P_{(E\setminus A)xB}(w) = (1-P_A).P_B(w) = (1-P_A)(v) = v$$

$$P_{Ax(F\setminus B)}(w) = P_A.(1-P_B)(w) = P_A(u) = u$$

Suppose now that the physical system is in the state w. Then there is at least one possible outcome $x \in A$ and one possible outcome $z \in E\setminus A$ for e. And there is at least one possible outcome $y \in B$ and one possible outcome $t \in F\setminus B$ for f. But, while (x,t) and (z,y) are possible outcomes for exf, the outcomes (x,y) and (z,t) are not possible. This shows that e and f are not separated experiments. If $P_A.P_B(H) \neq 0$ and $(1-P_A).(1-P_B)(H) \neq 0$ we proof in an analoguous way that e and f are not separated.

It is easy to see that it is the superposition principle that compels the existence of states such that e and f are not separated.

Consequence: If e and f are separated experiments on a physical system, then the experiment exf cannot be described by quantum mechanics without leading to contradictions. Hence quantum mechanics cannot describe separated physical systems without leading to contradictions.

As our following analysis will show, it is this shortcoming of quantum mechanics that is at the origin of the EPR paradox.

6. THE REASON WHY QUANTUM MECHANICS CANNOT DESCRIBE SEPARATED SYSTEMS

We shall try to explain briefly in this section the study that can be found in Refs. (2), (3) and (7), where a description of separated systems is given in a more general theory than quantum mechanics. This theory is an elaboration of Piron's appraoch to quantum logic. (1) It is more general than classical mechanics and quantum mechanics, because we can define a set of axioms such that , when these axioms are satisfied, the theory reduces to classical mechanics or to quantum mechanics. (2, 3, 7) The mathematical framework used in this theory is that of a complete lattice, where the elements of the lattice represent the properties of the physical system. The axioms to reduce the theory to quantum mechanics are formulated on this lattice of properties. The first two axioms introduce the mathematical stucture of an orthocomplementation on the lattice. The third axiom makes the lattice atomic. The fourth axiom makes the lattice weakly modular and the fifth axiom is equivalent to the covering law (or makes the lattice semi modular). What we found is that axiom 4 and axiom 5 are never satisfied for the description of separated systems. Let us repeat the main theorem of Refs. (2) and (3).

Theorem 4: Suppose that S is a physical system composed of two separated physical systems S_1 and S_2. If the lattice of properties of S satisfies axiom 4 (is weakly modular) or if the lattice of properties of S satisfies axiom 5 (the covering law), then S_1 or S_2 is a classical system for which the lattice of properties is a Boolean lattice.

The lattice of properties of a system described by quantum mechanics satisfies both axioms 4 and 5, but is never a Boolean lattice. So, from this theorem it follows that if S_1 and S_2 are both described by quantum mechanics, then S can never be described by quantum mechanics. Or, in other words, quantum mechanics cannot describe separated quantum systems. This failure of quantum mechanics to describe separated systems is really due to the mathematical structure of quantum mechanics and does not depend on the interpretation. The covering law (axiom 5) is the axiom that introduces the vector space structure for the set of states. Hence the set of states of the system composed of two separated systems will not have a vector-space structure. As a consequence, the superposition principle will not be valid. The weak modularity (axiom 4) more or less corresponds to the possibility of representing observables by operators. Indeed, for example, a von Neumann algebra has a weak modular lattice of projections. Hence, also in the algebraic approach, separated systems will not be describable. The theory used in Refs. (2), (3) and (7) allows the description of separated systems, since none of our axioms have to be satisfied in this theory.

7. NONSEPARATION, LOCALISATION IN SEPARATED REGIONS OF SPACE, SPACELIKE SEPARATED CORRELATIONS, AND THE EPR EXPERIMENTS

In most of the discussions about the EPR paradox, different concepts are often confused. We already made a distinction between the EPR reasoning and the effect predicted by quantum mechanics that we called the Schrödinger effect, and which is paradoxical in the case of separated systems. Most of the discussions of the EPR paradox concern this Schrödinger effect. Also experiments performed in connection with the EPR paradox are about this Schrödinger effect. (9) In connection with this Schrödinger effect and the EPR experiments we also should distinguish different concepts. For two physical systems S_1 and S_2 the following in-principle-different situations are possible. S_1 and S_2 can be separated or nonseparated. We explained already in detail what is meant with separation and nonseparation. S_1 and S_2 can be localized in separated regions of space. For the particles in the singlet-spin state of example 1 no experiment has been made till now to see whether these two particles are localized in separated regions of space. Quantum mechanics predicts however that they are not, because a singlet state is not a state of one particle localized in one region of space and another particle localized in a separated region of space. The question about separability could be formulated now as follows: Are two systems that are localized in separated regions of space separated?

Separability is not violated by the two vessels of water of example 2, since before the experiment the water is not localized in separated regions of space. Whether the two spin-1/2 particles of example 1 can be put in such a state that they are localized in separated regions of space and still violate Bell's inequalities is a question that has not been answered. But we can imagine that an attempt to localize the particles in separated regions of space will also separate them, and hence Bell inequalities are not violated anymore. In both examples, the experiments, used to detect the nonseparation of the systems separate the systems and localize them in separated regions of space. Often the two particles in the singlet-spin state are called "correlated systems." If by correlation is meant that the two systems are in a state such that certain outcomes of different experiments are correlated, then it is not the correlation which is specific for the two particles in the singlet-spin state. Indeed, also separated systems can be put in a state that produces correlation for different measurements. Let us see this again for an example.

Fig. 3

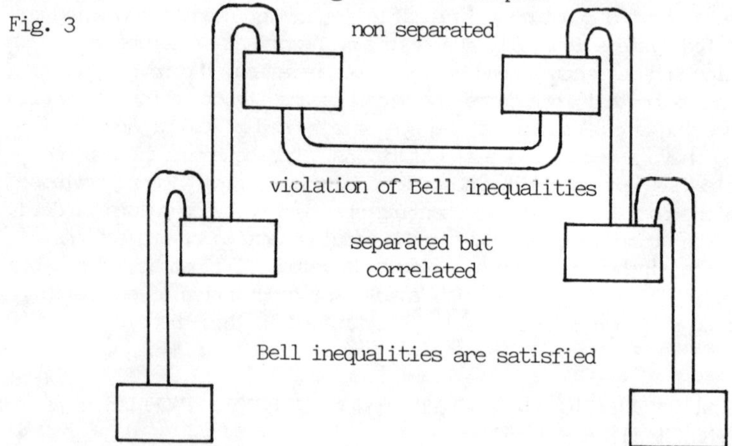

When the two vessels (Fig. 3) are connected by a tube, they are nonseparated systems. If we make the experiments described in example 2, the Bell inequalities are violated, as shown in Sec. 4. After the experiment, all the water is in the reference vessels. In this state the water is localized in two separated regions of space. If we make the same experiment, we will find that the outcomes are still strictly correlated. But now Bell inequalities are not violated anymore. So, in this case the experiments are separated. Hence the first experiment separates the water. The same process happens with the two particles in the singlet-spin state. Two such particles in the singlet-spin state are nonseparated systems, as follows from the quantum mechanical description of these particles and as is also tested now by experiments. (9)

Fig. 4

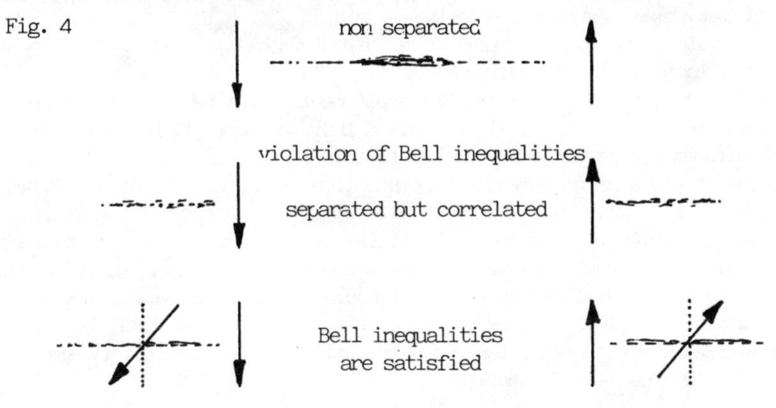

After the first experiment that measures the polarization, the two particles are separated but they are still correlated. Hence also here the experiment separates the particles. That the particles are really separated after the first polarization measurement can be tested by making a second polarization measurement in another direction. By means of this second polarization measurement it will not be possible to violate Bell inequalities.

The correlations that are found in the EPR experiments are tested by means of coincidence experiments. Hence the two events that correspond to the detection of the correlation are spacelike separated events. Intuitively one could think that such spacelike separated events that are correlated are in contradiction with relativity theory. This would indeed be so if these correlations could be used to transmit a signal. This seems not to be the case, but still a number of people find these spacelike separated correlated events the most mysterious part of the EPR paradox. Let us show that also the example with the two vessels produces such spacelike separated correlations. Indeed, for the two particles in the singlet-spin state we say that a coincidence experiment is made if the two particles are detected simultaneously. Hence there is a coincidence if the experiment at both sides ends simultaneously. Let us now look at the experiments for the two vessels of water in example 1. The end of an experiment occurs when the water stops flowing. This happens always simultaneously on both sides. Hence the events that corresponds to the detection of the correlation are spacelike separated events. Clearly there is no problem with relativity theory and there is even no mystery at all.

There is a number of other analogies we can remark on between the experiments with the two vessels and the experiments with the particles in the singlet-spin state:

(1) In both cases the results do not depend on the distance.

(2) It does not seem to be possible to transmit a signal by means of the produced correlations.

(3) There is no exchange of energy corresponding to the correlations.

(4) If we try to refine the description of the two vessels of water connected by a tube by introducing additional variables (often called hidden variables), then clearly some of these additional variables will be nonlocal. This is so because the variables describing the water in the tube are not variables of the water in the vessels.

(5) The systems are not localized in separated regions of space.

(6) The coincidences are simultaneous.

Sometimes it is claimed that the EPR experiments force us to change fundamental conceptions of the world. It is claimed that, because of the existence of such simultaneous correlations, one has to abandon the principle of separability or one has to abandon a realistic conception of the world. Since the two vessels of water also give us a situation with simultaneous correlations, the philosophical reasoning applied to the particles in the singlet state that leads to these conclusions can also be made on the two vessels of water. The principle of separability is not violated by the two vessels of water, as we already explained. It is also clear that we do not have to abandon a realistic conception of the world because of the simultaneous correlations encountered in the experiments with the two vessels of water.

8. THE EPR REASONING

Since we showed that separated systems cannot be described by quantum mechanics, the EPR reasoning is still valid, but, it becomes a reasoning *ad absurdum*. Indeed, EPR suppose that quantum mechanics can describe separated physical systems. Let us analyse again their reasoning knowing this result.

EPR consider the following two sentences:

(1) Quantum mechanics is not complete.

(2) Physical quantities that are not compatible cannot have simultaneous reality.

Obviously these two sentences cannot both be wrong. Indeed if two noncompatible quantities can have simultaneous reality, then quantum mechanics is not complete, because the wave function cannot describe these elements of reality. So we have one of the three cases:

A. (1) false, (2) true;

B. (1) true, (2) false;

C. (1) true, (2) true;

Once EPR come to this conclusion, they consider the situation of two separated systems S_1 and S_2. By applying quantum mechanics to describe these two separated systems and using the Schrödinger effect they can show that it is possible to attach simultaneously elements of reality to noncompatible quantities. We explained this in sec. 2. Hence (1) is true. So, what EPR show is the following:

Quantum mechanics correctly describes separated systems.
⇒ Quantities that are not compatible can have simultaneous reality.
⇒ Quantum mechanics is not complete.

From this EPR can conclude that quantum mechanics does not correctly describe separated systems or quantum mechanics is not complete. EPR mention in the beginning of their paper that they suppose quantum mechanics to be correct, and then they can indeed conclude that quantum mechanics is not complete. If one supposes quantum mechanics to give a correct description of separated systems, this reasoning of EPR indicates which are the missing elements of reality in quantum mechanics. These are elements of reality corresponding to noncommuting observables. This made many people think that it should be possible to solve the problem by introducing classical hidden variables, which would take into account these missing elements of reality, and in this way complete quantum mechanics. If we take into account that the quantum mechanical description of separated systems is wrong, then the EPR reasoning is still valid and the conclusion is correct. But, because the premise is false, it does not indicate the missing elements of reality. Indeed, the statement "Quantum mechanics is not complete because quantities that are not compatible can have simultaneous reality" is not a true statement, because it is only a statement used in an *ad absurdum* reasoning. I want to draw attention to this fact, because often people are not interested in the conclusion of EPR, namely that quantum mechanics is incomplete if it is correct, but rather in the conclusion that quantities that are not compatible can have simultaneous reality if quantum mechanics is correct. And it is this conclusion that leads directly to the thought that, to avoid the EPR problem, we have to build a theory that takes into account the fact that noncompatible physical quantities can have simultaneous reality. Met us summarize all this and see that there is no paradox left. There are two possible situations for two systems:

First situation: The two systems are separated. Then quantum mechanics does not give a correct description of this situation. Correcting quantum mechanics does not happen by adding states to the subsystems, but by taking states away from the compound system (see Refs. (2), (3), and (7)).

Second situation: The two systems are not separated. In this case, as we explained in Sec. 2, it is not possible to hold the EPR reasoning. Indeed, if we try to attribute an element of reality to one of the systems S_1 by making an experiment on the other systems S_2, then, because the experiment on the system S_2 changes the state of S_1, the element of reality of S_1 is created by the experiment on S_2. Hence we cannot continue the EPR reasoning and say that the element of reality was already there before we made the experiment on the system S_2. This step is however neccesary if we want to show that S_1 can have two elements of reality corresponding to noncompatible observables.

So, as one can see, the EPR paper touches on a major shortcoming of quantum mechanics, namely its incapacity to describe separated systems. Since the EPR reasoning is a reasoning *ad absurdum*, it however does not indicate a way to solve the problem.

The condition of completeness of a theory was put forward by EPR in the following way: "A theory is complete if every element of reality has a counterpart in the theory". Clearly EPR did not mean that a theory should describe all possible elements of reality of the physical system at once. Indeed, if this was what they meant, then of course every theory is incomplete, because a theory only gives a model for the physical system and this model describes a well defined set of elements of reality of the physical system. Therefore we would like to put this criterium of completeness in a different way. We prefer to say: "A theory is complete if it can describe every possible set of elements of reality of the physical system without leading to contradictions". This criterium should be satisfied by any reasonable physical theory. It means in fact that the theory is flexible enough to provide a model for any well defined set of elements of reality of the physical system. This is not the case for quantum mechanics, because quantum mechanics cannot provide us with a model for the description of separated physical systems. In Refs. (14) we explicitely construct the elements of reality that cannot be represented by quantum mechanics. In Refs. (3) and (6) we give an experimental example of observables that have not simultaneous reality. Even for macroscopic systems it is possible to find observables of this kind, as we show in Ref. (7). This demonstrates that of the three cases A, B and C it is C which is the correct one.

9. CONCLUSION

The shortcoming of quantum mechanics of not being able to describe separated systems is due to the fact that the mathematical structure of quantum mechanics always compels the existence of states that should not be there in the case of separated systems. It is the superposition principle that is at the origin of these states. Now, one could think in the following way: Why not simply drop these states and proceed with quantum mechanics? This is indeed the solution that has been adopted for the problem in connection with superselection rules. But for the case of separated systems this solution is not possible because the unwanted superpositions are not superpositions between states of different orthogonal subspaces, as is the case for superselection rules. If one drops the unwanted states, one is left with a set of states that cannot be embedded in a Hilbert space such that the mathematical formalism of linear operators remains applicable. Things are even worse, for not only are there too many states in the quantum mechanical description of separated systems, there are also missing projection operators (higher-dimensional elements of the lattice of closed subspaces). This is shown explicitely in Ref. (14). We must also remark that there is another logicaly possible way out of the dilemma. It is possible to pretend that separated systems do not exist, so we do not have to try describing them with quantum mechanics. Often people defend this solution. This is I think because they have a wrong intuitive image of what separated systems are. When two systems are separated, this does not mean that there is no interaction between the two systems. No, in general there is an interaction, and by means of this interaction the dynamical change of the state of one system is influenced by the dynamical change of the state of the other system. In classical mechanics for example, almost all two-body problems are problems of separated bodies (e.g. the Kepler problem). It is the analogies of such situations for quantum systems that cannot be treated by quantum mechanics. Two systems are not separated when an experiment on one system changes the state of the other system. For two classical bodies this would for example be the case when they are connected by a rigid rod. But as we know, two bodies connected by a rigid rod are treated as one body. To provide a still better intuitive feeling of what separated systems are, we note the following: One system is separated from the rest of the universe, but one system is not separated from the measuring apparatus during a measurement. Hence it is the whole range of interesting situations for two separated systems with interactions between them that cannot be treated in quantum mechanics. The inability of quantum mechanics to describe such separated systems with interaction between them can also be remarked on with reference to their dynamics. Let us show this. The evolution of a system from one instant to another one is described by an automorphism of the set of its states. In quantum mechanics such automorphisms are represented by unitary or anti-unitary operators of the Hilbert space describing the system. In classical mechanics they are represented by permutations of the state space describing the system. If one consider in classical mechanics a system S, described in a state space X, consisting of two systems S_1 and S_2, described in state space X_1 and X_2, then $X = X_1 \times X_2$. It is easy to see that a permutation of X cannot always be decomposed in the product of a permutation of X_1 and a permutation of X_2. Moreover it is these nonproduct permutations of X that make it possible to describe an evolution with interactions between S_1 and S_2. On the other hand, if one has in quantum mechanics a system S, described in a Hilbert space H, composed of two systems S_1 and S_2, described in Hilbert spaces H_1 and H_2, then $H = H_1 \otimes H_2$. If one should now try to give a description of the evolution of two separated systems in this tensor-product Hilbert space, unitary transformations that conserve product states must be considered. But a unitary transformation U that conserves product states is of the form $U_1 \times U_2$, and with these product unitary transformations one is not able to describe interactions between the two systems. So, something very peculiar happens in quantum mechanics. To describe a system composed of separated systems with interactions between them, it is sufficient to have a state space consisting of states that are products of subsystem states, but one needs automorphisms of this state space that are not products of automorphisms of the subsystem state spaces. In quantum mechanics exactly the reverse happens. For the state space one always has these nonproduct states, but for the automorphisms one always finds products.

REFERENCES

1. C. Piron, *Foundations of quantum physics* (W.A Benjamin, Menlo Park, Calif., 1976).
2. D. Aerts, "The one and the many. Towards a unification of the quantum and the classical description of one and many physical entities". Doctoral thesis, Vrije Universiteit Brussel, TENA (1981).
3. D. Aerts, *Found. Phys. 12*, 1131 (1982).
4. A. Einstein, B. Podolsky, and N. Rosen, *Phys. Rev. 47*, 777 (1935).
5. E. Schrödinger, *Proc. Cambr. Phil. Soc. 31*, 555 (1935).
6. D. Aerts, *J. Math. Phys. 24*, 2441 (1983).
7. D. Aerts, "The description of one and many physical systems", *Proceedings of the 25e Cours de Perfectionnement de l'Association Vaudoise des Chercheurs en Physique: Les Fondements de la Mécanique Quantique*, Montana, march 1983.
8. D. Bohm, *Quantum Theory* (Prentice-Hall, Englewood Cliffs, J., 1951).
9. A. Aspect, P. Grangier, and G. Roger, *Phys. Rev. Lett. 43* , 91, (1982);

49, 1808 (1982).
S.J. Freedman and J.F. Clauser, *Phys. Rev. Lett. 28* , 938 (1972).
R.A. Holt and F.M. Pipkin, preprint, Harvard University (1973).
J.F. Clauser, *Phys. Rev. Lett. 36*, 1223 (1976).
E.S. Fry and R.C. Thompson, *Phys. Rev. Lett. 37*, 465 (1976).
M. Faraci et al., *Lett. Nuovo Cimento 9*, 607 (1974).
M. Lamechi-Rachti and W. Mittig, *Phys. Rev. 14* , 2543 (1976).
10. D. Foulis, C. Piron and C. Randall, *Found. Phys. 13* , 813 (1983).
11. J.S. Bell, *Physics 1*, 195 (1964).
12. J.F. Clauser and A. Shimony, *Rep. Prog. Phys. 41*, 1883 (1978).
13. D. Aerts, *Lett. Nuovo Cimento 34* 107 (1982).
14. D. Aerts, "The Missing Elements of Reality in the Description of Quantum Mechanics of the EPR Paradox Situation," Preprint, TENA, Brussels.

DOES THE BELL INEQUALITY HOLD FOR ALL LOCAL THEORIES ?

Thomas D. Angelidis

University College London
London WC1E 6BT, England

ABSTRACT

The claim that the Bell inequality $D(\underline{a},\underline{b}) \leq 2$ is valid for *all* local theories is criticised. The criticism is based on the author's proof that this universality claim is *incompatible* with the conservation law of angular momentum. There exists an infinity of counterexamples to the universality claim made for $D(\underline{a},\underline{b}) \leq 2$. So, Bell offers *no valid proof of the non-locality of the quantum formalism*; consequently the experiments culminating in the Aspect experiment with the switches cannot be said to refute Einsteinian locality.

1. INTRODUCTION

Einstein, after formulating his principle of locality (1) (and even before (2)), advanced the argument that if the (unproved) Copenhagen interpretation of the quantum formalism (QF) is true, then there must be action at a distance or, as it is now called, non-locality. So, Einstein first drew *"the Einsteinian alternative"* (3): Either the Copenhagen interpretation or Einsteinian locality is false. 1

According to Bell (4,5) any triple combination $(\rho(\lambda), A(\underline{a},\lambda), B(\underline{b},\lambda))$ 2 of certain functions that satisfy the Bell premises (explained here in Sec. 3) constitutes a formulation of a local theory; and he states that:

$$D(\underline{a},\underline{b}) \leq 2 < [D(\underline{a},\underline{b})]_{QF}. \tag{1}$$

Here

G. Tarozzi and A. van der Merwe (eds.), Open Questions in Quantum Physics, 51–62.

$$D(\underline{a},\underline{b}) \leq 2, \tag{2}$$

where $D(\underline{a},\underline{b}) \equiv |E(\underline{a}_1,\underline{b}_1) - E(\underline{a}_1,\underline{b}_2)| + |E(\underline{a}_2,\underline{b}_1) + E(\underline{a}_2,\underline{b}_2)|$ and $E(\underline{a},\underline{b}) = \int_\Lambda A(\underline{a},\lambda)B(\underline{b},\lambda)\rho(\lambda)d\lambda$ is the expectation value for any product $A(\underline{a},\lambda)B(\underline{b},\lambda)$, is the more general (5) Bell inequality which is *not empirically testable*. 3 And

$$2 < [D(\underline{a},\underline{b})]_{QF} \tag{3}$$

is the inequality that is valid, for some unit vectors $(\underline{a},\underline{b})$, for the predictions of the quantum formalism, where $[D(\underline{a},\underline{b})]_{QF} \equiv |C| (|\underline{a}_1 \cdot \underline{b}_1 - \underline{a}_1 \cdot \underline{b}_2| + |\underline{a}_2 \cdot \underline{b}_1 + \underline{a}_2 \cdot \underline{b}_2|)$ and $\underline{a}.\underline{b}$ denotes the dot product of any two vectors. The coefficient C ($|C| \leq 1$) denotes a measure of efficiency of the detectors, polarizers, etc. For example, for a nearly ideal experiment (that is, $|C| \simeq 1$) and four coplanar unit vectors $\underline{a}_1,\underline{b}_1,\underline{a}_2,\underline{b}_2$ successively making an angle $\phi = \pi/4$ with each other, inequality (3) is satisfied because $2 < 2\sqrt{2} \simeq 2.828$.

What is challenged here is the usual interpretation that identifies (1) and (2) with

$$[D(\underline{a},\underline{b})]_{LT} \leq 2 < [D(\underline{a},\underline{b})]_{QF} \tag{1a}$$

$$[D(\underline{a},\underline{b})]_{LT} \leq 2 \tag{2a}$$

respectively, where $[D(\underline{a},\underline{b})]_{LT} \leq 2$ is the inequality claimed (4,5) to be valid, for *all* $(\underline{a},\underline{b})$'s, for the predictions of *all* local theories (LT). My criticism is: If (1) and (2) could *indeed* be interpreted as (1a) and (2a), then the quantum formalism would *indeed* be shown to be a non-local theory. *But if (2) cannot be identified with (2a), then the quantum formalism could be a local theory.*

Interpreting (1) and (2) as (1a) and (2a) *would be permissible only if what I shall call below the universality claim (UC) were actually proved true*. The universality claim was raised by Bell (5) who put it as follows: *"no local... theory can reproduce all the experimental predictions of quantum mechanics."* 4 Jammer's (2) statement, namely that "Bell proved that no correlation function P_{ab} ($\equiv E(\underline{a},\underline{b})$) for *any* ρ, A or B can reproduce the quantum-mechanical expectations for *all* $\underline{a}$ and $\underline{b}$," raises the same universality claim.

The universality claim (UC) could be formulated as follows:

(UC) The left side of $D(\underline{a},\underline{b}) \leq 2$ covers *all* local theories (ρ,A,B)

or, more precisely, as the *identity* $D(\underline{a},\underline{b}) \equiv [D(\underline{a},\underline{b})]_{LT}$. The immediate consequence of (UC) is that:

(NL) The quantum formalism (QF) is not a local theory.

Clearly, if (UC)[5] is true, (NL) must be true; and so the quantum formalism, whose predictions give $2 < [D(\underline{a},\underline{b})]_{QF}$, would thereby be proved to be non-local. Thus, experiments like Aspect's (6,7) would, in principle, be *crucial* for deciding between locality and non-locality.

So, (UC) is crucially important. To prove (UC) true, the identity $D(\underline{a},\underline{b}) \equiv [D(\underline{a},\underline{b})]_{LT}$ *must be shown to be valid for all arbitrary* (8) $\rho(\lambda)$'s, $A(\underline{a},\lambda)$'s, $B(\underline{b},\lambda)$'s and *all* their combinations as covered according to (UC).[6]

I propose here to defend two theses. The *first* is that (UC) has nowhere been proved. My *second* thesis, which actually contains the first, is that the universality claim (UC) *can be disproved: there exists an infinity of counterexamples to (UC) because (UC) is incompatible with the conservation law of angular momentum*.

The existence is admitted here of a class of local models that satisfy the Bell premises (as will be shown in Sec. 4), and therefore $D(\underline{a},\underline{b}) \leq 2$. But my criticism of the identity $D(\underline{a},\underline{b}) \equiv [D(\underline{a},\underline{b})]_{LT}$ is the following. *First, no model of this class*, although it establishes the consistency of the Bell premises, *can be generalized consistently with the conservation law of angular momentum for all* $\rho(\lambda)$*'s*. So, $D(\underline{a},\underline{b}) \leq 2$ is valid only for a class of local models of *measure zero*, and therefore $D(\underline{a},\underline{b}) \not\equiv [D(\underline{a},\underline{b})]_{LT}$. This is why there exists an infinity of counterexamples to (UC). This will be shown in Sec. 5.

Second, if we were to *assume*, mistakenly, that $D(\underline{a},\underline{b}) \equiv [D(\underline{a},\underline{b})]_{LT}$ is true, then we are authorized to choose *all* those local theories that *can* be generalized *consistently* with the conservation law of angular momentum for *all* $\rho(\lambda)$'s; these local theories *must necessarily be covered* according to (UC). Then, as will be shown in Sec. 5, my criticism is that *almost all* (that is, except for a class of measure zero) the functions $A(\underline{a},\lambda)$, $B(\underline{b},\lambda)$ of this class must satisfy the *constraint* that $A(\underline{a},\lambda) \equiv A(\lambda)$ and $B(\underline{b},\lambda) \equiv B(\lambda)$. As a consequence of this absurd constraint[7], the expectation value $E(\underline{a},\underline{b}) \equiv E$ is a constant independent of (a,b); therefore, different inequalities must hold, namely

$$D \leq 2, \tag{2b}$$

where $D(\underline{a},\underline{b}) \equiv D \equiv 2|E|$ is also a constant independent of $(\underline{a},\underline{b})$, is the inequality that must replace (2). Since, by definition, $|A(\underline{a},\lambda)| \leq 1$ and $|B(\underline{b},\lambda)| \leq 1$, inequality (2b) is a *tautology*, because $2|E| \leq 2$ is always satisfied. And

$$2 < [D]_{QF} \tag{3a}$$

where $[D]_{QF} \equiv 2|[E(\underline{a},\underline{b})]_{QF}| \equiv 2|C\underline{a}.\underline{b}|$, is the inequality that must replace (3). But (3a) is *not* satisfied for any unit vectors $(\underline{a},\underline{b})$:

even for a nearly ideal experiment ($|C|\simeq 1$) and $\underline{a}$ and $\underline{b}$ parallel, $2 < 2|1| = 2$ is *not* greater than 2 (no number is greater than itself). So, $[D]_{QF} < 2$ *contrary* to the almost universal belief in (NL).

In other words, my argument here is an *ad-absurdum* disproof of (UC), and in deriving my absurd formula (5) I show that my assumption (A2) of Sec. 2 specifies a class of counterexamples that refute (UC) (That I have done so, and not only shown the existence of counterexamples, was pointed out to me by Popper (9)). However, it should be pointed out again that all this does *not* show the inconsistency of the Bell premises (which I will show to be consistent in Sec. 4) but only the inconsistency of the Bell premises with (UC).

Incidentally, the universality claim made for the Clauser-Horne (10,11) inequality is also invalid. (12) It seems that this crucial point has been overlooked in the debate whether or not (NL) is true.

2. CAN THE UNIVERSALITY CLAIM (UC) BE DISPROVED ?

Yes, even a weaker universality claim than (UC) can be disproved. Of course, the *non*-validity of a weaker universality claim than (UC) *logically entails* the *non*-validity of the stronger universality claim (UC). In this section I shall formulate a weaker universality claim and then, in Sec. 5, disprove it.

Since the identity $D(\underline{a},\underline{b}) \equiv [D(\underline{a},\underline{b})]_{LT}$ must be shown to be valid for *all* arbitrary $\rho(\lambda)$'s, $A(\underline{a},\lambda)$'s, $B(\underline{b},\lambda)$'s and *all* their combinations, clearly $D(\underline{a},\underline{b}) \equiv [D(\underline{a},\underline{b})]_{LT}$ must also be shown to be valid for the following two *weaker assumptions:*

(A1) the conservation law of angular momentum is valid for *all* $\rho(\lambda)$'s,

(A2) the triples (ρ,A,B) belong to $L^2(\Lambda)$, the real Hilbert space of square-integrable functions defined almost everywhere on the domain Λ.

Of course, $L^2(\Lambda)$ covers continuous and non-continuous or generalized functions, (13) and it is similar to the Hilbert space of functions of the quantum formalism. 8,9

3. THE BELL PREMISES

Except (B3), the Bell (5) paper gives the following premises claimed to be sufficient for deriving $D(\underline{a},\underline{b}) \leq 2$; (B3) expresses the conservation of the z-component of the total angular momentum. Bell (4) considers only conservation of spin in every direction expressed by $A(\underline{a},\lambda) = -B(\underline{a},\lambda)$. But, as will be argued in Sec. 4, (B3) is more fundamental than $A(\underline{a},\lambda) = -B(\underline{a},\lambda)$. The *consistency* of

(B3) with (B1) and (B2) is established in Sec. 4.

(B1) Bell defines Einsteinian locality by the requirement that the functions $A(\underline{a},\lambda)$ do *not* depend on $\underline{b}$ *nor* $B(\underline{b},\lambda)$ on $\underline{a}$.

(B2) Bell assumes that *all* $\rho(\lambda)$'s are *not* conditional on $(\underline{a},\underline{b})$. That is $\rho(\lambda|\underline{a},\underline{b})= \rho(\lambda)$ for *all* $\rho(\lambda)$'s.

(B3) The conservation of the z-component of the total angular momentum can be formulated by the condition that the expectation values $\int_\Lambda A(\underline{a},\lambda)\rho(\lambda)d\lambda \equiv A(\underline{a}) \equiv A=$ constant, and $\int_\Lambda B(\underline{b},\lambda)\rho(\lambda)d\lambda \equiv B(\underline{b}) \equiv B=$ constant, must be constants independent of $(\underline{a},\underline{b})$.

4. CONSERVATION OF ANGULAR MOMENTUM

The conservation of the z-component of the total angular momentum assumes the following form. Take, say $A(\underline{a}) \equiv A=$ constant for *all* $\underline{a}$'s. Then, for every pair $(\underline{a}_1,\underline{a}_2)$, $A(\underline{a}_1)= A(\underline{a}_2)= A=$ constant, and therefore the *difference* $A(\underline{a}_1)- A(\underline{a}_2)= 0$ or

$$\int_\Lambda G(\underline{a}_1,\underline{a}_2,\lambda)\rho(\lambda)d\lambda= 0 \qquad (4)$$

where $G(\underline{a}_1,\underline{a}_2,\lambda) \equiv A(\underline{a}_1,\lambda)- A(\underline{a}_2,\lambda)$, with a similar formula holding for parameter $\underline{b}$. 10

The *consistency* of formula (4) with (B1) and (B2) can be established by giving a (formal) model that satisfies (4) and $\int_\Lambda \rho(\lambda)d\lambda= 1$. Take

$$\rho(\lambda)= n\exp(-n\lambda)$$

and

$$G(\underline{a}_1,\underline{a}_2,\lambda) \equiv H(\underline{a}_1,\underline{a}_2)[1-(n+m)\exp(-m\lambda)/n] \neq 0,$$

where $H(\underline{a}_1,\underline{a}_2)$ is a function of $(\underline{a}_1,\underline{a}_2)$ and (n,m) are arbitrary constants such that $|G(\underline{a}_1,\underline{a}_2,\lambda)| \leq 2$. Let Λ be the interval $[0,+\infty)$. Then,

$$\int_\Lambda \rho(\lambda)d\lambda= \int_0^{+\infty} n\exp(-n\lambda)d\lambda= -[\exp(-n\lambda)]_0^{+\infty}= 1$$

and

$$\int_\Lambda G(\underline{a}_1,\underline{a}_2,\lambda)\rho(\lambda)d\lambda= \int_0^{+\infty} H(\underline{a}_1,\underline{a}_2)[1-(n+m)\exp(-m\lambda)/n]n\exp(-n\lambda)d\lambda$$

$$= H(\underline{a}_1,\underline{a}_2)- H(\underline{a}_1,\underline{a}_2)= 0$$

This model will be used in Sec. 5. 11

Bell (14) also established the consistency of formula (4) with (B1) and (B2) by showing that (4) and $\int_\Lambda \rho(\lambda)d\lambda= 1$ are satisfied by taking $\rho(\underline{\lambda})= [\int_\Lambda d\underline{\lambda}]^{-1}=$ constant and $A(\underline{a},\underline{\lambda})= \mathrm{sign}(\underline{a}.\underline{\lambda})$ (this model has already been shown by Bell (4) to satisfy (B1) and (B2)), and therefore $G(\underline{a}_1,\underline{a}_2,\underline{\lambda}) \equiv \mathrm{sign}(\underline{a}_1.\underline{\lambda})- \mathrm{sign}(\underline{a}_2.\underline{\lambda}) \neq 0$. Then, $\int_\Lambda G(\underline{a}_1,\underline{a}_2,\underline{\lambda})\rho(\underline{\lambda})d\underline{\lambda}= 0$ because $\int_\Lambda \mathrm{sign}(\underline{a}_1.\underline{\lambda})d\underline{\lambda}= \int_\Lambda \mathrm{sign}(\underline{a}_2.\underline{\lambda})$

$d\underline{\lambda}= 0$.

Bell (4) considers only conservation of spin in every direction $\underline{a}$ defined by $A(\underline{a},\lambda)= -B(\underline{a},\lambda)$. But the following arguments show that the conservation of the total angular momentum is more fundamental than $A(\underline{a},\lambda)= -B(\underline{a},\lambda)$. It should also be noted that $A(\underline{a},\lambda)= -B(\underline{a},\lambda)$ implies $E(\underline{a},\underline{b})= -1$, a consequence criticised by Clauser *et al*. (15) and acknowledged by Bell. (5)

Wigner (16) shows that the conservation of the z-component of the total angular momentum imposes a limitation, in principle, on the measurability of the spin x-component of a particle. So, Bell's perfect correlation $A(\underline{a},\lambda)= -B(\underline{a},\lambda)$ between spin measurements in *every* direction $\underline{a}$ does not obtain even in principle if $\underline{a}= \underline{x}$.

Berestetskii *et al*. (17) argue that in relativistic quantum theory the orbital angular momentum $\underline{l}$ and the spin $\underline{s}$ of particles are *not* separately conserved. Only the total angular momentum $\underline{j}= \underline{l}+\underline{s}$ is conserved. The component of the spin in any fixed direction is therefore *not* conserved. Only the component of the spin $\underline{s}.\underline{n}$ in the direction $\underline{n}$ $(= \underline{k}/|\underline{k}|)$ of the particle's momentum $\underline{k}$ *is conserved*. That is, since $\underline{l}= \underline{r} \times \underline{k}$, only then $\underline{s}.\underline{n}$ is equal to the conserved quantity $\underline{j}.\underline{n}$ known as the *helicity* of the particle.

In addition, if a particle has zero rest mass (such as a photon, antineutrino, etc.), then there is always a *distinctive* direction of space, the direction of the particle's momentum $\underline{k}$, because for such a particle there is *no* rest frame. In such a case there is clearly *no* symmetry along every direction of space, but only axial symmetry about the direction $\underline{k}$, and therefore only the helicity of such a particle is conserved.

Thus, Bell's perfect correlation $A(\underline{a},\lambda)= -B(\underline{a},\lambda)$ between spin measurements in *every* direction $\underline{a}$ and for *all* sources is clearly less fundamental than formula (4), if not objectionable (for $\underline{a}\neq \underline{k}$) in the context of relativistic quantum theory. And Bell (5) states that his theorem "is in no way restricted to the context of nonrelativistic wave mechanics."

Two-body decay experiments support the above arguments from relativistic quantum theory. Take the decay $\pi^- \rightarrow \mu^- + \bar{\nu}$ in the rest frame of the spin-zero π^- meson into two spin-$\frac{1}{2}$ particles, a μ^- meson and a $\bar{\nu}$ antineutrino. Consider the component of the angular momentum along the direction $\underline{k}_\mu$ of the momentum of μ^- (this is the two-body decay direction and may be taken as the z-axis). The spin component of μ^- along $\underline{k}_\mu$ is completely determined, *irrespective of μ^-'s total angular momentum*, by the helicity of $\bar{\nu}$ along $\underline{k}_\mu$, and this has been well supported by experiments. (18) The reason is that the zero rest mass $\bar{\nu}$ has *no* spin component orthogonal to $\underline{k}_\mu$ (the spin of the antineutrino is *always* parallel to its momentum), but the accompanying μ^- may have a spin component orthogonal to $\underline{k}_\mu$. In other words, if $\underline{k}= \underline{z}$, only $\underline{j}^2$ and the z-component of the total angular momentum are conserved in such two body decay experiments.

5. THE DISPROOF OF THE UNIVERSALITY CLAIM (UC)

In this section, (UC) is disproved by disproving $D(\underline{a},\underline{b}) \equiv [D(\underline{a},\underline{b})]_{LT}$ for the weaker assumptions (A1) and (A2). 9

The disproof requires the theorems (T1) and (T2) which are proved in the Appendix.

The assumptions required for the proofs of these theorems are all satisfied. This can be shown as follows. First, the assumption that (ρ,A,B) belong to $L^2(\Lambda)$; 12 for fixed $(\underline{a},\underline{b})$ there is always in $L^2(\Lambda)$ an equivalent function to G which depends only on λ. Second, the inner product $(y|z)= 0$ in theorem (T2) is identical with $\int_\Lambda G(\underline{a}_1,\underline{a}_2,\lambda)\rho(\lambda)d\lambda= 0= (G|\rho)$; for any pair of fixed $(\underline{a}_1,\underline{a}_2)$, set $G(\underline{a}_1,\underline{a}_2,\lambda) \equiv y(\lambda)$ and, for every $\rho(\lambda) \geqq 0$, set $\rho(\lambda)= z(\lambda)$. Then, according to theorem (T2), 13 $G(\underline{a}_1,\underline{a}_2,\lambda) \equiv A(\underline{a}_1,\lambda)- A(\underline{a}_2,\lambda) = 0$ is the *constraint* imposed on the functions $A(\underline{a},\lambda)$. But $A(\underline{a}_1,\lambda)= A(\underline{a}_2,\lambda)$ implies that the functions $A(\underline{a},\lambda)$ must be independent of $\underline{a}$, and similarly $B(\underline{b},\lambda)$ of $\underline{b}$. So, if $D(\underline{a},\underline{b}) \equiv [D(\underline{a},\underline{b})]_{LT}$ *is assumed* true for the weaker assumptions (A1) and (A2), then

$$A(\underline{a},\lambda) \equiv A(\lambda), \quad B(\underline{b},\lambda) \equiv B(\lambda). \tag{5}$$

The *immediate consequences* of the constraint (5) are that: (C1) Any local theory or model, whose functions A, B depend on $\underline{a}$, $\underline{b}$ respectively, is *excluded* in the sense that *no* such model, although it establishes the consistency of the Bell premises (as shown in Sec. 4), *can be generalized consistently with the conservation law of angular momentum for all* $\rho(\lambda)$*'s*. So, $D(\underline{a},\underline{b}) \leqq 2$ is valid only for a class of local models of *measure zero*, and therefore $D(\underline{a},\underline{b}) \not\equiv [D(\underline{a},\underline{b})]_{LT}$.

To illustrate (C1), take my formal model (of Sec. 4) whose functions A, B depend on $\underline{a}$, $\underline{b}$ respectively and which established the consistency of the Bell premises with (4). From the set of *all* $\rho(\lambda)$'s, choose another $\rho(\lambda)$, say $\rho(\lambda)= 1/2\pi$, and let Λ be the interval $[0,2\pi]$ so that $\int_0^{2\pi} d\lambda/2\pi= 1$. Then,

$$\int_\Lambda G(\underline{a}_1,\underline{a}_2,\lambda)\rho(\lambda)d\lambda= H(\underline{a}_1,\underline{a}_2)[1-(n+m)(1-\exp(-2\pi m)/2\pi nm] \neq 0$$

for some (n,m) *contrary to the required* $\int_\Lambda G(\underline{a}_1,\underline{a}_2,\lambda)\rho(\lambda)d\lambda= 0$. So, this model cannot be generalized consistently with the conservation law of angular momentum for *all* $\rho(\lambda)$'s because, together with $\rho(\lambda)= 1/2\pi$, it does not satisfy this conservation law.

(C2) Instead of (2) a *different inequality* (2b) must hold because $D(\underline{a},\underline{b}) \equiv D \equiv 2|E|$, where the expectation value E is a constant independent of $(\underline{a},\underline{b})$. Then, $2< [D]_{QF} \equiv 2|[E(\underline{a},\underline{b})]_{QF}| \equiv 2|C\underline{a}.\underline{b}|$ is the inequality (3a) that the predictions of the quantum formalism, namely $[E(\underline{a},\underline{b})]_{QF}= C\underline{a}.\underline{b}$, must satisfy. But, as discussed in the introduction, inequality (3a) is *not* satisfied for any pair of unit vectors $(\underline{a},\underline{b})$. So, $[D]_{QF} \leqq 2$, and therefore *neither (UC) nor (NL) is true.*

(C3) The Bell definition of locality, namely that the functions $A(\underline{a},\lambda)$ and $B(\underline{b},\lambda)$ do not depend on $\underline{a}$ and $\underline{b}$, respectively *entails* the formulae $A(\underline{a},\lambda)\equiv A(\lambda)$ and $B(\underline{b},\lambda)\equiv B(\lambda)$. But these formulae *make no physical sense* because, given an initial state λ, the results A and B of measurements performed on the components s_1 and s_2 of a two-component system (s_1+s_2) must depend on the directions $(\underline{a},\underline{b})$ of the polarizers.

So (C1), (C2) and (C3) are some of the consequences of the absurd formula (5) which has been derived from formula (4) with the help of Theorem (T2). But what has been added to formula (4), that is, quite simply, (UC) which, if it is assumed true, authorises the use of Theorem (T2): my argument here is an *ad-absurdum* disproof of (UC).

In other words, the Bell theory, which includes (UC) among its premises, is *inconsistent*. Also the Bell definition of Einsteinian locality *is too strong*, in the sense that *almost all* the local theories that satisfy the definition *clash with the conservation law of angular momentum*. Einsteinian locality remains essentially a *weaker* notion than the Bell definition of it.

6. CONCLUSIONS

Only if one could prove the universality claim (UC) true, only then would the quantum formalism be proved to be non-local. But (UC) has been disproved.

Again, only if (UC) was proved true, only then would the experiments (6,11) culminating in the recent Aspect *et al.* (7) experiment with the switches at all be able to refute Einsteinian locality, because all other arguments in favour of the belief in (NL), such as the instantaneous collapse of the wave packet etc., are merely part of some (unproved) interpretation of the quantum formalism and *not* really part of the quantum formalism itself.

In consequence of these findings and the absence of other proofs, the present author believes that the quantum formalism could be *interpreted* as a local theory. That is, Einsteinian locality could be compatible with the quantum formalism *although not compatible with its Copenhagen interpretation.* The Einsteinian alternative remains intact and, as Einstein (1) put it, there is nothing (except the disproved universality claim (UC)) that theoretically speaks for action at a distance or non-locality. Thus, for Einstein, a crucial reason for accepting locality would be that the burden of proof still rests on the shoulders of those who advocate non-locality.

ACKNOWLEDGEMENTS

I am most grateful to Professor Sir Karl Popper for his invaluable help and warm encouragement, to Professor Clive W. Kilmister for his help with the Appendix, and to Professor Franco Selleri for his kind invitation to speak at the Bari conference.

APPENDIX

Here two theorems are formulated and proved. $L^2(c,d)$ denotes the real Hilbert space of square-integrable functions defined almost everywhere on $c \leqq t \leqq d$. The inner product $(y|x) = \int_c^d y(t)x(t)dt$ defined for every pair of functions of $L^2(c,d)$, is strictly positive, (19) that is $(y|y) = 0$ implies that $y(t) = 0$ in $[c,d]$.

Theorem (T1). If $y(t) \varepsilon L^2(c,d)$ and if $(y|x) = 0$ for all $x(t) \varepsilon L^2(c,d)$, then $y(t) = 0$ in $[c,d]$.

Proof: It is sufficient to take $x(t) = y(t)$. Then, the strictly positive inner product $(y|y) = 0$ implies that $y(t) = 0$ in $[c,d]$.

Theorem (T2). If $y(t) \varepsilon L^2(c,d)$ and if $(y|z) = 0$ for all $z(t) \geqq 0$ $\varepsilon L^2(c,d)$, then $y(t) = 0$ in $[c,d]$.

Proof: For every $x(t) \varepsilon L^2(c,d)$, there is always a pair (z_1, z_2) of non-negative functions in $L^2(c,d)$, defined as: (1) $z_1(t) = x(t)$ for $x(t) \geqq 0$ and $z_1(t) = 0$ for $x(t) < 0$, and (2) $z_2(t) = -x(t)$ for $x(t) \leqq 0$ and $z_2(t) = 0$ for $x(t) > 0$. Therefore, for every $x(t)$, $(y|x) = (y|z_1 - z_2) = (y|z_1) - (y|z_2) = 0 - 0 = 0$, since, by assumption, each of the inner products must be zero. Then, by theorem (T1), $y(t) = 0$ in $[c,d]$.

NOTES

1. Einstein never claimed that the quantum formalism was non-local, but he did claim that the Copenhagen interpretation was non-local. Thus, Einstein clearly *did not identify* the quantum formalism with its Copenhagen interpretation. So, Einstein drew this distinction in order to use his locality principle against the Copenhagen interpretation and *not against the quantum formalism*.
2. The variable λ denotes the emission state of a two-component system (s_1+s_2). The distribution function of the λ's is $\rho(\lambda)$. The functions $A(\underline{a},\lambda)$ and $B(\underline{b},\lambda)$ denote respectively results of measurements performed on s_1 and s_2 and $|A(\underline{a},\lambda)| \leqq 1$ and $|B(\underline{b},\lambda)| \leqq 1$. The parameters $(\underline{a},\underline{b})$ are taken to be unit vectors denoting polarizer orientations.

3. $D(\underline{a},\underline{b}) \leq 2$ requires the normalization condition $\int_\Lambda \rho(\lambda)d\lambda = 1$ for its derivation. So, $D(\underline{a},\underline{b}) \leq 2$ can be made empirically testable only if the total number N of source emissions is counted. But in practice "event ready" detectors depolarize particles, and so the number N of *undisturbed* particles is impossible to count. Only the Clauser-Horne (10) inequality $S(\phi) \leq 1$ is claimed to be empirically testable.
4. Bell raises the universality claim in the form of a negation of an existential statement, e.g., "There is no perpetual motion machine." But the negation of an existential statement is always equivalent to a universal statement, and *vice versa*. For a further explanation of this point see K.R. Popper, *The Logic of Scientific Discovery* (Hutchinson, London, 1972), pp. 68-70.
5. My formulation of (UC) is *weaker* than the Bell universality claim that *"no local... theory can reproduce all the... predictions of quantum mechanics"* because there exists so far *no proof* that the functions (ρ,A,B) are *all* the conceivable functions that satisfy the fairly wide conception of Einsteinian locality and *fully cover* the Bell claim.
6. Clearly, by the word *"... arbitrary..."*, Bell (8) does not mean that the functions (ρ,A,B) do not satisfy any constraints: for they must obviously satisfy the premises of Bell's theorem. Bell's word refers to the fact that he has not anywhere specified any particular class of functions to which the triples (ρ,A,B) have to belong.
7. The (absurd) constraint $A(\underline{a},\lambda) \equiv A(\lambda)$ and $B(\underline{b},\lambda) \equiv B(\lambda)$ makes *no physical sense* because, given an initial state λ, the results A and B of measurements performed on the components s_1 and s_2 of a two-component system (s_1+s_2) must depend on the directions $(\underline{a},\underline{b})$ of the polarizers. This constraint can only be obtained from contradictory premises: the inconsistency of the Bell premises with (UC).
8. Assumptions (A1) and (A2) provide a *weaker* universality claim than (UC) because the class of arbitrary functions (8) covers both classes of continuous and non-continuous functions.
9. The *weaker* universality claim based on assumptions (A1) and (A2) covers *all* those local theories $(\rho,A,B) \varepsilon L^2(\Lambda)$ that *can be generalized consistently with the conservation law of angular momentum for all* $\rho(\lambda)$*'s*.
10. In case the name given here to my formula (4) might seem inappropriate, I am prepared to drop it and adopt any other more appropriate name. What is important, however, is the fact that if (UC) is *assumed* to be true, then we are authorized to choose *all* those local theories that *can be generalized consistently with formula (4) for all* $\rho(\lambda)$*'s* .
11. This model, with $\rho(\lambda) = n\exp(-n\lambda)$, clearly shows that *it is not necessary to assume that* $\rho(\lambda)$ is uniform or isotropic for formula (4) to hold; even if such an assumption were necessary, it would simply be covered by (UC). So, no such "additional"

assumption is made anywhere in my argument.

12. The authority to choose $L^2(\Lambda)$, that is to stipulate (A2), is not based on anything "arbitrary", but based on three facts: first, the fact that (UC) entails my (A2) provided (A2) is not excluded by the Bell premises. Secondly, on the fact that my (A2) is not excluded by the Bell premises: Bell has *not* anywhere specified (8), 6 any particular class to which the triples (ρ,A,B) have to belong. Thirdly, it would be physically unreasonable to adopt *ad-hoc* an assumption that excludes (A2).
13. If (UC) is true, then we are authorized to choose *all* those local theories that satisfy formula (4) for *all* $\rho(\lambda)$'s, and therefore we are authorized to invoke Theorem (T2) since the assumptions required for the proof of this theorem are all satisfied. 9

REFERENCES

1. A. Einstein, *Dialectica* 2, 320 (1948).
2. M. Jammer, *The Philosophy of Quantum Mechanics* (John Wiley, New York, 1974), pp. 115-119, 307.
3. K.R. Popper, *Quantum Theory and the Schism in Physics* (Hutchinson, London, 1982 and Rowan & Littlefield, Totowa, New Jersey, 1982), pp. 20-22.
4. J.S. Bell, *Physics (N.Y.)* 1, 195 (1964).
5. J.S. Bell, in *Foundations of Quantum Mechanics*, B. d'Espagnat, ed., (Academic Press, New York, 1971), p. 178.
6. A. Aspect, P. Grangier, and G. Roger, *Phys. Rev. Lett.* 47, 460 (1981); *Phys. Rev. Lett.* 49, 91 (1982).
7. A. Aspect, J. Dalibard, and G. Roger, *Phys. Rev. Lett.* 49, 1804 (1982).
8. J.S. Bell, private communication (23 June, 1982); Bell's own explicit universality claim, namely that "Bell's theorem *is* valid for arbitrary $\rho(\lambda)$," clearly shows that Bell has *not* specified any particular class to which the functions (ρ,A,B) have to belong. In other words, the class to which (ρ,A,B) belong is assumed by Bell to be *arbitrary*.
9. K.R. Popper, private communication.
10. J.F. Clauser and M.A. Horne, *Phys. Rev. D* 10, 526 (1974).
11. J.F. Clauser and A. Shimony, *Rep. Prog. Phys.* 41, 1881 (1978).
12. Th. D. Angelidis, *Phys. Rev. Lett.* 51, 1819 (1983).
13. R.D. Richtmyer, *Principles of Advanced Mathematical Physics* (Springer-Verlag, New York, 1978), pp. 4-8, 21.
14. J.S. Bell, private communication (4 June, 1982).
15. J.F. Clauser, M.A. Horne, A. Shimony, and R.A. Holt, *Phys. Rev. Lett.* 23, 880 (1969).
16. E.P. Wigner, *Z. Physik* 133, 101 (1952).

17. V.B. Berestetskii, E.M. Lifshitz, and L.P. Pitaevskii, *Relativistic Quantum Theory*, Part 1 (Pergamon Press, Oxford, 1971), pp. 22, 47.
18. G. Backenstoss, B.D. Hyams, G. Knop, P.C. Marin, and U. Stierlin, *Phys. Rev. Lett.* 6, 415 (1961); M. Bardon, P. Franzini, and J. Lee, *Phys. Rev. Lett.* 7, 23, (1961).
19. Y. Choquet, C. Dewitt, and M. Dillard, *Analysis, Manifolds and Physics* (North-Holland, Amsterdam, 1977), p. 11.

IS IT POSSIBLE TO SAVE CAUSALITY AND LOCALITY IN QUANTUM MECHANICS?

Eftichios Bitsakis

Department of Philosophy, University of Ioannina
Department of Physics, University of Athens, Greece

ABSTRACT

The physical meaning of EPR experiment is first analysed. By generalizing the EPR criterion of reality to include potential and real states in quantum mechanics, the author investigates the possibility to save causality and locality in microphysics.

1. INTRODUCTION

The argumentation of Einstein, Podolsky, and Rosen (EPR) that quantum mechanics is not a complete theory, is based on two assumptions: (1) the assumption of the existence of an objective reality, independent of any observation (realistic principle); (2) the assumption concerning the local character of physical interactions, the finite velocity of which implies the actual separability of two physical systems separated by a space-like interval.

According to Einstein, the realistic principle can be extrapolated to the microscopic level. For him, "there is something such as a 'real state' of a physical system, existing objectively, independently of any observation".(1) So a microphysical system is characterized by a number of elements of reality, accessible to measurement: "If, without in any way disturbing a system, we can predict with certainty (i.e., with probability equal to unity) the value of a physical quantity, then there exists an element of physical reality corresponding to this physical quantity".(2) (We shall discuss in Sec. 3 the meaning of this statement of EPR.)

The principle of locality, on the other hand, and its implication, physical separability, lead to the conclusion that a measurement on a quantum system A cannot affect a system B if the two systems are separated by a space-like interval. So an element of reality of B cannot be created by a measurement on A. The local

G. Tarozzi and A. van der Merwe (eds.), Open Questions in Quantum Physics, 63–73.

character of physical interactions implies also that an element of reality of A cannot be generated by any future experiment (rejection of retroactive causality).

In their well-known mental experiment, EPR concluded that the existing quantum mechanical description of physical reality, given by the wave function, is not complete. The starting point of their demonstration was locality (and separability). By contrast, Niels Bohr's assertion that quantum mechanical description is complete, was based on the principle of non-separability of the two physical systems A and B and, more generally, of the physical system, the instrument, and the observer. This principle, however, which is the starting point of the whole Copenhagen interpretation (CI) presupposes the existence of interactions with infinite velocity and consequently contradicts the principle of relativity. Bohr and the Copenhagen school have never assumed a clear position on this crucial point.

2. THE PHYSICAL MEANING OF THE EPR EXPERIMENT

Let us discuss now the EPR experiment, as it was illustrated by Bohm and Aharonov.(3)

Consider a system of total spin zero, consisting of two particles, each of spin one-half. The state vector of the system, according to Bohm-Aharonov, is

$$|\Psi\rangle = \frac{1}{\sqrt{2}}\left\{|u^+\rangle\ |v^-\rangle - |u^-\rangle\ |v^+\rangle\right\} \qquad (1)$$

The two particles are then separated by a method that does not affect the total spin. When the particles are sufficiently far apart, so that they cease to interact, we measure, say, the 3-component of the spin of A. If we find the value $+\frac{1}{2}$, we can predict with certainty that the corresponding spin component of B has the value $-\frac{1}{2}$, and vice versa.

Had the system been a classical one, there would have been no difficulty, because all components of the spin of each particle should have a well-defined value at each instant of time. But quantum mechanichs cannot explain the above correlation of the spins of A and B. It is true that if we accept the principle of non-separability (Bohr), the difficulty vanishes. In that case, however, we must postulate the existence of interactions with infinite velocity, which violate the principle of relativity and are outside the scope of today's physics. So, the search for a local and causal explanation seemed more natural to many physicists and philosophers.

We can now put the question: What is the physical meaning of (1)? For Bohm and Aharonov, Eq.(1) represents the state of the composite system before the separation of the two particles. But during that time the two particles interact, so that their state vectors cannot be factorized. The factorized state vectors

$$|u^+> \ |v^-> \quad \text{and} \quad |u^- > |u^+>$$

of (1) denote particles that do not interact and, consequently, are separated. So, (1) cannot represent the composite system (A+B). We can also suppose that (1) represent the spin state after the particles are separated. (In fact, Bohm and Aharonov formulated this point of view in another paper.(4)) But this assumption is wrong, because the final statistical ensemble is a mixture, while (1) represents a pure state. The mixture

$$|u^+>|v^-> \quad , \quad |u^->|v^+>$$

is not physically equivalent with (1), as it concerns the correlations of measurements of the $\vec{OX}$ components.(5)

We can, however, consider (1) as a pure state (a superposition of states), but not in the formal spirit of the orthodox school. We can treat it as a measure of the potentialities of the statistical ensemble under given experimental conditions.

Before the separation, the composite system constitutes a pure state. So, the elements of reality of A and B are correlated. But the state variables of the composite system are in a random fluctuation, and the same assertion is valid for the microscopic part of the separating apparatus. This "fine structure" of the states of the composite system and the microscopic part of the apparatus can, in principle, explain the realisation of two different potentialities,

$$|u^+>|v^-> \quad \text{and} \quad |u^->|v^+>$$

for a large number of individual measurements. In that case, the Hilbert space to which (1) belongs is not the space of real states but of potential states of the initial pure ensemble. It is a potential space, and the statistical ensemble can be treated as a potential coherent mixture. So the state vector is not a superposition of states, but the measure of the potentialities of the statistical ensemble under the given external conditions.(6) If, after the separation of the particles we measure the value of the spin of one of them, we are able to make a prediction about the value of the corresponding component of the other. The common past of the two particles can in principle explain this correlation.

The preceding interpretation is a deterministic and a local one. In its frame there is no EPR paradox but lack of physical knowledge. In fact, the possibility to predict the state of B from a knowledge of the state of A does not presuppose any interaction between A and B. So, the fact that A realises the state $|u^+>$ does not mean that B realises the state $|v^->$ and vice versa. The certainty of this prediction means that B has the potentiality to realise that state only if we make a measurement on it. Thus the results of measurement are correlated, but the measurements themselves are not. The correlations were established during the time of interaction of the two particles. As J.S. Bell puts it, "Correlations between physical events in different space-time regions should be explicable in terms of physical events in the overlap of the backward light cones."(7)

In the frame of the preceding interpretation, we can also understand the physical difference between

$$|\Psi> = \frac{1}{\sqrt{2}}\{|u^+> \; |v^-> - |u^-> \; |v^+>\} \text{and } |u^+> |v^- >, \; |u^-> |v^+>$$

The first expression is a measure of the potentialities of a pure state (of a potential coherent mixture). The two state vectors, on the other hand, represent a mixture. The pure state was transformed into a mixture by the action of the separating apparatus. The elements of reality of the separated systems A and B are correlated and can realise different, but mutually correlated, states of spin.

The irreversible character of the process of separation and registration is another indication of the objective character of the transformations involved in EPR experiment.

It is well known that the state vector is invariant under rotation of the coordinate system. So, the total spin is always zero, independently of the direction of measurement. The physical meaning of this invariance is that the spin is a potentiality realised during the measurement. The correlation of the actualised values is an expression of the correlation of the elements of reality of A and B, established during the time interval $\Delta\tau$, necessary for the separation of the constituents of the global system. The process of separation is non-linear and is not described by the actual theory. Thus the only expression that we can write a posteriori, on the basis of the experimental data, is (1). The EPR paradox emphasises the impossibility of the actual quantum mechanics to describe the "reduction" of the wave packet, that is the creation of new elements of reality during the measurement.(8) Accordingly, this famous problem, which constitutes the crucial test for the rejection of the CI, constitutes also the crucial point for the understanding of EPR experiment.

3. POTENTIAL AND REAL ELEMENTS IN QUANTUM MECHANICS

In their paper, EPR established the following criterion of reality: "If, without in any way disturbing a system, we can predict with certainty (i.e., with probability equal to unity) the value of a physical quantity, then there exists an element of physical reality corresponding to this physical quantity."(9)

The validity of the above criterion is obvious for classical systems, because for that class of systems it is accepted that the measurement does not generate new elements of reality. Moreover, although in Newtonian mechanics the velocity of interactions is taken as infinite, Newtonian physics accepts tacitly the separability of classical systems. But in quantum mechanics the measurement can generate new elements of reality. We must therefore distinguish between the classical and quantal criterion of reality.

In fact, what is the meaning of the expression: "without in any way disturbing"? We can suppose that the perturbation caused by the apparatus is "very small." This expression, however, is vague. Franco Selleri affirms that the EPR reality criterion applies in fact only to those very particular situations in which the outcome of an act of measurement is predictable with certainty. (10) However, a probability of prediction equal to unity is a necessary, but not always sufficient, condition for the application of the EPR criterion in its classical form.

In quantum mechanics, we must distinguish between actual and potential properties (or elements of reality). Thus we must distinguish between the cases for which the element of reality is actual before the measurement (actual state) and the cases for which the element is realised during the measurement.

In the case of an unidimensional Hilbert space, the probability of prediction is equal to unity. This "sharp state" can exist before the measurement. In that case the instrument does not disturb the system. However, there are instances where the only possible state is not actual and is realised during the measurement. In that case, the interaction of (S+A) creates new elements of reality. Also, in the case of a "superposition" of states, the elements of reality are potential elements and are realised during the measurement. We have in that case many potentialities, realised according a probabilistic distribution. The so-called "reduction of the wave packet" is an irreversible process which generates new elements of reality.

If the property $a \in L$ is real at the end of a measurement, the answer is "yes". The measurement is ideal if the property was real before the measurement. If the property $a \in L$ is real after, but not before the measurement, we can speak of a <u>measurement of the first kind</u>. In that case, the potential property is actualised in an objective and irreversible manner. So, the initial state is transformed into an eigenstate of the system.

We can now proceed to a <u>first generalisation of the EPR criterion of reality</u>, to include also measurements of the first kind, that is measurements actualising potential elements of reality. The case of EPR belongs to that kind of measurement, because the elements of reality caracterising the eigenstate of the spin are actualised during the measurement. We can also make a <u>second generalisation</u> of the EPR criterion of reality, to include the case of creation of more than one state:

$\{\Psi_1, \Psi_2, \ldots, \Psi_n\}$ with corresponding probabilities $\{P_1, P_2 \ldots, P_n\}$

In this case, if we can predict with a certain probability P_i the existence of an element of reality λ_i, we can say that this element is actualised during the measurement. The Bohm-Aharonov example can be treated within the frame of this generalised criterion of reality.

In the case of the EPR experiment, the initial elements of reality are determined by the interaction of the particles during the time of their common history. These correlated elements are potential elements for the new set of elements realised during the process of separation of the two particles. In this manner, the elements realised by the separated particles during the measurement of spin are correlated.

4. LOCALITY AND THE VIOLATION OF BELL'S INEQUALITIES

Until now we have not taken deliberately into account the existence of Bell's inequalities and their violation by experiment, which is almost a practical certainty. We will try now to see if causality and locality can be saved, after taking into account the situation created by the theorem of J.S. Bell.

As it is well known, according to Bell, the validity of causality and locality in microphysics can, in certain special cases, be in contradiction with the statistical predictions of quantum mechanichs.(11) Thus the validity of locality and causality can be tested. However, the great majority of experiments confirmed the predictions of quantum mechanics. Experience seems to contradict the existence of local hidden variables. Many escapes from the impasse created by the violation of Bell's inequalities have been proposed:

1. For a number of physicists, the violation of the inequalities of Bell is a decisive argument in favour of non-separability. For them experience seems to favor the old idea of N. Bohr. But non-separability presupposes the existence of interactions with infinite velocity, and this is a fundamental handicap of this solution. There is more: The CI gives a subjectivist interpretation of measurement. The transformation of quantum systems is considered as an instantaneous and non-causal "reduction of the wave packet." This reduction is however impossible in the frame of this interpretation. One must also accept the positivist "principle" of non-existence of unobervable physical quantities and, more generally, the anti-realistic attitude of the Copenhagen School. The contradictions of the CI obliged many physicists <u>to support non-separability from a realistic point of view</u>.

2. A second solution was proposed by O. Costa de Beauregard. For him, separability and objectivity are illusions related to our pragmatic approaches. Costa de Beauregard accepts the possibility of superluminal transmission of energy (or information). In his scheme, positive energy can be propagated towards the past and it is possible to influence the future via the past. Retroactive causality is possible in that paradoxical interpretation, which violates the relativistic order of physical events. Thus there is no "ontological

priority" of the cause relative to the effect, and a direct space-like connection is possible. For Costa de Beauregard, parapsychology, telepathy, and magic are all possible. (12)

3. Bohm and Hiley proposed a solution that retains realism and causality, but rejects separability. According to these authors, the non-locality is an essential novel feature of quantum mechanics. Bohm and Hiley use the "quantum potential" to explain the quantum properties of matter. Non-separability leads to the new notion of unbroken wholeness for the entire universe. One can object that the instantaneous propagation of the quantum potential violates the principle of relativity. However, Bohm and Hiley argue that this type of interaction does not necessarily imply the transmission of energy (of a signal).(13) But in what way can an interaction, which does not transport energy, generate observable effects?

4. Another model for solving the contradiction between locality and the violation of the inequalities of Bell, was proposed by J. P. Vigier. Vigier also rejects locality in order to save realism and causality. But his scheme is not a non-local one in the strict sense of the term, because it presupposes the existence not of instantaneous, but of superluminal interactions. The starting point of Vigier is the ether of Dirac, or the subquantum vacuum which he considers as a "milieu" capable of transmitting collective movements with a superluminal velocity. The principle of relativity is not violated, according to Vigier, because these stochastic movements do not transport energy. (14)

If the preceding scheme corresponds to quantum reality, we could affirm that we have here a model of generalized causality and locality. This idea is a very interesting one, and it is in conformity with the views of Einstein, De Broglie, and Dirac. But the existence of these superluminal, non-instantaneous interactions is also contradicted by the violation of Bell's inequalities. We can also imagine a complete synchronisation of the measurements on the two particles of the EPR experiment. In that case we need instantaneous interactions in order to explain the correlations observed. According to Vigier these "ondes de phase" do not violate the principle of relativity, because they do not carry energy. But in that case one needs a concrete explanation of the mechanism by which observable phenomena are produced without energy exchange.

5. IS IT POSSIBLE TO SAVE CAUSALITY AND LOCALITY?

In accordance with the view developed in the preceding sections, there is no EPR paradox, but a lack of knowledge concerning the physical processes of correlations of the quantum systems. Anyway, according to the generally accepted interpretation, the paradox exists only for local theories. If we accept the non-separability, there is no paradox. So, non-local hidden-variable theories (of the type of D. Bohm) reproduce the statistical predictions of quantum mechanics. But if we want to reject the metaphysical postulates of

CI and the impasse of this interpretation, we must find a way out of the situation created by the violation of the inequalities of Bell.

It is well known that many authors criticised these inequalities.(15) Their arguments were however not conclusive. Other inequalities were also formulated, which contradict quantum mechanichs.(16) But the contradiction between causality and locality persists.

It is also well known that there are some experiments that violate the quantum mechanical predictions while verifying Bell's inequalities. However, the great majority of experiments violates Bell's inequalities, and the Aspect experiment is considered by the majority of the specialists as conclusive. All the same, the long distance and the high accuracy obtained constitute a quantitative improvement, but not a new decisive argument. Thus we can claim that "long distance" does not necessarily imply an exchange of information between the two particles. Even if the results of measurement are correlated, the measurements can be considered mutually independent. The fact that the two photons are emitted by the same atom at the same moment is a nearly tangible indication that the correlations observed resulted from the actualisation of potential elements established during the time of interaction of the two particles. However, the existing situation demands a radically different idea in order to retain causality plus locality in microphysics:

1. One interesting idea is formulated by Selleri, who uses the old idea of Einstein concerning the existence of "ghost waves" guiding the photon. All energy-momentum is carried by the particle, which is embedded in an objectively real undulatory phenomenon, the empty wave. The latter does not carry energy-momentum. But, although it is not a carrier of energy-momentum, the empty wave can stimulate emission. According to Selleri, Cozzini's experiment is consistent with the empty-wave idea. Also, a detailed analysis of the other Pisa experiments leads to the same conclusion. The collected evidence is strong enough to motivate us towards the search for experiments which could prove (rather than indicate) the existence of empty waves. This intuitive picture contradicts non-locality. So the confirmation of the existence of "ghost waves" could afford a mechanism for a causal and local explication of the correlation of EPR type.

2. Another idea should be the following: The dynamical form of determinism is not the only one in physics that characterizes the electromagnetic theory and the relativistic theory of gravitation. There is also the classical (or mechanical or Laplacian) form, which characterizes mechanical phenomena. There are, finally, the statistical forms (classical and quantum statistical determinism). (17) So one can ask the question: Must a local hidden-variable theory necessarily be deterministic in the classical (dynamic) sense? The theories presupposed by Bell's inequalities are deterministic in the dynamic sense. It is not however possible to exclude a priori the possibility of a probabilistic hidden-variable theory. As Selleri

notes, from the time of the discovery of Bell's inequality, several probabilistic proofs of the inequalities have been proposed, "but they all contain some features which can hardly allow one to consider the problem as satisfactorily solved." Selleri himself, from a generalized reality criterion, deduces an inequality which is violated by [illegible] mechanics, exclusively from the reality criterion and separability. This new inequality can be tested on the basis of local probabilistic models. (18)

A probabilistic theory of local hidden variables would be the most general case that accepts as a limiting case the dynamic-deterministic model. According to this point of view, the probabilistic law is the most general form of physical law. The status of probabilistic law is not only epistemological. Probabilistic law also has its ontological status, because probability does not mean lack of causality and determination, but a more complex determination of the effect.

3. Another way out of the actual impasse would be to show that the physical premises of Bell's inequalities are not consistent with the real physical processes taking place in the EPR experiment. So, L.de Broglie affirms that Bell considers two electrons which are separated and at the same time are transported by the same "wave train." However, these two hypotheses are contradictory. (19) In a subsequent paper, de Broglie, Lochak, Beswick, and Vassalo-Pareira affirmed that Bell defines a classical probabilistic scheme and calculates the mean values of the results of measurements on the basis of these classical probabilities. But this classical scheme is incompatible with quantum mechanics. (20)

It is well known that classical physics respects the inequalities of Bell, which are violated only in exceptional cases by quantum mechanical predictions. But the principle of separability is tacitly admitted by classical physics. Here two separate measurements are considered mutually independent, and it is taken as a matter of fact that measurement does not create new elements of reality. In EPR experiments, by contrast, the variables of A and B are correlated because of the common past history of the two systems. Thus one can think that the inequalities of Bell must be violated by quantum mechanical experiments if, during the derivation of these inequalities, the two sets of variables λ_i and λ_j corresponding to A and B were considered as independent.

Bell calculates the mean value P $(\hat{a}, \hat{b})$ on the basis of

$$p(\hat{a},\hat{b}) = \int d\lambda\ \rho(\lambda)A(\hat{a},\lambda)B(\hat{b},\lambda)$$

The locality condition requires that A does not depend on $\hat{b}$ nor B on $\hat{a}$. But what is the physical distribution of λ?

In his first paper, Bell defines a more complete specification of the state, by means of a parameter λ, which is treated as a single continuous parameter. Some might prefer, writes Bell, a formulation in which hidden variables fall into <u>two sets</u>, with A dependent on one and B on the other. However, this possibility is already contained in the above, according to Bell, since λ stands for any number of variables and the dependences thereon of A and B are unrestricted. (22) But how are these two abstract sets correlated? This is perhaps an interesting question, because if the formalism means that the set of variables λ_i and λ_j are separated, then the inequalities of Bell must be violated by quantum mechanics.

6. ON THE EPISTEMOLOGICAL STATUS OF SEPARABILITY AND NON-SEPARABILITY

A realistic and causal interpretation (and, eventually, formulation) of quantum mechanics must be based on two principles: realism and the validity of some form of determinism. Locality is not a necessary condition for a realistic and causal conception of quantum mechanics, because it is possible to imagine, and eventually to discover, more general forms of determination than the known relativistic-local ones. The interpretations of Bohm-Hiley and of Vigier are realistic and causal, without being local.

We can see, however, that non-locality (and non-separability) do not possess the same status as locality (and separability).

The primary concept in physical theories is interaction. Causality and determinism are derived concepts. But all known physical interactions have a finite velocity. So locality is a well-established characteristic of all known forms of interaction. It is therefore a well defined physical concept, verified by experimental evidence. The same is true for separability, which is an implication of locality.

Non-separability, on the contrary, presupposes instantaneous action at a distance. But this class of interactions is hypothetical and, moreover, it contradicts the experimentally established principle of relativity. Non-separability is therefore based on a hypothetical class of interactions, and so does not possess the status of a scientific concept.

Surely, one can object that the formalism of quantum mechanics is non-local. But conventional quantum mechanics is a non-relativistic approximation to the real relativistic nature of quantum systems and interactions. Thus this argument does not modify the status of the concept of non-separability.

The above distinction is a supplementary argument for the need to save causality and locality in quantum mechanics.

ACKNOWLEDGMENT

Work on the present paper has been realised in the framework of a program of the Group of Interdisciplinary Research, partially financed by Scientific Research and Technology Agency, Greece.

REFERENCES

1. A.Einstein, in *Louis de Broglie, physicien et penseur* (Albin Michel, Paris, 1952), p.7.

2. A. Einstein, B. Podolsky, and N. Rosen, *Phys. Rev. 47*,777 (1935).

3. D. Bohm, and Y. Aharonov, *Phys. Rev. 108*,1070 (1957).

4. D. Bohm and Y. Aharonov, *Nuovo Cim. XVII*, 964 (1960).

5. B. d'Espagnat, *Conceptions de la physique contemporaine* (Hermann, Paris, 1965), p.34.

6. E. Bitsakis, *Le problème du déterminisme en physiqye* (Thèse, Paris, 1976); *Physique et matérialisme* (Ed. Sociales, Paris, 1983).

7. J.S. Bell, *Epistem. Letters*, March 1976.

8. E. Bitsakis, *Ann. Fond. Louis de Broglie 5*, 263 (1980).

9. A. Einstein, B. Podolsky, and N. Rosen, *Phys. Rev. 47*, 777 (1975).

10. F. Selleri, "Quantum Reality as an Empirical Problem," Athens Meeting in Epistemology, 1982.

11. J.S. Bell, *Physics 1*, 195 (1964).

12. See *Epist. Letters*, May 1977, January 1978, and June 1979.

13. D. Bohm, and B. Hiley, in *Quantum Mechanics, a Half-Century Later*, J.L. Lopes and M. Paty, ed. (Reidel, Dordrecht,1977).

14. See: D. Bohm and J.-P. Vigier, *Phys. Rev. 109*, 882 (1958); J.-P. Vigier, *Epist. Letters*, Nov. 1980; N. Cufaro-Petroni and J.-P. Vigier, *Phys. Lett. 73*, 289 (1979); J.-P. Vigier *Lett. Nuovo Cim. 29*, 467 (1980); A. Garuccio and J.-P.Vigier, *Found. Phys. 10*, 797 (1980).

15. See: G. Lochak, in *Quantum Mechanics, a Half Century Later*, cited above; L. de Broglie *et al.*, *Cahiers Fund. Scientiae 55* (1976).

16. See Roy and Singh, *J. Phys. A. Math. Gen. 11*, 1167 (1978).

17. For a detailed analysis, see E. Bitsakis, *Physique et matérialisme*, cited above.

18. F. Selleri, *Found. Phys. 12*, 645 (1982).

19. L. de Broglie, *C.R. Acad. Sc. Paris*, Série B, *278*, 722,(1974).

20. L. de Broglie *et al.*, *Cahiers Fund. Scientiae 55* (1976).

21. J.S. Bell, in *Foundations of Quantum Mechanics* (Academic Press, New York, 1971).

22. J.S. Bell, *Physics* 1, 195 (1964).

EINSTEIN-PODOLSKY-ROSEN EXPERIMENTS AND MACROSCOPIC LOCALITY

K. Kraus

Physikalisches Institut
der Universität Würzburg
D-8700 Würzburg, FRG

ABSTRACT

Special relativity requires that (at least) macroscopic events in a space-time region I are not influenced by changes of the macroscopic conditions in another, spacelike separated region II. It is shown that previous attempts at deriving a contradiction between this postulate and the predictions of quantum mechanics for EPR experiments are inconclusive.

1. INTRODUCTION

According to special relativity, events in a certain space-time region I should not depend on anything that happens in another space-time region II if these two regions are spacelike separated. In particular, macroscopic events in region I (e.g., the readings of measuring instruments) should not be influenced if the macroscopic conditions (e.g., the settings of other measurings instruments) are deliberately changed in region II. When restricted, as in such examples, to the realm of macrophysics,

G. Tarozzi and A. van der Merwe (eds.), Open Questions in Quantum Physics, 75–86.

this fundamental postulate (or hypothesis) of relativity will briefly be called macroscopic locality.

We will not consider the extensions of this hypothesis to microphysics, as expressed, e.g., by the locality postulate of relativistic quantum field theory. Whether or not there might be superluminal influences on the microscopic level is thus not discussed here. Rather, we shall examine the question whether or not macroscopic locality is compatible with the particular correlations predicted by quantum mechanics for experiments of the Einstein-Podolsky-Rosen (EPR) type. [1] It has in fact been claimed in the literature [2] and also at this conference [3] that these correlations cannot be reconciled with macroscopic locality. On the other hand, the quantum mechanical predictions have been verified with high accuracy by recent EPR experiments with correlated photons. [4] If, therefore, the arguments of Refs. [2,3] were conclusive, then we would indeed be forced to abandon macroscopic locality, and thus to revise quite drastically our usual conceptions of space-time and causality. We want to prove, however, that the quantum mechanical description of EPR experiments is compatible with the postulate of macroscopic locality; in particular, the arguments presented in Refs. [2,3] will be reexamined, and shown to be inconclusive.

2. SPIN CORRELATIONS IN THE SINGLET STATE

For definiteness, we shall discuss the EPR experiment with pairs of spin-(1/2) particles invented by Bohm. [5] [1] In an experiment of this kind, pairs of spin-(1/2) particles are produced -- e.g., by suitable scattering or decay processes -- so that the total spin of each pair is zero and its two particles are moving away from each other in two (nearly) fixed and opposite directions. The preparation apparatus thus produces two oppositely directed particle beams -- beam 1 and beam 2, say -- such that each particle in beam 1 is paired with another particle in beam 2, and vice versa, and the total spin of each of these particle pairs is zero; i.e., all pairs are in the singlet spin state.

With two suitably placed and oriented Stern-Gerlach devices, then, the spin components $\sigma_a^{(1)}$ and $\sigma_b^{(2)}$ along two arbitrarily chosen spatial directions (unit vectors) $\underset{\sim}{a}$ and $\underset{\sim}{b}$ of the particles in beam 1 and beam 2, respectively, can be measured. The measured values for the i-th particle pair are denoted by a_i and b_i. If spin components are measured in units of $\hbar/2$, as natural for spin-(1/2) particles, both a_i and b_i are either +1 or -1. In quantum mechanics, the observables $\sigma_a^{(1)}$ and $\sigma_b^{(2)}$ are represented by commuting Hermitean operators on the two-particle state space of the pairs considered. The statistical correlations between the values measured for $\sigma_a^{(1)}$ and $\sigma_b^{(2)}$ on such pairs are described by the observable $\sigma_a^{(1)} \cdot \sigma_b^{(2)}$, with measured values given by $a_i b_i$ for the i-th particle pair.

For pairs in the singlet state, i.e., with zero total spin, quantum mechanics predicts the expectation values

$$< \sigma_a^{(1)} > \; = \; < \sigma_b^{(2)} > \; = \; 0 \tag{1}$$

and

$$< \sigma_a^{(1)} \cdot \sigma_b^{(2)} > \; = \; -\underset{\sim}{a} \cdot \underset{\sim}{b} \tag{2}$$

for the observables considered. This is easily verified if, as usual, one makes the additional assumptions that the pairs consist of distinguishable particles 1 and 2 and that the particle pairs produced by the source are selected such that particle 1 is always emitted into beam 1 and particle 2 into beam 2. (In this case, the observables $\sigma_a^{(1)}$ and $\sigma_b^{(2)}$ may be simply represented by $\underset{\sim}{a} \cdot \underset{\sim}{\sigma}^{(1)}$ and $\underset{\sim}{b} \cdot \underset{\sim}{\sigma}^{(2)}$ in terms of the spin vectors $\underset{\sim}{\sigma}^{(1)}$ and $\underset{\sim}{\sigma}^{(2)}$, with components given by the three Pauli matrices and operating on the spin state space of particles 1 and 2, respectively.) However, Eqs. (1) and (2) remain true also if the source emits both kinds of particles into each of the two beams or if the pairs consist of identical particles. There is little doubt that actual experiments would confirm Eqs. (1) and (2), if they could be performed with a precision comparable to that of the recent photon pair experiments. [4]

Imagine $\sigma_a^{(1)}$ and $\sigma_b^{(2)}$ to be measured for $N \gg 1$ pairs. This sequence of N (or rather, 2N) measurements with fixed orientations $\underset{\sim}{a}$ and $\underset{\sim}{b}$ of the two Stern-Gerlach devices is called a <u>run</u> $(\underset{\sim}{a},\underset{\sim}{b})$. It is convenient to collect the 2N measured values a_i and b_i $(i = 1...N)$ of such a run $(\underset{\sim}{a},\underset{\sim}{b})$ into two "result vectors" $\underset{\sim}{A}$ and $\underset{\sim}{B}$, with components $a_i/\sqrt{N}$ and $b_i/\sqrt{N}$, respectively. Since $a_i^2 = b_i^2 = 1$, these result vectors are N-dimensional unit vectors.

The predictions (1) and (2) of quantum mechanics can be reformulated as follows. Denote, for the run $(\underset{\sim}{a},\underset{\sim}{b})$ considered, by N_{++}, N_{+-}, N_{-+}, and N_{--} the numbers of pairs i with $a_i = b_i = +1$, $a_i = +1$ and $b_i = -1$, $a_i = -1$ and $b_i = +1$, and $a_i = b_i = -1$, respectively. Then, clearly,

$$N_{++} + N_{+-} + N_{-+} + N_{--} = N \ ,$$

whereas Eqs. (1) and (2) imply

$$\frac{1}{N}\sum_i a_i = \frac{1}{N}(N_{++} + N_{+-} - N_{-+} - N_{--}) = 0 \ ,$$

$$\frac{1}{N}\sum_i b_i = \frac{1}{N}(N_{++} - N_{+-} + N_{-+} - N_{--}) = 0 \ ,$$

and

$$\frac{1}{N}\sum_i a_i b_i = \frac{1}{N}(N_{++} - N_{+-} - N_{-+} + N_{--}) = -\underset{\sim}{a}\cdot\underset{\sim}{b} \ .$$

The last four equations yield

$$\left.\begin{aligned} N_{++} = N_{--} &= \frac{N}{4}(1 - \underset{\sim}{a}\cdot\underset{\sim}{b}) , \\ N_{+-} = N_{-+} &= \frac{N}{4}(1 + \underset{\sim}{a}\cdot\underset{\sim}{b}) . \end{aligned}\right\} \tag{3}$$

Equations (3) represent the quantum mechanical predictions (1) and (2) in a different but equivalent form.

In actual experiments, the numbers N_{++}, N_{+-}, etc. will deviate from the values predicted by (3), due to statistical fluctuations. Being of the order of magnitude of $\sqrt{N}$, however, these statistical errors become negligible for large enough N and will thus here be ignored. Equation (2), then, may be rewritten in the simple form

$$\underset{\sim}{A}\cdot\underset{\sim}{B} = -\underset{\sim}{a}\cdot\underset{\sim}{b} \tag{4}$$

in terms of the result vectors $\underset{\sim}{A}$ and $\underset{\sim}{B}$ of an arbitrary run $(\underset{\sim}{a},\underset{\sim}{b})$.

3. MACROSCOPIC LOCALITY AND SPIN CORRELATIONS

Being applied to two oppositely directed particle beams, the two Stern-Gerlach devices may in principle be separated very far from each other and from the particle source. It may thus be achieved that the spin components $\sigma_a^{(1)}$ and $\sigma_b^{(2)}$ of the two particles of a given pair are measured in two spacelike separated space-time regions. For sufficiently large separation, one may even achieve that, e.g., the Stern-Gerlach device applied to beam 2 can be deliberately reorientated along a new direction $\underset{\sim}{c} \neq \underset{\sim}{b}$ immediately before the arrival of the i-th particle of that beam, and that this reorientation occurs in a space-time region II which is still spacelike separated from the region I in which $\sigma_a^{(1)}$ is measured on the accompanying i-th particle in beam 1. The value a_i obtained in this measurement is registered by a macroscopic event -- e.g., the discharge of a counter, indicating whether the particle has been deflected "up" or "down" by the Stern-Gerlach magnet. Macroscopic locality thus implies that the particular value a_i measured in region I should be independent of the orientation of the other measuring instrument in region II; i.e., it should be the same regardless of whether $\sigma_b^{(2)}$ or some other spin component $\sigma_c^{(2)}$ of the other particle of the same pair i is measured.

This reasoning can be applied to an arbitrary pair i of the run $(\mathbf{a},\mathbf{b})$ considered. Macroscopic locality thus requires that the same values a_i would have been measured for $\sigma_a^{(1)}$, if instead of $(\mathbf{a},\mathbf{b})$ another run $(\mathbf{a},\mathbf{c})$ with $\mathbf{c} \neq \mathbf{b}$ had been performed with the given set of N particle pairs, i.e., if instead of $\sigma_b^{(2)}$ another spin component $\sigma_c^{(2)}$ had been measured on beam 2. (Note also that, at least in principle, the two measuring instruments could be so far separated from each other, that the choice of the orientation of the Stern-Gerlach device for beam 2 *and* $N \gg 1$ successive measurements of either $\sigma_b^{(2)}$ or $\sigma_c^{(2)}$ may be performed in a single space-time region II which is spacelike separated from the single space-time region I containing *all* N accompanying measurements of $\sigma_a^{(1)}$ in beam 1. In this situation, macroscopic locality applies directly to whole runs, rather than to measurements on individual pairs only.) Since the experimental situation is symmetric with respect to the two beams of particles, macroscopic locality also requires that the values b_i obtained for $\sigma_b^{(2)}$ should not be influenced if the orientation of the instrument applied to beam 1 is changed, and thus instead of $(\mathbf{a},\mathbf{b})$ another run $(\mathbf{d},\mathbf{b})$ with $\mathbf{d} \neq \mathbf{a}$ is performed.

Since in an *actual* run the orientations $\mathbf{a}$ and $\mathbf{b}$ are fixed, macroscopic locality cannot be tested directly by experiments. If the run $(\mathbf{a},\mathbf{b})$ is actually performed, this clearly excludes the actual performance of other runs with the same set of N particle pairs. One may at most *imagine* alternative runs, in which one of the orientations, $\underset{\sim}{a}$ or $\underset{\sim}{b}$, is changed into another one, $\underset{\sim}{d} \neq \underset{\sim}{a}$ or $\underset{\sim}{c} \neq \underset{\sim}{b}$, respectively. Due to the symmetry just mentioned, it suffices to discuss the second case, i.e., the hypothetical run $(\underset{\sim}{a},\underset{\sim}{c})$.

If quantum mechanics is correct, the results a_i and b_i $(i = 1...N)$ of the actual run $(\underset{\sim}{a},\underset{\sim}{b})$ will confirm the predictions of that theory, as expressed by Eqs. (3). According to macroscopic locality, the hypothetical run $(\underset{\sim}{a},\underset{\sim}{c})$ would yield the same results a_i for $\sigma_a^{(1)}$, accompanied by certain values c_i for the hypothetical measurements of $\sigma_c^{(2)}$. Macroscopic locality by itself does not yield any predictions for the latter. However, if this postulate is to be compatible with quantum mechanics, the hypothetical run $(\underset{\sim}{a},\underset{\sim}{c})$ must also confirm quantum mechanics since, instead of $(\underset{\sim}{a},\underset{\sim}{b})$, the run $(\underset{\sim}{a},\underset{\sim}{c})$ could have been actually performed as well. This imposes the following restrictions on the hypothetical measured values c_i: Defining numbers M_{++}, M_{+-}, etc. analogous to the numbers N_{++}, N_{+-}, etc. introduced above, with a_i unchanged but b_i replaced by c_i, we must have (cf. Eqs. (3))

$$\left.\begin{aligned} M_{++} = M_{--} &= \frac{N}{4}\,(1 - \underset{\sim}{a}\cdot\underset{\sim}{c})\;, \\ M_{+-} = M_{-+} &= \frac{N}{4}\,(1 + \underset{\sim}{a}\cdot\underset{\sim}{c})\;. \end{aligned}\right\} \tag{5}$$

In the situation considered, therefore, macroscopic locality is compatible with quantum mechanics if and only if the hypothetical measured values c_i can be <u>chosen</u> to satisfy, together with the <u>given</u> values a_i, these Eqs. (5).

The existence of such numbers c_i is easily shown, however. Denote by S_+ and S_- the subsets of pairs i with $a_i = +1$ and $a_i = -1$, respectively. By (4), or (1), both S_+ and S_- contain N/2 pairs. To satisfy (5), it suffices to select $(N/4)(1 - \underset{\sim}{a}\cdot\underset{\sim}{c})$ pairs from S_+ and $(N/4)(1 + \underset{\sim}{a}\cdot\underset{\sim}{c})$ pairs from S_-, to set $c_i = +1$ for these pairs, and $c_i = -1$ for the remaining ones. This is always possible, and in most cases the choice of the c_i's is still highly arbitrary.

From the "actual" run $(\underset{\sim}{a},\underset{\sim}{b})$ with result vectors $\underset{\sim}{A}$ and $\underset{\sim}{B}$, we may thus construct a hypothetical but possible run $(\underset{\sim}{a},\underset{\sim}{c})$, whose result vectors are $\underset{\sim}{A}$ and $\underset{\sim}{C}$, by macroscopic locality, and which along with the actual run also confirms quantum mechanics. This construction, then, may be continued as follows. As the run $(\underset{\sim}{a},\underset{\sim}{c})$ is possible at least, and therefore could have been actually performed, one may apply macroscopic locality again, e.g., by changing $\underset{\sim}{a}$ into $\underset{\sim}{d} \neq \underset{\sim}{a}$, and obtain another possible run $(\underset{\sim}{d},\underset{\sim}{c})$ with result vectors $\underset{\sim}{D}$ and $\underset{\sim}{C}$; etc. In each step of this construction, only one of the two directions of the previous run is changed, and the corresponding result vector is replaced by a new one, whereas the other result vector is left unchanged. If n possible runs are constructed in this way, they contain $n + 1$ independent result vectors, since the original run contains two, and each step introduces one additional result vector.

Again, macroscopic locality is compatible with quantum mechanics only if it is possible to require, in addition, that <u>all</u> runs constructed in this way satisfy the statistical predictions of quantum mechanics. But this requirement may indeed be satisfied by an appropriate choice of the new (hypothetical) result vectors introduced in each step. This has already been proved for the first step, the replacement of the run $(\underset{\sim}{a},\underset{\sim}{b})$ by $(\underset{\sim}{a},\underset{\sim}{c})$; and, obviously, the same argument may be applied to each one of the following steps as well. No contradiction between macroscopic locality and quantum mechanics can thus be derived in this way.

4. CRITICISM OF PREVIOUS ARGUMENTS

In view of this result, a critical reexamination of the arguments [2,3] leading to the opposite conclusion seems to be appropriate. In Refs. [2], four (actual, or supposedly possible) runs $(\underset{\sim}{a},\underset{\sim}{b})$, $(\underset{\sim}{b},\underset{\sim}{b})$, $(\underset{\sim}{a},\underset{\sim}{c})$, and $(\underset{\sim}{b},\underset{\sim}{c})$ are considered, with result vectors $\underset{\sim}{A}$ and $\underset{\sim}{B}$, $\underset{\sim}{B}'$ and $\underset{\sim}{B}$, $\underset{\sim}{A}$ and $\underset{\sim}{C}$, and $\underset{\sim}{B}'$ and $\underset{\sim}{C}$, respectively, and directions $\underset{\sim}{a}$, $\underset{\sim}{b}$, and $\underset{\sim}{c}$ as shown in Fig. 1. Quantum mechanics, cf. Eq. (4), predicts the relations

$$\underset{\sim}{A} \cdot \underset{\sim}{B} = -\underset{\sim}{a} \cdot \underset{\sim}{b} = 0 \ , \tag{6}$$

$$\underset{\sim}{B}' \cdot \underset{\sim}{B} = -\underset{\sim}{b} \cdot \underset{\sim}{b} = -1 \ , \tag{7}$$

$$\underset{\sim}{A} \cdot \underset{\sim}{C} = -\underset{\sim}{a} \cdot \underset{\sim}{c} = -\frac{1}{\sqrt{2}} \ , \tag{8}$$

$$\underset{\sim}{B}' \cdot \underset{\sim}{C} = -\underset{\sim}{b} \cdot \underset{\sim}{c} = \frac{1}{\sqrt{2}} \ . \tag{9}$$

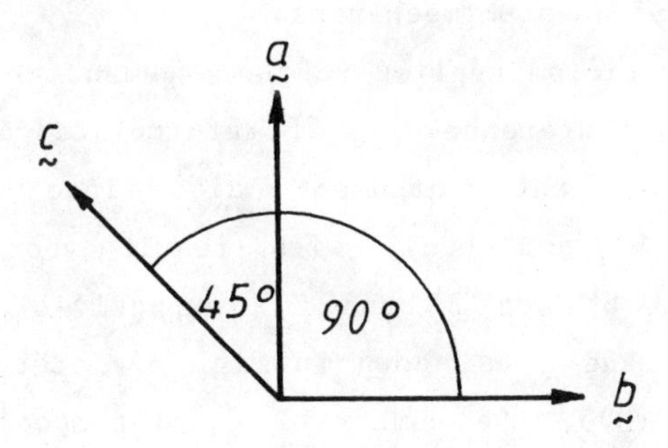

Fig. 1. Orientations of Stern-Gerlach devices as discussed in Refs. [2].

However, this set of equations cannot be satisfied by any choice of the four result (unit) vectors $\underset{\sim}{A}$, $\underset{\sim}{B}$, $\underset{\sim}{B}'$, and $\underset{\sim}{C}$. [2] A simple proof goes as follows. By (7), we have $\underset{\sim}{B}' = -\underset{\sim}{B}$ (i.e., $b_i' = -b_i$, the perfect anticorrelation predicted by quantum mechanics for runs of the type $(\underset{\sim}{b},\underset{\sim}{b})$). Inserting this into (9), and subtracting the resulting equation from (8), we get

$$(\underset{\sim}{A} + \underset{\sim}{B}) \cdot \underset{\sim}{C} = -\sqrt{2} \ . \tag{10}$$

By (6), $\underset{\sim}{A}$ and $\underset{\sim}{B}$ are orthogonal. Therefore $\underset{\sim}{A} + \underset{\sim}{B}$ has the length $\sqrt{2}$, and (10) implies $\underset{\sim}{A} + \underset{\sim}{B} = -\sqrt{2}\,\underset{\sim}{C}$ or, in components,

$$a_i + b_i = -\sqrt{2}\, c_i \ . \tag{11}$$

Since a_i, b_i, and c_i are all ± 1, however, Eq. (11) cannot be satisfied for any i.

However, this argument does not prove the incompatibility of macroscopic locality and quantum mechanics. Indeed, the argument involves <u>four</u> different runs with only <u>four</u> independent result vectors. The simultaneous possibility of four runs of this kind cannot be concluded from macroscopic locality. When applied to the case considered, the latter will always lead to <u>five</u>, rather than only four independent result vectors, which moreover, as shown above, may always be chosen in accordance with the statistical predictions of quantum mechanics.

The same criticism applies to the argument presented by C.W. Rietdijk at this conference. [3] If reformulated in the terminology used here, this argument also involves four runs $(\underset{\sim}{a},\underset{\sim}{b})$, $(\underset{\sim}{a},\underset{\sim}{c})$, $(\underset{\sim}{b},\underset{\sim}{b})$, and $(\underset{\sim}{b},\underset{\sim}{c})$, with result vectors $\underset{\sim}{A}$ and $\underset{\sim}{B}$, $\underset{\sim}{A}$ and $\underset{\sim}{C}$, $\underset{\sim}{B}'$ and $\underset{\sim}{B}$, and $\underset{\sim}{B}'$ and $\underset{\sim}{C}$, respectively, but now with directions $\underset{\sim}{a}$, $\underset{\sim}{b}$, and $\underset{\sim}{c}$ as shown in Fig. 2. According to quantum mechanics (Eqs. (3)), the number of equal components $a_i = b_i$ of the result vectors $\underset{\sim}{A}$ and $\underset{\sim}{B}$ of the run $(\underset{\sim}{a},\underset{\sim}{b})$ is

$$N_{++} + N_{--} = \frac{N}{2}(1 - \underset{\sim}{a}\cdot\underset{\sim}{b}) = \frac{N}{2}(1 - \cos\phi) = N \sin^2(\phi/2) \ .$$

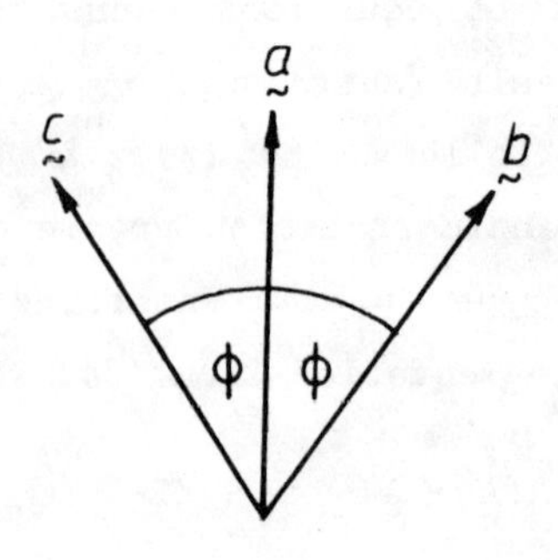

Fig. 2. Orientations of Stern-Gerlach devices as discussed in Ref. [3].

By symmetry, obviously, the same number of components are equal in the result vectors $\underset{\sim}{A}$ and $\underset{\sim}{C}$ of the run $(\underset{\sim}{a},\underset{\sim}{c})$. Therefore, both $\underset{\sim}{B}$ and $\underset{\sim}{C}$ differ from $-\underset{\sim}{A}$ in $N\cdot\sin^2(\phi/2)$ components, and thus $\underset{\sim}{B}$ and $\underset{\sim}{C}$ differ from each other in at most $2N\cdot\sin^2(\phi/2)$ components.

On the other hand, quantum mechanics predicts perfect anti-correlation, $b_i' = -b_i$, i.e., $\underset{\sim}{B}' = -\underset{\sim}{B}$, for the run $(\underset{\sim}{b},\underset{\sim}{b})$. For the run $(\underset{\sim}{b},\underset{\sim}{c})$, Eqs. (3) imply that

$$\frac{N}{2}(1 - \underset{\sim}{b}\cdot\underset{\sim}{c}) = \frac{N}{2}(1 - \cos 2\phi) = N \sin^2\phi$$

components of $\underset{\sim}{B}'$ and $\underset{\sim}{C}$ are equal. Accordingly, the result vectors $\underset{\sim}{B} = -\underset{\sim}{B}'$ and $\underset{\sim}{C}$ are different in $N\cdot\sin^2\phi$ components. For all angles ϕ smaller than $\pi/2$, however, this contradicts the statement derived in the preceding paragraph, since $\sin^2\phi > 2\sin^2(\phi/2)$ for such ϕ.

But again this contradiction does not imply that macroscopic locality and quantum mechanics are incompatible. Indeed, as in the previous case, the argument involves four different runs with only four different result vectors, whose simultaneously possibility does not follow from macroscopic locality. The inconsistencies derived in Refs. [2] and [3] thus refer to sets of runs with fewer independent result vectors -- i.e., with more tightly "interlocked" measured values -- than any set of runs that can be constructed from the hypothesis of macroscopic locality.

Strictly speaking, of course, the foregoing considerations do not prove the compatibility of quantum mechanics and macroscopic locality. They only demonstrate that previous attempts at proving the opposite are inconclusive. It is thus still conceivable that by other methods, applied perhaps to quantum systems quite different from the ones considered here, the desired contradiction with macroscopic locality could indeed be derived. The present author, however, does not consider this to be very likely.

ACKNOWLEDGEMENT

I want to express my hearty thanks to the organizers of the workshop, Profs. A. Garuccio and F. Selleri, for their generous hospitality during my stay at Bari, and to Dr. W. Petzold for his patient help with the preparation of this paper.

NOTES

1. Corresponding experiments with photon pairs, which are easier to perform in practice, [4] could be discussed along the same lines. This would only require a suitable reinterpretation and some inessential modifications of the explicit formulae (1), (2), and (3) used below.

2. One might also conceive actual runs, during which the orientations of the applied instruments are changed. Such runs, however, may be decomposed into "subruns" with fixed orientations, to which the following discussion applies.

REFERENCES

1. A. Einstein, B. Podolsky, and N. Rosen, Phys. Rev. **47**, 777 (1935). For a comprehensive review, see: F.J. Clauser and A. Shimony, Rep. Progr. Phys. **41**, 1881 (1978).

2. H.P. Stapp, Phys. Rev. D **3**, 1303 (1971); Nuovo Cim. **40** B, 191 (1977); Found. Phys. **9**, 1 (1979); P.H. Eberhard, Nuovo Cim. **38** B, 75 (1977).

3. C.W. Rietdijk, "On Nonlocal Influences," in this volume.

4. A. Aspect, P. Grangier, and G. Roger, Phys. Rev. Lett. **49**, 91 (1982).

5. D. Bohm, Quantum Theory (Prentice Hall, Englewood Cliffs, N.Y., 1951), Ch. XXII, pp. 614-623; D. Bohm and Y. Aharonov, Phys. Rev. **108**, 1070 (1957).

ON THE COMPATIBILITY OF LOCAL REALISM WITH ATOMIC CASCADE EXPERIMENTS

T.W. Marshall

Department of Mathematics, Manchester University
Manchester M13 9PL, U.K.

E. Santos

Departamento de Fisica, Universidad de Santander
Spain

F. Selleri

Dipartimento di Fisica, Università di Bari
INFN, Sezione di Bari, Italy

ABSTRACT

We present two criticisms of the reported violations of local realism in atomic cascade experiments. Firstly, we show that physically reasonable local-hidden variable models can reproduce within errors the results of those experiments. Secondly, we show that photon rescattering in the beam is important for the experiments reported to violate Bell's inequality.

1. INTRODUCTION

The seven experiments[1-7] performed with atomic cascade to study the contradiction between quantum mechanics and local realism, expressed, for example, by Bell's inequality,[8] are repored to agree well with quantum predictions and to violate up to 40 standard deviations the predictions of any local realistic

G. Tarozzi and A. van der Merwe (eds.), Open Questions in Quantum Physics, 87–101.

model.

The theoretical background against which local realistic models are being discussed stems from the following:

(i) Von Neumann's theorem,[9] along with similar theorems which seek to exclude the possibility of a causal completion of quantum mechanics, is now known to be completely ineffective,[10] so causal realistic theories are logically possible and should be sought.

(ii) All known interactions (gravitational, weak, electromagnetic and strong) decrease with distance, and hence *physical* connections between two atomic systems should go to zero as their mutual separation increases.

(iii) Most of the many formulations of wave-particle duality[11] imply that there is some sort of wave, associated with atomic emission, whose squared amplitude determines the probability of events (such as emissions, absorption, scattering, and ionization) occurring in other atomic systems.

We will show that, from these ideas, a consistently local and realistic picture of atomic cascades may be constructed, and that this picture fits the existing data as closely as the quantum mechanical model.[12] We will also show that important objections can be raised against the approach used in order to analize the experimental data.[13]

2. A CLASS OF LOCAL REALISTIC MODELS

In the experimental situation that we want to describe, it is assumed that suitably excited atoms decay through a cascade, emitting two light signals with different frequencies. Each of these light signals passes through a filter, a polarizer, and a system of lenses to arrive eventually at a detector. Assuming that each atom is excited, and decays, independently of the other atoms, the quantum mechanical model gives, as the probability for joint detection of a pair of signals, after they pass through two polarizers whose axes form an angle (a-b),

$$P_{12}(a-b) = \frac{1}{4}\eta_1\eta_2 f_1 g \left[\varepsilon_+^1\varepsilon_+^2 + \varepsilon_-^1\varepsilon_-^2 F\cos 2(a-b)\right] \qquad (1)$$

where $\eta_1 (\eta_2)$ is the quantum efficiency of the first (second) phototube, f_1 and g are geometrical factors which take account of the fact that the lenses do not cover all directions of emissions, and F is a factor, usually slightly less than unity, which takes into account certain depolarization effects. The efficiency of the polarizer i (i = 1,2) for light polarized parallel (perpendicular) to the polarization axis is $\varepsilon_M^i(\varepsilon_m^i)$, and

$$\varepsilon_\pm^i = \varepsilon_M^i \pm \varepsilon_m^i .$$

The sign + in Eq. (1) holds for the cascade 0-1-0 of calcium.

The probability of joint detection without polarizers is

$$P_o = \eta_1 \eta_2 f_1 g \qquad (2)$$

The ratio between the joint detection rate with polarizers inserted at a relative angle (a-b) and the rate without polarizers should equal the ratio of the probabilities, that is

$$\frac{R_{12}(a-b)}{R_o} = \frac{1}{4}\left[\varepsilon_+^1\varepsilon_+^2 + \varepsilon_-^1\varepsilon_-^2\, F\cos 2(a-b)\right]. \qquad (3)$$

This quantum mechanical prediction has been found to be consistent with the experimental data.

A class of local realistic theories called, by Clauser and Horne,[14] objective local theories (OLT), gives the joint detection probability as

$$P_{12}(a-b) = \int_\Lambda d\lambda\,\varrho(\lambda)\, P_1(a,\lambda)\, P_2(b,\lambda) \qquad (4)$$

where λ denotes the state of the pair of signals emitted, Λ is the space of these states, and $\varrho(\lambda)$ is the probability measure over this space. $P_1(a\lambda)\left[P_2(b\lambda)\right]$ is the probability that the first (second) signal is detected after passing through a polarizer with axis set at a (b). A model based on (4) was obtained by Clauser and Horne that gives probabilities exactly equal to those of the quantum mechanical model for all experiments so far performed. At the same time Clauser and Horne pointed out that their model contained what they regarded as a highly implausible feature, called "enhancement." They were able to show that OLTs which satisfy an additional no-enhancement hypothesis are all refuted by the experimental data in the atomic cascade experiments.

In what follows, we shall show that the implausible nature of the Clauser-Horne model stems rather from the artificial nature of the expressions they assume for P_1 and P_2 than from the enhancement feature. We shall do this by showing that a very plausible model with enhancement features exists, which, though not giving identical predictions to the quantum mechanical model, fits the experimental data very closely.

We shall use a single angular variable with uniform probability distribution, that is

$$\int_\Lambda \varrho(\lambda)d\lambda = \frac{1}{\pi}\int_0^\pi d\lambda . \tag{5}$$

Now, rotational and reflection symmetry requires that P_1 and P_2 be even functions of $2(a-\lambda)$ and $2(b-\lambda)$, so that they have the Fourier expansions

$$P_1(a\lambda) = \sum_{n=0}^{\infty} a_n \cos 2n(a-\lambda), \tag{6}$$

$$P_2(b\lambda) = \sum_{n=0}^{\infty} b_n \cos 2n(b-\lambda). \tag{7}$$

A model of this type with only two terms in each Fourier series ($a_i = b_i = 0$ for $i \geqslant 2$) is not possible if it is to give predictions close to the quantum predictions. For positivity requires that $a_0 \geqslant a_1$ and $b_0 \geqslant b_1$ and then (4), (5), (6) and (7) give

$$P_{12}(a, b) = \text{const.}\left[1 + k\cos 2(b-a)\right] \tag{8}$$

with $k = a_1 b_1/2a_0 b_0 \leqslant \frac{1}{2}$. This disagrees with the experimental data, which give values of k in the range from 0.8 to 0.9 . Clauser and Horne could reproduce the quantum mechanical prediction by assuming sharply asymmetrical expressions for P_1 and P_2 .

Taking, however, three terms in (6) and (7), we obtain

$$P_{12}(a, b) = c\left[1 + \frac{1}{2}\alpha_1\beta_1 \cos 2(b-a) + \frac{1}{2}\alpha_2\beta_2\cos 4(b-a)\right], \tag{9}$$

with $$c = a_0 b_0, \quad \alpha_i = \frac{a_i}{a_0}, \quad \beta_i = \frac{b_i}{b_0}, \quad (i=1,2). \tag{10}$$

Positivity of P_1 and P_2 now requires that α_1 and α_2 lie within the dashed region of Fig. 1, and similarly for β_1 and β_2 .

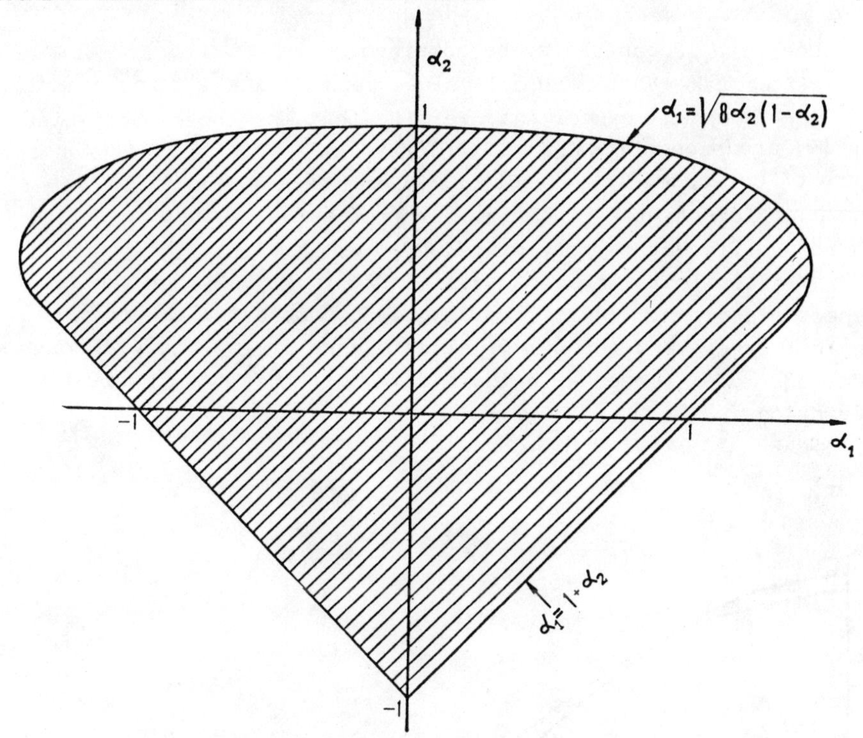

Fig. 1. The region in the α_1, α_2 plane where P (a λ), defined by (6), with only three terms retained in the Fourier series, is positive-definite.

No positive-definite P_1 and P_2 may be found for $|\alpha_1|, |\beta_1| > \sqrt{2}$. It is interesting that, if the limiting values are substitued in (9), they give $|\frac{1}{2}\alpha_1\beta_1| = 1$, which is precisely the maximum value allowed by quantum mechanics.

We now make the following choices:

$$c = \frac{1}{4}\varepsilon_+^1\varepsilon_+^2\eta_1\eta_2 f_1 g \;, \quad |\alpha_1| = \frac{\varepsilon^1}{\varepsilon_+^1}\sqrt{2F} \;, \quad |\beta_1| = \frac{\varepsilon^2}{\varepsilon_+^2}\sqrt{2F}\,. \tag{11}$$

The first two terms in (9) are now identical with the corresponding quantum mechanical terms, while it is possible to choose a very small additional term proportional to $\cos 4(b-a)$. We choose minimum values of α_2, β_2 allowed by the limits shown in Fig. 1, that is

$$\alpha_2 = \frac{1}{2}\left[1 - \sqrt{1 - \frac{1}{2}\alpha_1^2}\right], \quad \beta_2 = \frac{1}{2}\left[1 - \sqrt{1 - \frac{1}{2}\beta_1^2}\right], \tag{12}$$

whence $\alpha_2 \beta_2 /2$ can easily be obtained.

If we take the numerical values appropriate in, for example, the Fry-Thompson[4] experiment, we find that the three coefficients in (9) are proportional to 1, 0.84, 0.04 . This is a common feature of our model in respect of all the experiments: The first two coefficients take their quantum mechanical value, while the third coefficient is very small.

We illustrate the closeness of fit in Fig. 2 for the first Aspect experiment.[5] From the diagram accompanying the report on that experiment (no numerical data were given), it would appear that our model agrees with their data to within two standard deviations.

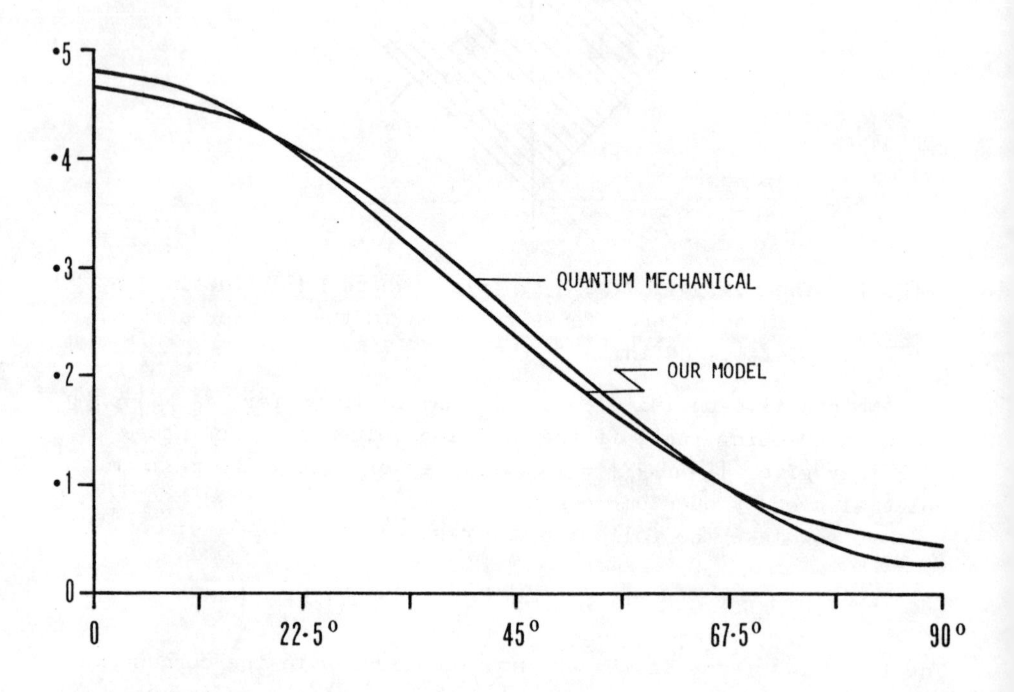

Fig. 2. Agreement of our model [Eqs.(9) to (12)] with the quantum mechanical predictions. The Orsay experimental data (not shown) agree with our model within two standard deviations.

A further feature to be stressed is that *in most cases only the angles* $22\frac{1}{2}^\circ$ *and* $67\frac{1}{2}^\circ$ *have been considered for checking Bell's inequality: Our predictions are identical to the quantum mechanical ones for these values* owing to the vanishing of $\cos 4(b-a)$ at these two angles.

The maximum deviation between our model and quantum mechanics corresponds to the extreme values of $\cos 4(b-a)$, namely at 0°, 45°, and 90°.

In the second experiment of Aspect *et al.*, the four quantities

$$R_{++}(a-b)\,, \quad R_{+-}(a-b)\,, \quad R_{-+}(a-b)\,, \quad R_{--}(a-b)$$

are measured at each angle, giving four coincidence counts. Our model gives the following correlation function

$$\begin{aligned} E(a-b) &= \frac{R_{++} - R_{+-} - R_{-+} + R_{--}}{R_{++} + R_{+-} + R_{-+} + R_{--}} \\ &= F\,\frac{\varepsilon_-^1\varepsilon_-^2}{\varepsilon_+^1\varepsilon_+^2}\cdot\frac{\cos 2(a-b)}{1+q\cos 4(a-b)}\,, \end{aligned} \tag{13}$$

with $q \simeq 0.078$. The maximum deviation from quantum mechanics is about 7% at $(a-b) = 0$. From the diagram published by Aspect *et al.* (again they give no numerical data.), this would seem to be contradicted by their data. We would however point out, firstly, that a closer fit to the data may be expected by the inclusion of extra terms in Eqs. (6) and (7), and, secondly, that important criticism may be made of their data analysis, as shown below.

We therefore believe further experimental effort should be put into this, with a view of obtaining a reliable upper bound for the term in $\cos 4(a-b)$. According to our realistic model, this term must have a certain non-zero *lower* bound, which can be read from Fig. 1.

According to the current quantum mechanical model, of course, this term is zero. Its value is obtained experimentally from the quantity

$$\left[\frac{1}{4}R(o) + \frac{1}{4}R(90°) - \frac{1}{2}R(45°)\right]\,.$$

We now say a little about the physical significance of our model. It is possible to view it simply as a counterexample which shows that, because of the low detection rate, atomic cascade experiments have not refuted local realism. However, Eq. (9) has such a simple structure that we are led to examine its possible physical significance. In fact, inserting (12) in (6) and (7) one obtains perfect squared expression for $P_1(a\lambda)$ and $P_2(b\lambda)$; for example

$$P_1(a\lambda) = a_o \left\{ \frac{1}{\sqrt{2}} \left[1+\sqrt{1-\frac{1}{2}\alpha_1^2}\right]^{1/2} + \left[1-\sqrt{1-\frac{1}{2}\alpha_1^2}\right]^{1/2} \cos 2(\lambda - a) \right\}^2 .$$

If we confine our attention to the ideal case where F = 1, corresponding to infinitely small lens apertures, then (6), (7), (11), and (12) are equivalent to the expressions

$$\begin{cases} P_1(a\lambda) = \dfrac{\eta_1 f_1}{2\sqrt{2}} \left[\varrho_M^1 + \varrho_m^1 + \sqrt{2}(\varrho_M^1 - \varrho_m^1) \cos 2(\lambda - a)\right]^2 , \\ P_2(b\lambda) = \dfrac{\eta_2 g}{2\sqrt{2}} \left[\varrho_M^2 + \varrho_m^2 + \sqrt{2}(\varrho_M^2 - \varrho_m^2) \cos 2(\lambda - b)\right]^2 , \end{cases} \tag{14}$$

where $\varrho_M^1 = \sqrt{\varepsilon_M^1}$, and so on.

A natural interpretation, based on preserving the concept of the photon, is that f_1 is the probability of a photon entering the right-hand lens system, $\varepsilon_+^1/2$ is the probability that it is transmitted by the polarizer , and

$$\eta_1(\lambda) = \eta_1 \frac{1}{2\varepsilon_+^1} \left[\varrho_M^1 + \varrho_m^1 + \sqrt{2}(\varrho_M^1 - \varrho_m^1) \cos 2(\lambda - a)\right]^2 \tag{15}$$

is the probability that it be detected by the counter. As is natural, the average value of $\eta_1(\lambda)$ is simply η_1, as is easy to check. But, equally naturally, once we admit of the possibility that hidden variables may play a role at the point of detection; there are some values of λ giving detection rates higher than η_1 and other values giving rates lower than η_1. It is clear to us that the ban on "enhancement" proposed by Clauser and Horne is an entirely artificial restriction on hidden-variable theories. It is a small wonder that, by imposing such a ban, one comes to the conclusion that hidden variable theories cannot explain the experimental data!

3. PHOTON RESCATTERING IN THE BEAM

In deriving inequalities of Bell's type, one begins by defining four probabilities associated with the two-photon decay of the atom: Both photons are detected, the first photon is detected and the second is not, the second photon is detected and the first is not, and no photon is detected. In more sophisticated experiments, like the second one of Aspect,[6] there are two possible places where each photon can be detected and there are nine, instead of four, probabilities associated with each atom. These probabilities have a meaning only if each atomic decay is independent; if it is not so, one must consider probabilities for more complex collective events such as: "Two photons of atom 1 are detected and two photons of atom 2 are also detected, etc.." In such a case the probabilities of individual events are not well-defined and the inequalities cannot be derived.

After these considerations it should be clear that a reliable test of Bell's type inequalities is only possible if collective effects are small. Next we estimate the (resonant) rescattering of the second photon of the atomic cascade due to absorption and re-emission by atoms in the lowest level of the cascade (which corresponds to the ground state for all performed experiments, the one by Holt and Pipkin[2] excepted). Our calculations show that *resonant rescattering in the beam is very important*.[13]

Consider the ratio γ between the half-width of the atomic beam, L/2, and the mean free path ℓ of the second photon in the atomic beam. The value of ℓ^{-1} equals the product of the atomic density n_o times the cross-section σ for absorption. For a rough estimate we identify σ with its unitary limit ($\sigma_r \simeq \lambda_o^2$, λ_o being the wavelength of the photons). It is well known that such a unitary limit is approached in resonance absorption. Therefore,

$$\gamma \equiv \frac{L}{2\ell} \simeq \frac{1}{2} L n_o \sigma_r \ . \tag{16}$$

Only if γ is much smaller than unity can we be sure that no radiation trapping exists for the second photon of the cascade within the atomic beam. Not all the published reports contain enough information for the calculation of γ, but the most important ones do. Some such data are given in the following table.

Experiment	Density n_o	Wavelength λ_o	Beam dimension L	Parameter γ
Freedman-Clauser[1]	10^{10} cm^{-3}	$4.2 \cdot 10^{-5}$cm	4 mm	3.5
Holt-Pipkin[2]	4.10^{8} cm^{-3}	$4.5 \cdot 10^{-5}$cm	.5 mm	0.02
Aspect[7]	8.10^{10} cm^{-3}	$4.2 \cdot 10^{-5}$cm	1 mm	7.1

The low n_o value in the case of Holt and Pipkin is due to the fact that the *lowest* level of the cascade (which is the only level capable of giving rise to the resonant scattering) is not the ground-state level.

The value of γ is large for the Freedman-Clauser experiment, and indeed some radiation trapping is reported in Freedman's thesis.[15] Also large is the γ value for Aspect's experiment, and we see no justification for the claim made in Aspect's thesis[16] that rescattering is very small in the Orsay experiments.

One must however consider the Doppler effect, which is not negligible in the atomic beam: Its contribution reduces σ in (1), since only a fraction of the atoms can actually give rise to resonant scattering if the photon Doppler shift is taken into account.[17] The velocity distribution of the calcium atoms has been measured[18] in conditions similar to those of the Orsay experiment. From the results of Ref. 18 we deduce that a Maxwellian distribution

$$F(v) = \frac{2v^3}{\alpha^4} e^{-v^2/\alpha^2} \tag{17}$$

is able to provide a reasonable fit to the velocity distribution of the beam of the Orsay experiment if α is chosen to satisfy the relation

$$\langle v \rangle = \int_0^\infty dv \cdot v \cdot F(v) = \frac{3\sqrt{\pi}}{4} \alpha . \tag{18}$$

For $n = 8 \cdot 10^{10}$ atoms/cm^3, Aspect shows[16] that one should take

$$\langle v \rangle = .70 \cdot 10^5 \text{ cm/s} . \tag{19}$$

The value of α can then be deduced from (18).

The average *relative* velocity η of two atoms in the beam is

$$\eta = \int_0^\infty dv \int_0^\infty dv' \cdot |v' - v| \cdot F(v) \cdot F(v') .$$

A direct calculation using (17) gives $\eta = 0.54\,\alpha$, whence, using (18) and (19),

$$\eta = 2.84 \cdot 10^4 \text{ cm/s} .$$

A photon emitted by atom A is therefore expected to be seen Doppler-shifted according to the formula

$$\nu^1 = \nu \left(1 - \frac{\eta}{c} \cos \vartheta\right)$$

by an atom B moving in the beam with velocity η relative to atom A, if ν is the unshifted frequency and ϑ the relative angle between the photon momentum and the B-atom momentum. The percental frequency shift is therefore, for the average atom,

$$\frac{\delta \nu}{\nu} = \frac{\eta}{c} \left|\cos \vartheta\right| . \tag{20}$$

The intrinsic width $\Delta \nu$ of the J = 1 level giving rise to the second photon of the cascade in calcium can be deduced from its lifetime, $\tau = 5$ nanoseconds. One gets

$$\Delta \nu \simeq (2\pi\tau)^{-1} \simeq 3.2 \cdot 10^7 \text{ sec}^{-1} .$$

The frequency corresponding to $\lambda = 4227$ Å is $\nu_0 = .71 \cdot 10^{15} \text{ sec}^{-1}$. Therefore,

$$\frac{\Delta \nu}{\nu_0} \simeq 4.51 \cdot 10^{-8} .$$

The cross section for photon-atom scattering can be represented near resonance by the Breit-Wigner formula

$$\sigma(\nu) \simeq \frac{\left(\frac{c}{\nu_0}\right)^2 \left(\frac{1}{2}\Delta \nu\right)^2}{(\nu - \nu_0)^2 + \left(\frac{1}{2}\Delta \nu\right)^2} ,$$

whence one deduces the $\sigma(\nu)$ values of the following table, where also the corresponding $\gamma(\nu)$'s, calculated from (16), are shown

	$\sigma(\nu)$	$\gamma(\nu)$
$\nu = \nu_o$	σ_r	7.15
$\nu = \nu_o \pm \Delta\nu$	$\frac{1}{5}\sigma_r$	1.43
$\nu = \nu_o \pm 2\Delta\nu$	$\frac{1}{17}\sigma_r$	.42
$\nu = \nu_o \pm 3\Delta\nu$	$\frac{1}{37}\sigma_r$	.19

Remembering that γ gives the average number of re-scatterings in the beam, we see that re-scattering is very important in the $\nu_o - \Delta\nu \leqslant \nu \leqslant \nu_o + \Delta\nu$ range (where $\gamma \simeq 4.3$ on the average) and important in the ranges $\nu_o - 2\Delta\nu \leqslant \nu \leqslant \nu_o - \Delta\nu$ and $\nu_o + \Delta\nu \leqslant \nu \leqslant \nu_o + 2\Delta\nu$ (where $\gamma \simeq .9$ on the average). Limiting the considered frequency interval to $\nu_o \pm 2\Delta\nu$, one has, from (20),

$$|\cos\vartheta| \leqslant 2\,\frac{c}{\eta}\,\frac{\Delta\nu}{\nu} \simeq .095$$

Therefore, only photons propagating within the angles

$$84^\circ.5 \leqslant \vartheta \leqslant 95^\circ.5 \qquad (21)$$

can undergo resonant re-scattering.

Next we estimate the fraction of the 4227 Å photons that emerge *outside* the angular interval (21) and are observed by the counter ("clean" photons). The source of photon pairs is a cylinder of length 1 mm and thickness 60 μ, just in the middle of the beam, where laser light generates atomic excitation followed by pratically immediate de-excitation. The photons emitted by the source hitting the focalizing lens (see Fig. 3) move pratically within a cone of half-opening angle 30°.

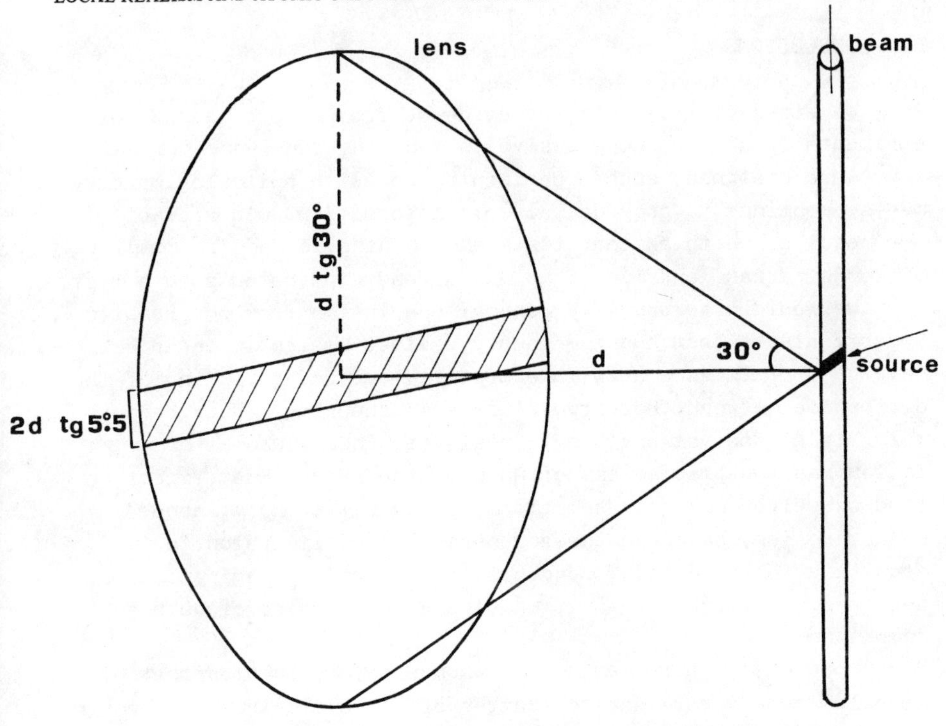

Fig. 3. Photons emitted by the source and moving towards the hatched area of the lens are practically certain to undergo at least one rescattering in the beam (Orsay experiments).

From the full sample of such photons we must subtract those which satisfy (21), that is, photons emitted in the region included between the planes passing through the source's symmetry axis (we neglect the small thickness of the source) and forming angles of 84°.5 and 95°.5 with the beam's direction. The fraction of "clean" photons is therefore given by

$$\Gamma = (A_1 - A_2)/A_1 ,$$

where, if d is the beam-to-lens distance, A_1 is the area of the circle having radius d tg 30°, and A_2 the area of the rectangle with basis 2d tg 30° and height 2d tg 5°.5 . It is a simple matter to show that $\Gamma \simeq 0.79$.

We conclude that about 21% of the photons with wavelength 4227 Å emitted by the source and travelling towards the lens undergo rescattering in the beam.

5. CONCLUSIONS

We stress that, while the evidence from atomic cascade experiments is so far inconclusive in deciding the issue for which they were designed, such experiments can, with suitable improvements, continue to provide valuable information. In view of our results, we think that tests should also be made in completely different areas, and one of us has already indicated such a test.[19]

We would draw one very general conclusion from our results. Experiments to test quantum theory against realistic theories have been of immense help in sharpening our understanding of the difference between these two classes of theories. But, because there is not as yet any single realistic theory whose formulation has the preciseness of quantum theory, we must expect to find ourselves making assumptions, often unwittingly, about such theories, based on three generations of immersion in and acculturation by quantum concepts. Any serious examination of the realist alternative must involve a challenging of such assumptions.

Apparently there was a time when everyone assumed that plane-polarized γ rays had to scatter off free electrons according to the Klein-Nishina formula.[20] This assumption caused us for a long time to think that the Bell inequalities could be tested by observing the scattering of pairs of annihilation photons. And this was almost universally assumed, even though nobody had ever seen a plane polarized γ ray, nor even had a proposal for producing one. We now see that the no-enhancement assumption is another thing of this sort.

Furthermore, if there are correlations in the emission, the rate of accidental coincidences in the no-delay channel should be smaller than in the delay channel -- where certaintly there are no correlations.[21] In such a case the usual background subtraction would not be correct. The Orsay experiments combine the highest atomic densities with laser excitation and a large detection rate, facts which imply a high rate of accidental coincidences. If this is appropriate for decreasing the statistical errors, it makes the experiments unreliable as tests of Bell's inequality.

REFERENCES

1. S.J. Freedman and J.F. Clauser, *Phys. Rev. Lett. 28,* 938 (1972).
2. R.A. Holt and F.M. Pipkin, Harvard University preprint (1974).
3. J.F. Clauser, *Phys. Rev. Lett. 36,* 1223 (1976).
4. E.S. Fry and R.C. Thompson, *Phys. Rev. Lett, 37,* 465 (1976).
5. A. Aspect, P. Grangier, and G. Roger, *Phys. Rev. Lett. 47,* 460 (1981).
6. A. Aspect, P. Grangier, and G. Roger, *Phys. Rev. Lett. 49,* 91 (1982).
7. A. Aspect, J. Dalibard, and G. Roger, *Phys. Rev. Lett. 49,* 1804 (1982).
8. J.S. Bell, *Physics 1,* 195 (1965).
9. J. von Neumann, *Mathematische Grundlagen der Quantenmechanik* (Springer, Berlin, 1932).
10. F.J. Belinfante, *A Survey of Hidden-Variable Theories* (Pergamon Press, Oxford, 1973).
11. F. Selleri, *Found. Phys. 12,* 1087 (1982).
12. T.W. Marshall, E. Santos, and F. Selleri, *Phys. Lett. 98A,* 5 (1983); T.W. Marshall, *Phys. Lett. 99A,* 163 (1983).
13. T.W. Marshall, E. Santos, and F. Selleri, *Lett. Nuovo Cimento 38,* 417 (1983).
14. J.F. Clauser and M.A. Horne, *Phys. Rev. D10,* 526 (1974). That the probabilistic theories factorable for fixed λ do not exhaust all conceivable probabilistic local theories has been shown by F. Selleri, *Found. Phys. 12,* 645 (1982).
15. S.J. Freedman, Ph. D. Thesis, Berkeley (1972).
16. A. Aspect, These, Université de Paris-Sud, Centre d'Orsay, No. 2674 (1983).
17. F. Selleri, "Photon scattering with Doppler-shift in the atomic-cascade experimental tests of Bell inequalities.", *Nuovo Cim. Lett.*, to be published.
18. G. Giusfredi, P. Minguzzi, F. Strumia, and M. Toselli, *Z. Physik A274,* 279 (1975).
19. F. Selleri, *Nuovo Cim. Lett. 36,* 521 (1983).
20. T.W. Marshall, *Phys. Lett. 79A,* 147 (1980).
21. E. Santos, Universidad de Santander, preprint No. FT/AS/83/6. See also E. Santos' contribution to these proceedings.

IS THE EINSTEIN-PODOLSKY-ROSEN PARADOX DEMANDED BY QUANTUM MECHANICS?

O. Piccioni, P. Bowles, C. Enscoe, R. Garland, and W. Mehlhop
University of California, San Diego, La Jolla, California

ABSTRACT

At variance with the more popular opinion that the EPR paradox is a necessary consequence of quantum mechanics, we argue that: (1) The assumption that two distant particles can be in a singlet state implies actions at a distance simply because of preemptive considerations of classical mechanics. There is no need to discuss concepts of reality, Bell's inequalities, or "interpretations of quantum mechanics." (2) The original EPR example, always cited as unquestionable, is not a valid one. (3) Our approach decreases the importance of measuring Bell's inequalities, emphasizing the importance of experiments where such inequalities cannot be defined.

1. INTRODUCTION

The EPR paradox is described by most writers using Bohm's model of two, in general not identical, fermions. Assuming that the two particles are initially interacting in a singlet state and that they are later separated by an interaction that "does not disturb their spins," the writer takes it as obvious that the state of singlet will be "conserved" when the particles are at a distance from each other.[1]

Such reasoning would be justified if each particle had a defined spin state according to quantum mechanics (QM). But the singlet state does not assign a defined QM state to each of the particles, so they cannot very well "remain" in "their" state. Yet the described reasoning is often accepted without discussion, perhaps because it is obviously correct for the spins of classical mechanics (CM) or because of the

G. Tarozzi and A. van der Merwe (eds.), Open Questions in Quantum Physics, 103–118.

wrong interpretation that the singlet state is in effect equivalent to an incoherent mixture of defined QM states, so that "when" one spin is up "then" the other is down. However, the proof that the singlet state at a distance can exist at all, as a final or as an initial state, should precede any other consideration on the subject, such as general demonstrations of the development of a state into another, which imply the proof of existence.

Importantly, if it can be shown that an original singlet state cannot be modified by the interaction which produces the separation of the particles and if that state indeed exists as an original state, but it is impossible as a final state, we should conclude that the transition is simply forbidden. Indeed, we cannot be sure that an interaction to separate the particles leaving unchanged the relationship between their spins must exist.

One could think, as an alternative, that the separation of two particles interacting in a singlet state is part of an act of measurement of the individual spins, because before the separation a measurement of the state of one individual spin cannot be made without altering the other state, while afterward it is the total spin that cannot be measured directly, as distinguished from it being inferred from the sum of the two measurements of each particle. Such a notion is not in conflict with the concept of measurement: "The inapplicability of QM to some part of the measurement process has to be postulated or admitted," according to E. Wigner.[2]

2. THE FAMOUS EXAMPLE GIVEN BY EPR

EPR considered two particles A1 and A2 in the unidimensional state,

$$\Psi(A1,A2) = \int_{-\infty}^{+\infty} \exp\left[ik\left(x_1 - x_2 + x_0\right)\right] dk, \qquad (1)$$

which establishes that, for each eigenvalue x of x_1, the variable x_2 has the eigenvalue $x + x_0$, so that A1 and A2 are always at the fixed distance x_0. Similarly, the momentum p_2 is always accurately equal to $(-p_1)$. Note that $(x_2 - x_1)$ and $(p_1 + p_2)$ are commuting operators. We can measure x_1 accurately and determine x_2 with arbitrary accuracy, or measure p_1 and determine accurately p_2, in either case without disturbing A_2. Therefore, EPR wrote, the accurate values for both x_2 and p_2 are "elements of reality" at the same time. By contrast, QM can only describe a state where only one of them has an accurate value.

The fact that (1) represents a permanent state without any transitory perturbation makes the EPR example appear beyond any doubt, to the point that A. Pais[3] asserts that the two particles remain permanently a fixed distance apart, yet they collide with each other, a contradiction which is not at all contained in the EPR paper of 1935.

A serious objection is, however, that the state (1) cannot be the solution of a Schroedinger equation without a potential, because the second derivative of a delta function is not a delta function. Thus the Schroedinger equation of which the EPR state is the solution must have a potential (with infinities) which is a function of the relative coordinate x_2-x_1. Therefore, while the mathematics of EPR is correct, the verbal statement that measuring one particle does not disturb the other one is unwarranted. Rather, the EPR state represents two particles which are held by a rigid rod of negligible weight. Measuring one particle does indeed disturb the other.

After making the observation described above, we have searched the literature but have not found such comments published anywhere. An observation producing a similar consequence as ours was privately voiced by Epstein to Einstein.[4] Epstein apparently took back his statement because he never published it and we have never seen it cited in journal articles.

Notably, in his reply to EPR,[5] Bohr said, in the form of a side remark before coming to his substantive argumentation against EPR, that their discussion "would hardly seem suited to affect the soundness of quantum mechanical description, which is based on a coherent mathematical formalism covering automatically any procedure of measurement like that indicated." He was correct. It was the interpretation of the mathematics that was wrong.

However, Bohr accepted the existence of the paradox, and justified it on the basis that one could only predict one quantity or the other, but not both. Certainly, if the consequence of the paradox were a prediction of both quantities, the contradiction would be more evident, but the lack of such an extreme does not authorize us to ignore it altogether.

The explanation of Bohr consisted of the dictum that QM does not treat separately one part of the apparatus from the others, in particular the object to be measured from the object which performs the measurement. He wrote later[6] that the "feature of wholeness typical of proper quantum phenomena" justified the fact that the two particles would never be separated. As noted by several authors since 1935, the only justification for the argument was the very fact that without such "wholeness" the argumentation of EPR would be right. Let us note that the best meaning to be given to Bohr's wholeness is that a mechanism of action at distance exists in nature.

3. ACTION AT A DISTANCE

Several authors write that QM is "nonlocal" without explaining the meaning of that ambiguous word. The waves of the ocean are nonlocal in the commonly accepted sense, but perturbing the ocean here does not produce an immediate perturbation a mile away, as we would need for the EPR paradox. To see this, let us start from a simple observation:

Suppose that a device produces objects which leave the device and enter an instrument of macroscopic dimensions and provoke the instrument to produce a reaction either "up" or "down," a reaction which consists of a macroscopic, observable and tangible phenomenon, such as a counter increasing its number by one. Suppose that the objects come out of the device one at a time and that the device can only produce objects of a small number of types, so that it will be necessary to assign a probability for an object to provoke the reaction "up" or "down." Suppose further that we are sure that the objects, after being at some distance from the device, are no longer acted upon by any interaction. The following statements seem to us inevitable and applicable when the objects are at a large distance from the device:

(A) The objects' probability for registering "up" or "down" can no longer be altered, simply because of the principle of sufficient cause on which our scientific knowledge reposes. We could not have confidence in our instruments without it. "Our reliance on simple, non quantum-mechanical, everyday observations is perhaps nowhere as evident in the epistemology of quantum mechanics as in our assumption of the knowledge of the apparata used for measurements."[7]

(B) The assertion (A) is simply made on the basis of classical physics, preemptively with respect to the knowledge of the nature of the objects, whether of large or of microscopic dimensions.

(C) It is out of order to invoke QM for a dictum on (A) because QM is the science of the microscopic world and of its quantization properties, not of the domain of large distances. Surely QM did, unexpectedly, extend the wave properties to "heavy" bodies like electrons, but the propagation of those waves is simply inherited from classical physics.

Denying (A) is tautologically equivalent to assuming an action at a distance. It is important to note that such an assumption could not be reserved as a monopoly of the EPR effect for the convenience of explaining it. If the minute interaction of a nucleonic magnetic moment with the field of a Stern Gerlach apparatus can produce an action at a distance, the burden is on us to explain why is it not possible to produce that action more intensely with more powerful apparatuses, so as to study its nature.

Thus, the argument[8] that an action at a distance is acceptable for the EPR effect because such effect cannot be used for communications of messages is not clearly strong.

The letter of V. F. Weisskopf[9] is also the only publication known to us which details the logic behind the assumption of an action at a distance. We will discuss it because of the authority of the writer and because many physicists seem to accept the notions described in it. The letter, written as a comment to an article by d'Espagnat,[10] reads, at the relevant point:

> The third premise, that of "separability" or "locality," is surely not fulfilled. It postulates that there should be no connection between measurements on two separated protons except those measurements that could be considered arranged before the protons separate. Such unexpected connections do indeed occur, because the quantum state extends from one proton to the other even when the protons have separated. Such extended quantum states are nothing unusual. In the famous experiment of a beam of electrons passing through two slits of a diaphragm and forming interference patterns on a screen the quantum state extends over a region that includes the two slits. In principle this distance could be as large as one pleases.
>
> The spatial extension of the wave function of the two protons A and B in d'Espagnat's example is to be understood as follows. The wave function contains correlations between the spins of A and B: when the spin of A is up, then B is down, and vice versa. This is the case whatever the distance is between A and B.

The argument does not seem to flow easily. First, it is true that there is nothing surprising about an extended QM wave. A proton wave diffracts exactly like a radio wave, allowing for the difference in wave length, and in principle the "slits" can be very far apart. However, the diffraction pattern is observable only where the waves, emerging from the slits, converge on the screen. There is of course no interference between the amplitudes at a given time at points distant from each other. Thus, the extension of the state of each proton is not an argument for the property that, though separate, the protons are still somehow connected.

Second, and most important, the EPR proton amplitudes are not at all necessarily very extended or even overlapping with each other. In fact, the position of the protons can be observed without perturbing their spin states (as in the proton experiment of M. Lamehi-Rachti and W. Mittig[11]), and their trajectories as a function of time can be accurately known, so it is quite imaginable to have each proton separately completely enclosed in a heavily shielded cavity of small dimensions, containing a Stern-Gerlach apparatus, while the distance between the two cavities, thus between the two wave packets, is very large. Moreover, if the two fermions are not identical, it is possible to know that A is "here" and B is "there." The state of A does not extend at all to B. A corresponding situation obtains when the wave packets of two identical particles do not overlap: Their symmetry properties are inconsequential except for the effect of previous overlapping.[12] Thus the popular argument of the "extended waves" is not compelling.

Another approach to be found in the literature[13] is to emphasize that "it is not surprizing that the wave function collapse may involve nonlocal processes." E. Wigner characterized the notion of "collapse" as "very picturesque but not very informative.[14]" Let us consider an example. Suppose we produce a one-particle plane wave which is divided by some device into two wave packets, each enclosed in a separate shielded cavity. We can assume that our particle cannot be created or destroyed, and it will still be described by the Schroedinger equation. If we were to open the cavities we could of course prove the existence of two separate wave packets by letting them come to a common region to display interferences like in the example of the two slits.

Clearly, we can only know in which cavity the particle is if we make an observation in at least one of the cavities. If our particle is found in cavity A, we might say that the wave packet in B collapses instantaneously. The question is at what point such a statement is more than tautology, that is whether: (1) only the measurement in A causes the presence or absence of the particle in B (it is wrong to say that B was empty before) or (2) such measurement merely informed us that B was empty. The "collapse" thinking, as we understand it, has it that, since the possibility (2) cannot be proved, it cannot be accepted and alternative (1) must be the correct statement.

We observe that (1) cannot be proved either. Moreover, given our premises that the particle is indestructible, again classical preemptive principles compel us to accept (2), namely, that once a body is trapped in a cavity it cannot escape. In addition, it must be emphasized that (2) is entirely compatible with the existence of two packets of probability amplitudes, as long as we resist the temptation to assign a "real" physical attribute, such as energy, to such packets.

Thus, it is not correct to use the meaning that the word has in other contexts in order to introduce the paradox.

4. DISCUSSION OF THE EPR EFFECT IN THE BOHM MODEL

We suppose that a source periodically produces pairs of fermions, with one fermion going to the right and the other to the left at a distance from each other such that they are no longer interacting with each other in any way. We want to prove wrong the assumption that the pair of distant fermions could be in the singlet state s = ud-du, where the term ud means that the right-going fermion has a $+\frac{1}{2}$ component with respect to the z axis and the left-going fermion has a $-\frac{1}{2}$ component in the same direction; analogously for du. The state so defined is well known to properly describe fermions at a close distance. Most importantly s=ud-du is also valid with respect to any other axis. We want to show that instead, for a pair of distant fermions, if the state is correctly described by s with respect to z, it cannot be described by s with respect to x or y, and therefore two distant fermions cannot be in a singlet state. We start by assuming that one will observe the results predicted by s when s is interpreted along z.

We suppose that both the right-going (R) and the left-going (L) fermions arrive at their respective Stern-Gerlach (SG) apparatus oriented along z. Then s predicts that for 50% of the time the state will be projected as ud, and for the other times as du; that is, it will give either the result (+ -) (omitting the factors $\frac{1}{2}$) or (- +). Again we emphasize that the results consist in a macroscopic change in a macroscopic apparatus because of some "objects" (whose nature is at the moment irrelevant) sent to the SG's by a macroscopic device. Given the repeatability of the phenomenon, we clearly deal with a device that allows us to study the probabilities for R and L to register (+) or (-). Applying the principle of sufficient cause, again from a classical point of view, we must assume that after an object has left the source its probabilities cannot change.

If the source produced only one type of all identical pairs, each object should have equal probability of giving (+) or (-). The combinations (++) and (--), which are not observed, would then be inevitable. Thus the source must produce two types of pairs. (The inspection of the state s indicates that it is idle to consider more than two types.) Clearly, no formula can fit the results other than the choice of one type in which R has a probability 1 for registering (+), while L has a probability 1 for giving (-), and another type with inverted probabilities. In both types the probabilities are always either unity or zero.

We now ask what is the consequence of that classical statement when we know that the objects are fermions, constrained by the uncertainty principle, according to the Pauli theory with its spinors and matrices.

Since the only way one can produce a spinor with certainty that it registers (+) is to prepare it in the corresponding eigenstate, the first type must consist of a pair where R is in the (+) eigenstate respect to z and L in the (-) eigenstate, with the reverse for the second type. Here the word "eigenstate" has its ordinary meaning, namely, that a particle leaves the source prepared in that state and remains in it until the SG measures it and possibly perturbs it. (Note that the state s has no such property, because each particle has equal probability of registering (+) or (-) until the SG, formally represented by a projection operator, projects it into one of the two terms ud and du.)

Clearly, our forced description implies that the source will still produce pairs in the eigenstates along z even if the SG's are turned toward x or y, so that there will be no correlation observed in that case. On the contrary, the state s predicts the same perfect correlation for the x or y orientation of the SG's. Thus a simple and preemptive classical consideration seems to rule out the singlet state as a possible state for two distant fermions, once we reject the invention of actions at a distance.

Note that we ignore the concept of reality of the spins. Our ingredients are: (1) that the statistical properties impose the recognition of eigenstates, (2) that the source alone can prepare or change eigenstates, (3) that the source is a device that cannot be changed by a passive observer.

The EPR reliance on reality is thus reduced to reliance on the stability of a macroscopic object.

It is also interesting to analyze formally the structure of the state s.

(1) The particle R, in the cavity at right, is represented by a wave packet with the spin state (+) superimposed on a packet with the same space time factor but with the spin state (-) (again with respect to z).

In order to understand that the two packets do not combine to form a pure state, for instance an eigenstate in x, we remind ourselves that the two R packets are respectively associated with two orthogonal spin states for the particle L. However, the fact that the packets R do not overlap with the L packets forces us to reject any formalism that manifestly implies that the result of operating on the state R depends upon the existence of the state L.

(2) Similarly paradoxical is the feature that the choice of one of the two terms of s is made by one of the two particles, but must hold for both, even though the particles are completely separated. If they enter the SG's at the same time the two choices cannot always be compatible. This point gave rise to the idea of "superluminal" communication.

(3) Also amusing is the need to formally connect the partners of a pair, for instance a proton and a neutron, when their separation is orders of magnitude larger than the distance of each of them from millions of other identical particles close

by and when no parameter can establish a formal connection. The two-body togetherness of Bohr's concept now becomes the confusion of an unruly crowd, and the invention of an action at a distance cannot help without a creative addition of a powerful discriminating tool.

The points we have touched upon are not equivalent to asking the question "how does QM operate?" which is in fact the question motivating the debatable notion of "collapse". We have merely discussed statistical predictions for the occurrence of macroscopic phenomena in the light of the indisposable principle of sufficient cause.

5. POSSIBLE CORRELATIONS WITHOUT ACTIONS AT A DISTANCE

If the existence of actions at a distance is ruled out, we can easily narrow down the possible correlations for an EPR experiment according to QM or CM. When assuming, contrary to overwhelming evidence, that CM can describe elementary spins, we expect to reproduce, in their essence, the results of the beautiful work of Bell, possibly with some loss of generality but with the advantage of a clear physical insight.

First, in Sections 3 and 4 we concluded that the probability for each of the two objects to register "up" or "down" is determined after the object leaves the producing device and must remain constant until they enter the respective detectors. Inescapably, then, the probability for having one of the four combinations "up,up" "up,down," etc. is the product of separate probabilities for the left and right object. Each of the two types of productions assumed in Section 4 contributes an additive term (as in Section 4, we choose not to insist on the fact that a source of singlet state pairs must produce only one type of all identical events). The probability of a certain correlation, say "up,down," is then

$$(P1R(up))\ (P1L(down)) + (P2R((up))\ (P2L(down)),$$

where (P1R(up)), for instance, is the probability for type 1 to be produced (namely 0.5) times the probability that for that type the object R will register "up." This formula is the same as the starting expression of Bell,[15] with the simplification of limiting the number of terms to only two because no more than two are necessary for all the EPR experiments that we know.

Clearly, Bell's relations hold for QM and CM (we did not distinguish between the two). Moreover, the singlet state must violate the relations because, as we have shown, it is not reducible to the above formula and it needs actions at a distance. Conversely, it is clear that, if we admit actions at a distance, we invalidate at the same time our statement against singlet states for free particles as well as Bell's relations.

We thus see no justification for the often repeated statement that "Bell's relations prove that QM is "nonlocal."[16] Actually, the point that Bohr's concepts implied actions at a distance was already made by Einstein.[17] Apparently, as late as 1951, sixteen years after the EPR letter, Einstein still retained his opinion on the subject of the paradox.

It is surprizing to see that the literature still contains discussions of whether Bell's relations do apply to QM,[18] and whether they apply to "all local theories."[19] as well as discussions of "the different concepts of locality."[20]

To discuss now the possible correlations, we shall use the model of the two cascade photons of the atomic experiments of Clauser, Fry, and Aspect.[21]

In perfect duplication of the singlet state, the planes of polarization of the two photons, moving in the opposite directions along the z axis, are supposed to be in a EPR state, that is, the state xy-yx,[22] where the first letter denotes the plane of the right going photon and the second denotes the plane of the left photon for both terms. Again, the formula remains identically the same when we rotate the system xyz around the z axis, which feature produces the paradox.

Each of the two photons is detected by a plane polarization analyzer followed by a photomultiplier, and the planes of the analyzer are placed at various angles α with each other during the experiment. The final result is the probability for a coincidence between the two detectors versus α.

Constrained by the conclusion of Section 4, we cannot accept the EPR state for the photons once they are out of their source. Instead, we must assume that, in order to exhibit a correlation, each photon must somehow carry with it the information minimally necessary for such correlation, that is, a well-defined plane containing the z axis. Whether or not that plane is the plane of the vector potential A, we shall call it the plane of polarization.

It is clearly no limitation to assume that the planes of the two photons are orthogonal. Moreover, we are constrained by the physics of the atom levels to assume symmetry around the z axis. We are thus led to the model proposed (and rejected) by Furry[23]: The planes of the photons are determined, they cannot change after the photons have left the source, from where they emerge with equal probability at any azimuth around the z axis and always orthogonal to each other.

The only choice we have is the response of the polarimeter-detector, that is the probability $p(\theta)$ for detecting the photon when the angle between the polarimeter plane and the polarization plane is θ. To be consistent with our knowledge about light, $p(\theta)$ must be symmetric with respect to $\theta = 90°$ and must have a period of 180°.

It is also reasonable to assume that $p(\theta)$ is a decreasing function of θ and equal to 1 at $\theta = 0°$, though none of the conclusions would change on lifting such a restriction.

Importantly, there is no need to clarify whether the detector response is determined by QM or by CM.

The probability of detecting one unpolarized photon is the average of $p(\theta)$ with respect to θ. The probability for double coincidence in two detectors, with the polarimeter planes at the angle α with each other, of two photons with orthogonal polarization planes, averaged over their azimuthal angle around the z axis, is

$$(1/90)\int_0^{90} p(\theta)\, p(90 - \alpha - \theta)\, d\theta \tag{2}$$

In Fig. 1 are shown some possible responses to be used as the basis of our discussion. In Fig. 2 are plotted the corresponding double coincidence probabilities.

The curve QM in Fig. 1 is the ordinary $\cos^2$ dependence common to QM and to classical electromagnetism. Its average for one photon efficiency is $\frac{1}{2}$ and its corresponding coincidence probability, computed with our formula (2) above, is QM of Fig. 2.

With the same response QM, a coincidence curve EPR is computed on the basis of the EPR state xy-yx instead of formula (2). The rotational invariance of the EPR formula has the convenient consequence of allowing us to make the simplifying assumption that the polarization plane of one of the photons (our choice) is half the time parallel and half the time orthogonal to the plane of the detector polarizer for that photon, regardless of the angle α. The result is obviously $0.5\,(\sin\alpha)^2$.

The EPR curve is also quite close to the result of the experiments, if the current interpretation is correct.

The important difference between the coincidence curve QM and EPR is that QM never reaches one half or zero, as it obviously should not, because of the averaging with respect to θ.

Our limitations are clear: The integral of $p(\vartheta)$ should be one half for $\alpha=90°$ and zero for $\alpha=0°$, in order to agree with EPR. The last condition implies that $p(\vartheta) = 0$ for $\vartheta \geqq 45°$, because only then one of the two factors $p(\vartheta)$ or $p(90-\vartheta)$ is zero for any ϑ. Only the step function SF, first examined by Bell a long time ago,[24] satisfies both conditions at 0° and 90°. However, the discontinuity of SF makes it unacceptable on general grounds and for the specific reason that, since the angular momentum operator is the derivative of the wave function with respect to the angle, SF represents the effect of a photon of infinite spin.

The coincidence probability SF of Fig. 2, derived as "the best one can do without actions at a distance" is, not accidentally, also the "best allowed" by Bell's inequalities. In fact, the violation of the inequalities by EPR is a number representing the maximum deviation of the $\sin^2$ curve EPR from the straight line SF of Fig. 2, that is,

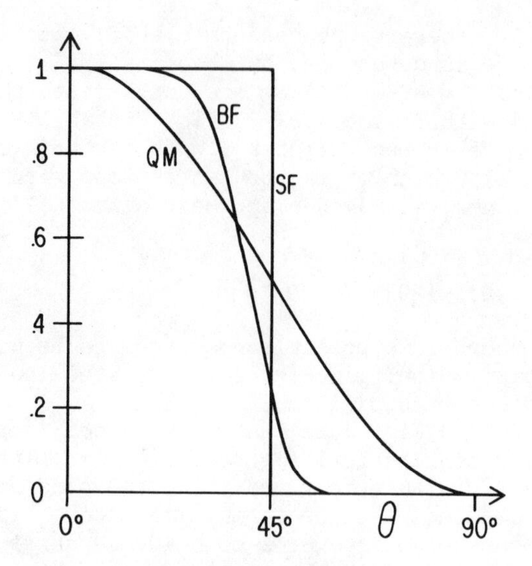

Fig. 1. Examples of possible forms of response of the analyzer. Plotted is the probability of detection versus the angle between the polarization plane and the plane of the polarimeter-detector. QM is the $\cos^2$ dependence. SF is a step function. BF is the best fit to the EPR curve of Fig. 2.

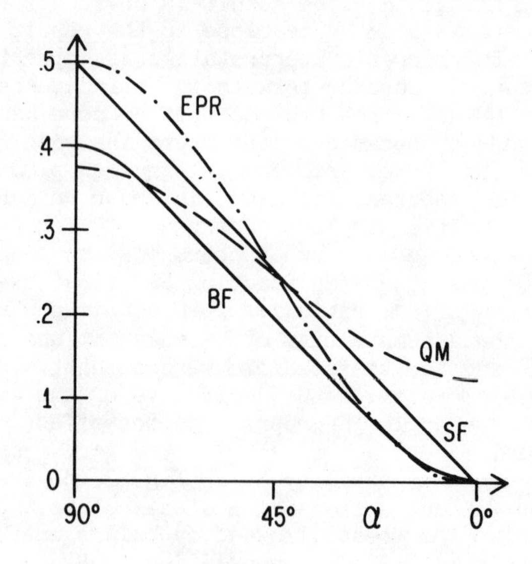

Fig. 2. Probability of coincidence in both analyzers for the various responses of Fig. 1. EPR is computed with an EPR state and QM of Fig. 1. QM, SF, BF are computed with the Furry model and the corresponding response of Fig. 1.

(EPR(67.5) - EPR(22.5))-(SF(67.5)-SF(22.5))= .10

We show in Fig. 1 the response BF, which is less offensive than SF. It has a single photon probability less than one half which allows for a good agreement with EPR at 0°, but not at 90°. Like other authors,[25] we have produced solutions which agree with EPR at least approximately, by appropriate inventions about the transmission of the analyzers or similar "liberal" methods. Though such inventions are not, in our opinion, as rejectable as actions at a distance, we do not see much point in elaborating on them.

CONCLUSIONS

We emphasize again that, coherently with our considerations of Sections 3 and 4, we had no need to clarify whether QM or CM determines the responses of Fig. 1, since the label QM simply stands for $\cos^2$.

Figures 1 and 2 seem to clearly exhibit the roots of the EPR problem. There seems to be no room for elaborate discussions on probabilities,[26] nor for suggestions of "a superdeterminism, in which the choice of the experimenters is not effectively free. Some tight connection from their common past binds the results in one region to the choice of experiment in the other."[27] Such notion would almost imply mental limitations for experimentalists (only), thus decreasing our confidence in the data.

The problem, we think, needs above all the recognition that a problem exists, and it is very encouraging that many authors[28] evidence such a recognition by the very fact that they work on the subject, with a variety of approaches.

Above all, the issue does not appear to be the compatibility of QM with "objective reality,"[29] because it is a contradiction between statistical predictions for macroscopic results and the principle of sufficient cause.

Though this contribution is not meant to include an adequate discussion of the experimental results, we note that the data agree well with the hypothetical state EPR, and they certainly do not appear in need of better statistical accuracy.

However, we better remind ourselves that oftentimes before experimental data have temporarily supported a wrong theory (for instance that "somehow" the Yukawa particle, once produced, would not interact strongly.)

Moreover, the "completeness" of the experiments is perhaps questionable, in the sense that none of the experiments can reproduce Bohm's description. The experiments with the photons from the annihilation of positrons,[30] as well as the work with low energy protons,[31] are not made with the equivalent of a Stern-Gerlach analyzer. Also, their interpretation must rely on the accuracy of important corrections to the observed scattering distributions.

The experiments with the atomic photons[32] use the analyzers but do not have the important feature that the detection of a photon at right implies with certainty the existence of a photon going to the analyzer at the left, so that the definition of the probabilities is less direct than desirable when dealing with results which seem to indicate miraculous actions at a distance. We note that the miracle has been elevated to a higher level by the experiment of Aspect,[33] which rejects actions at a distance unless we accept that such actions propagate much faster than light.

A point emerging from our discussion is that satisfaction of Bell's inequality is not necessarily the proper standard for the EPR effect.

This consideration is important for experimental work, because the accuracy needed to distinguish SF from EPR in Fig. 2 is not easy to achieve, while a determination of a probability of 0% at 0° is much easier.

Similarly, the experimental work on the subject of neutral kaons should be pursued with adequate vigor, independently of the possibility of its reduction to Bell's inequalities.

Claims that those inequalities (a beautiful work by Bell) only hold for CM (hidden variables) and not for QM, or that they prove that QM is "nonlocal," while QM "by itself" predicts the EPR experiment without actions at a distance, are, as we hope to have shown, a wrong basis for research on the subject.

The question is simply whether actions at a distance and the EPR paradox do exist in nature. The two phenomena are tied together, no matter which names are used to describe them.

ACKNOWLEDGMENTS

We are grateful to many colleagues for their generous donation of time spent in oral or written discussions, which have contributed greatly to this article. We especially thank E. Merzbacher, R. Feynman, U. Fano, E. Fry, F. Selleri, N. Kroll, H. Suhl, D. Wong, W. Thompson, J. Bell, P. Noyes, and P. Eberhard.

We also acknowledge the competent help of the staff of the scientific library of UCSD and of our secretary, Ms. P. Fisher.

REFERENCES

1. C. D. Cantrell and M. O. Scully, *Phys. Reports 43,* 499-508 (1978).
2. E. Wigner, in B. d'Espagnat, ed., *Foundations of Quantum Mechanics* (Academic Press, New York, 1971), p. 125.

3. A. Pais, *Revs. Mod. Phys. 51,* 861 (1979).
4. M. Jammer, *The Philosophy of Quantum Mechanics* (Wiley, New York, 1974), pp. 232-234.
5. N. Bohr, *Phys. Rev. 48,* 696 (1935).
6. M. Jammer, loc. cit., p. 197.
7. E. Wigner, loc. cit., pp. 18-19.
8. V. F. Weisskopf, *Scientific American,* May, 1980, p. 8.
9. Ibid.
10. B. d'Espagnat, *Scientific American,* Nov., 1979, p. 158.
11. M. Lamehi-Rachti and W. Mittig, *Phys. Rev. D14,* 2543 (1976).
12. L. Schiff, *Quantum Mechanics* (McGraw-Hill, 1968), 3rd ed., p. 368.
13. P. H. Eberhard, *Phys. Rev. Lett. 20,* 1476 (1982).
14. E. Wigner, loc. cit., p. 14.
15. B. d'Espagnat, loc. cit., p. 178.
16. H. Stapp,*Phys. Rev. Lett. 49,* 1470 (1982).
17. A. Einstein, in P. A. Schilpp, ed., *Albert Einstein Philosopher-Scientist* (Tudor Publishing Co., New York, 1951), pp. 681-682.
18. P. Eberhard, *Nuovo Cimento 38B,* 75 (1977).
19. T. Angelidis, *Phys. Rev. Lett. 51,* 1819 (1983).
20. P. Eberhard, *Nuovo Cimento 46B,* 392 (1978).
21. S. J. Freedman and J. F. Clauser, *Phys. Rev. Lett. 28,* 938 (1972).
    E. S. Fry and R. C. Thompson, *Phys. Rev. Lett. 37,* 465 (1976).
    A. Aspect, P. Grangier, and G. Roger, *Phys. Rev. Lett. 47,* 91 (1981).
22. J. F. Clauser and A. Shimony, *Rep. Prog. Phys. 41,* 1881-1927 (1978).
    F. Pipkin, *Advances in Atomic Molecular Physics 14,* 281 (1978).
    F. Selleri and G. Tarozzi, *La Rivista del Nuovo Cimento 4,* 1 (1981).
23. W. H. Furry, *Phys. Rev. 49,* 393 (1936); *Phys. Rev. 49,* 476 (1936).
24. J. S. Bell, *Physica 1,* 195 (1965).
25. T. W. Marshall, E. Santos, and F. Selleri, *Phys. Lett. 98A,* 5 (1983).
26. A. Fine, *Phys. Rev. Lett. 48,* 291 (1982).
    I. Pitowsky, *Phys. Rev. Lett. 48,* 1299 (1982).
27. H. Stapp, *Phys. Rev. Lett. 49,* 1473 (1982).
28. See F. Selleri and G. Tarozzi, loc. cit., for complete references.
29. A. Pais, loc. cit.
30. C. S. Wu and I. Shaknov, *Phys. Rev. 77,* 136 (1950);
    L. R. Kasday, in B. d'Espagnat, ed., op. cit., p. 195;
    M. Bruno, M. d'Agostino, and C. Maroni, *Nuovo Cimento 40B,* 142 (1977).
31. M. Lamehi-Rachti and W. Mittig, loc. cit.

32. S. J. Freedman and J. F. Clauser, loc. cit.;
E. S. Fry and R. C. Thompson, loc. cit.;
A. Aspect, P. Grangier, and G. Roger, loc. cit.
33. A. Aspect, loc. cit.

INFINITE WAVE RESOLUTION OF THE EPR PARADOX

R.D. Prosser

Central Electricity Generating Board
15 Newgate Street
London EC1A 7AU
U.K.

ABSTRACT

An infinite wave model of the photon is used to show that for light incident on a glass polarizer the transmitted and reflected photons originate in different parts of the beam. The presence of a second polarizer induces reflection and transmission zones in the incident beam. The position of these zones is affected by the orientation of the second polarizer which can thus determine whether a particular photon is transmitted or reflected at the first polarizer. It is this influence of the remote polarizer, specifically excluded by EPR, that resolves the paradox.

1. INTRODUCTION

On the occasion of the award of his Nobel prize in 1954 (1) Max Born gave a lecture on the statistical interpretation of quantum mechanics, which concluded with the following remark:

> The lesson to be learned from the story I have told of the origin of quantum mechanics is that, presumably, a refinement of mathematical methods will not suffice to produce a satisfactory theory, but that somewhere in our doctrine there lurks a concept not justified by any experience, which will have to be eliminated in order to clear the way.

I would like to suggest to you today that one concept that is commonly believed but that is not entirely justified by

G. Tarozzi and A. van der Merwe (eds.), Open Questions in Quantum Physics, 119–127.

experience concerns what happens when light passes through a glass plate, and I shall show how this possible misconception could lie at the heart of the historic paradox of Einstein Podolsky and Rosen.

Everyone has observed that when light falls on glass part of it is transmitted and part of it is reflected. But it is commonly believed that those photons that constitute the reflection could have come from any part of the beam. Such an idea, in the theoretical analysis of the EPR experiment, seems to be implied by the notion that a photon has a probability of 50% that it will pass through a particular polarizer. It is this idea that I wish to challenge with the aid of an infinite wave model of the photon.

2. PHOTON BEHAVIOUR AT A GLASS INTERFACE

In this infinite wave model, that is described more fully elsewhere (2), the photon is considered to consist of a Fourier superposition of infinite waves, such that these sum to zero except over a small region of space where the photon is manifest. The infinite waves are physically real everywhere and are required to satisfy the usual electromagnetic boundary conditions at the boundaries of the experimental apparatus. This has the effect of modifying the trajectory of the manifest aspect of the photon which may be far from any boundary. Only particular forms of this superposition are stable in time (3), but the only features that need concern us now are that this wave packet must be elliptically polarized and that it follows the direction of the Poynting vector of energy flow that is appropriate for the experimental arrangement. We shall consider the motion of such a photon wave packet in an EPR type of experiment with two glass polarizers, but for simplicity we shall assume that the polarizers consists of only one glass interface (Fig. 1).

Three waves are involved. A circularly polarized incident wave F_1 falls on the first polarizer; the boundary conditions here require that a plane polarized reflected wave F_2 results. The unmanifest components of the photon extend also to the second polarizer, and to satisfy the boundary conditions here we require a plane polarized reverse reflected wave F_3 that travels towards the second polarizer. This wave is unmanifest. The propagation functions for the three waves and their associated electric and magnetic components can be derived from electromagnetic theory as follows.

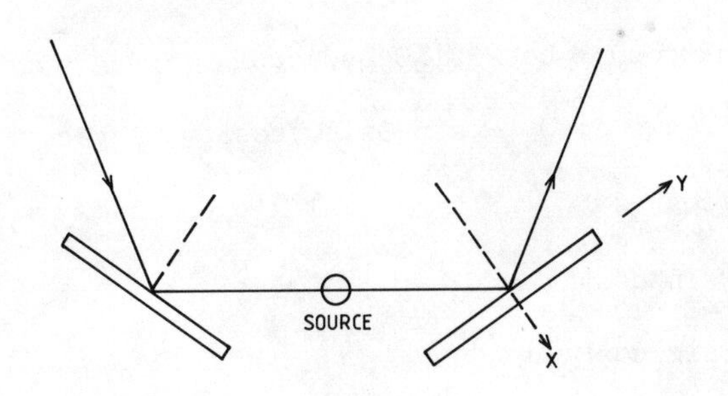

Fig. 1. EPR experiment with glass plates

We have:

$$F_1 = \omega t - k(y\sin\Theta + x\cos\Theta),$$

$$F_2 = \omega t - k(y\sin\Theta - x\cos\Theta),$$

$$F_3 = \omega t + ky(\cos2\Theta\sin\Theta + \sin2\Theta\cos\Theta\cos\emptyset) + kx(\cos2\Theta\cos\Theta - \sin2\Theta\sin\Theta\cos\emptyset) + kz(\sin2\Theta\sin\emptyset) - \emptyset,$$

$\emptyset = 57^\circ$; $\emptyset$=angle between transmission axes of polarizers.

Incident wave (circularly polarized):

$$ZH_{1x} = A_1\sin\Theta\cos F_1, \qquad E_{1x} = -A_1\sin\Theta\sin F_1,$$

$$ZH_{1y} = A_1\cos\Theta\cos F_1, \qquad E_{1y} = A_1\cos\Theta\sin F_1,$$

$$E_{1z} = A_1\cos F_1, \qquad ZH_{1z} = A_1\sin F_1.$$

Reflected wave at first polarizer

$$ZH_{2x} = A_2\sin\Theta\cos F_2,$$

$$ZH_{2y} = A_2\cos\Theta\cos F_2,$$

$$E_{2z} = A_2\cos F_2.$$

Reverse reflected wave at second polarizer:

$$ZH_{3x} = -A_3(\sin2\Theta\cos\Theta + \cos2\Theta\sin\Theta\cos\emptyset)\cos F_3 = -A_3 g_1 \cos F_3,$$

$$ZH_{3y} = A_3(\cos2\Theta\cos\Theta\cos\emptyset - \sin2\Theta\sin\Theta)\cos F_3 = A_3 g_2 \cos F_3,$$

$$E_{3z} = A_3\cos\emptyset\cos F_3,$$

$$E_{3x} = -A_3\sin\Theta\sin\emptyset\cos F_3,$$

$$E_{3y} = A_3\cos\Theta\sin\emptyset\cos F_3,$$

$$ZH_{3z} = -A_3\cos2\Theta\sin\emptyset\cos F_3.$$

In the above, Z represents the impedance of free space.

We assume that we are dealing with broad Gaussian beams, so that one may write

$$A_1 = \exp(-p_1^2/2\sigma^2), \qquad A_2 = \frac{n_2^2 - 1}{n^2 + 1} \exp(-p_1^2/2\sigma^2).$$

Here p_1 represents the perpendicular distance from any point (x,y,z) to the axis of the incident beam, and p_2 represents the corresponding distance for the wave reflected from the first polarizer.

We thus have, for the plane $z=0$,

$$p_1 = \frac{y-mx}{(1+m^2)^{\frac{1}{2}}}, \qquad p_2 = \frac{y+mx}{(1+m^2)^{\frac{1}{2}}}, \qquad m = \tan\Theta,$$

and a similar expression can be derived for p_3. The refractive index n is chosen to be 1.54, so that the Brewster angle Θ is 57° exactly.

The components of the Poynting Vector $S = E \times H$ are given by

$$S_x = E_y H_z - E_z H_y, \; S_y = E_z H_x - E_x H_z, \; S_z = E_x H_y - E_y H_x,$$

where $E_x = E_{1x} + E_{2x} + E_{3x}$, etc. We thus obtain:

$$ZS_x = (A_1^2 - A_2^2\cos^2F_2)\cos\Theta - A_1A_3((g_2 - \cos\Theta)\cos\Theta\cos F_1\cos F_3$$
$$-\cos\Theta\sin\emptyset(1-\cos\Theta)\sin F_1\cos F_3) - A_2A_3(g_2+\cos\Theta)\cos F_2\cos F_3$$
$$-A_3^2\cos\emptyset\cos^2F_3,$$

$$ZS_y = (A_1^2 + 2A_1A_2\cos F_1\cos F_2 + A_2^2\cos^2 F_2)\sin\Theta$$
$$+A_3((\sin\Theta - g_1)(A_1\cos F_1\cos F_3 + A_2\cos F_2\cos F_3) - A_1\sin\Theta\cos 2\Theta\sin\Theta \sin F_1\cos F_3$$
$$-A_3^2 g_1\cos^2 F_3,$$

$$ZS_z = -2A_1^2\sin\Theta\cos\Theta\sin F_1\cos F_1 - 2A_1A_2\cos\Theta\sin\Theta\sin F_1\cos F_2$$
$$+A_1A_3(g_1\cos\Theta - g_2\sin\Theta)\sin F_1\cos F_3$$
$$-A_2A_3(\sin\Theta\cos\Theta\sin\emptyset)\cos F_2\cos F_3$$
$$-A_3^2\sin\emptyset(g_1^2\cos\Theta - g_2^2\sin\Theta)\cos^2 F_3.$$

To simplify these expressions, we take the time averaged values of the trigonometrical products. Thus

$$\langle\cos^2 F_2\rangle = \tfrac{1}{2},$$

$$\langle\cos F_1\cos F_2\rangle = \langle\cos(F_1+F_2) + \cos(F_1-F_2)\rangle = \langle\cos(F_1-F_2)\rangle,$$

since the term in F_1+F_2 contains $2\omega t$ and is zero on average.

Hence

$$\langle\cos F_1\cos F_2\rangle = \tfrac{1}{2}\cos(k(a_1x+by+cz)+\emptyset) = \tfrac{1}{2}\cos(G_1+\emptyset),$$

$$\langle\cos F_2\cos F_3\rangle = \tfrac{1}{2}\cos(k(a_2x+by+cz)+\emptyset) = \tfrac{1}{2}\cos(G_2+\emptyset),$$

$$\langle\sin F_1\cos F_3\rangle = \tfrac{1}{2}\sin(k(a_1x+by+cz)+\emptyset) = \tfrac{1}{2}\sin(G_1+\emptyset),$$

where
$$a_1 = \sin 2\Theta\sin\Theta\cos\emptyset - \cos\Theta(\cos 2\Theta+1),$$
$$a_2 = \sin 2\Theta\sin\Theta\cos\emptyset - \cos\Theta(\cos 2\Theta-1),$$
$$b = \sin 2\Theta\cos\Theta\cos\emptyset - \sin\Theta(\cos 2\Theta+1),$$
$$c = \sin 2\Theta\sin\emptyset.$$

Substitution into the energy flow equations leads to

$$2ZS_x = (2A_1^2 - A_2^2)\cos - A_1A_3((g_2-\cos\Theta)\cos\emptyset\cos(G_1+\emptyset) + \cos\Theta\sin\emptyset(1-\cos 2\Theta)\sin(G_1+\emptyset))$$
$$-A_2A_3((g_2+\cos\Theta)\cos\emptyset\cos(G_2+\emptyset)) - A_3^2 g_2,$$

$$2ZS_y = 2(A_1^2 - A_1A_2\cos(kx\cos\Theta) + \tfrac{1}{2}A_2^2)\sin\Theta$$
$$+A_3((\sin\Theta - g_1)(A_1\cos(G_1-\emptyset) + A_2\cos(G_2+\emptyset) - A_1\sin\Theta\cos2\Theta\sin\emptyset\sin(G_1+\emptyset)$$
$$-A_3^2 g_1,$$

$$2ZS_z = 2A_1A_2\cos\Theta\sin\Theta\sin(2k\cos\Theta x) + A_1A_3(g_1\cos\Theta - g_2\sin\Theta)\sin(G_1+\emptyset)$$
$$- A_2A_3\sin\Theta\cos\Theta\sin\emptyset\cos(G_2+\emptyset) - A_3^2\sin\emptyset(g_1^2\cos\Theta + g_2^2\sin\Theta).$$

These equations represent the photon trajectories. From any point in the source plane a trajectory can be constructed. If the trajectory passes through the glass, the photon is transmitted. We can thus determine which photons are transmitted and which reflected.

The trajectories for a single interface are shown in Fig.2.

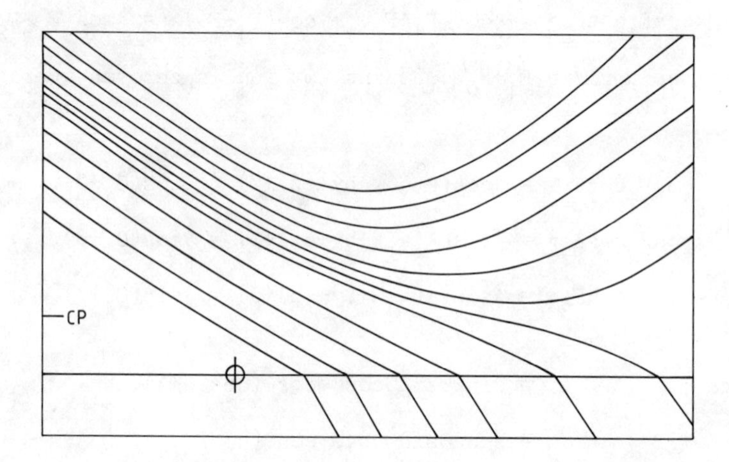

Fig. 2. Photon trajectories

Notice that all reflected trajectories are on one side of the beam. We conclude that all photons that are reflected come from one side of the beam only.

3. SITUATION WITH TWO POLARIZERS

The effect of a second polarizer is to induce undulations in the trajectories for the single polarizer. This may be understood by further consideration of the equation for S_x. The maximum deviation of the transmitted rays from a linear trajectory can be seen to occur in a critical plane designated as CP in Fig. 2. The gradient S_x/S_y of the flow lines is shown in Fig. 3 for this plane.

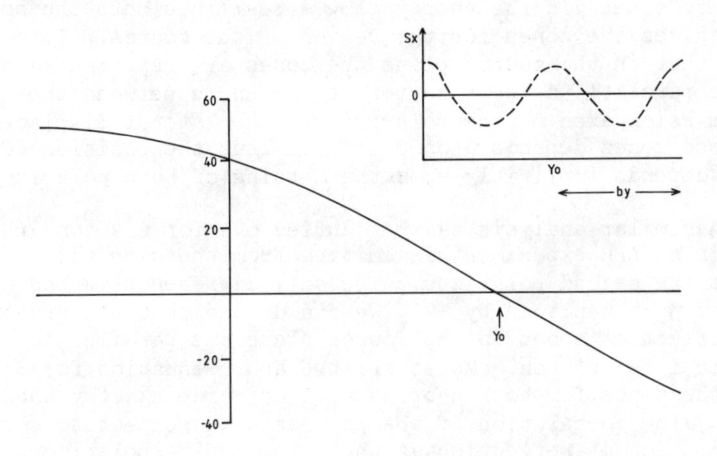

Fig. 3. Streamline gradient at critical plane

For value of y greater than y_o, S_x is negative and no transmission occurs in this area for the single polarizer. However, if we include the terms in A_3 that arise from the second polarizer, S_x becomes alternately positive and negative in the vincinity of the critical ray that passes through y_o. There are, therefore, transmission and reflection zones in this region which are bounded by the lines $S_x = 0$ given by

$$0 = (g_2-\cos\Theta)\cos(G_1+\emptyset)+\cos\Theta\sin\emptyset(1-\cos2\Theta)\sin(G_1+\emptyset)$$

$$- 2(g_2+\cos\Theta)\cos\emptyset\cos(G_2+\emptyset),$$

where we have neglected the small term in A_3^2. This equation can be written

$$C\sin(G_1+\emptyset+u) = 0, \quad C = \text{constant},$$

if we define
$$\tan u = \frac{r+s\cos q}{t-s\sin q},$$

where $r=(g_2-\cos\Theta)\cos\emptyset$, $s=\sqrt{2}(g_2+\cos\Theta)\cos\emptyset$, $t=(1-\cos\Theta)\cos\Theta\sin\emptyset$, and $q=(2k\cos\Theta)x_c$, x_c being the coordinate of the critical plane $x=x_c$ at which the trajectories of the transmitted rays attain their maximum deviation. From its definition, it follows that u can be written $u=\emptyset-\frac{1}{2}\pi+e$, where e is a nonlinear perturbation term which is zero at $\emptyset=\frac{1}{2}\pi,\pi$. Thus the condition for $S_x = 0$ is finally $k(a_1x+by+cz)+2\emptyset+e=(n+1)\pi/2$, where n is an integer. This represents a family of planes. The intersection of these planes with the critical plane $x=x_c$ delimits the transmission and reflection zones in that plane. The projection of these zones via the energy flow streamlines onto the source plane gives the zones for the region of the source. It can be shown that in the source plane the zones are represented by strips parallel to the bisector of the angle between the transmission axes of the polarizers. The lateral displacement of these zones depends upon $\emptyset$ and e. Thus the position of the zones depends implicitly upon the setting of both polarizers.

A similar analysis can be carried out for a second photon that in an EPR experiment is emitted from the same atom and enters the second polarizer. The only difference in the equations is that $\emptyset$ is replaced by $-\emptyset$. We can deduce that the transmission and reflection zones in the source plane are parallel to those for the first photon. Moreover, the above equation for S_x shows that the zones for both photons will coincide exactly when $\emptyset = 0$. The correlation of the photons with respect to transmission or reflection at their respective polarizers will thus be +1. When $\emptyset = \frac{1}{2}\pi$, the zones will be exactly out of phase so that the correlation is -1. The situation with regard to the intermediate angles is complicated and can only be calculated on the basis of an assumption concerning the exact position of the critical plane, and this depends critically upon the beam width. We are thus not able to prove that the correlation function is a function of $\cos 2\emptyset$, as required by quantum mechanics, but we can say that it should be a nonlinear periodic function of $2\emptyset$, equal to +1 at $\emptyset=0$ and -1 at $\emptyset=\frac{1}{2}\pi$.

The locality assumption in the EPR argument implies that the behaviour of the photon at one polarizer cannot be affected by the setting of the remote polarizer. But in our model we have shown that the position of the reflection and transmission zones at one polarizer is affected by the setting of the remote polarizer, and it is this that resolves the paradox. However, our explanation of the correlation can apply only in the vicinity of the critical ray, where the interference terms have a decisive effect, and this leads to the following experimental test for our hypothesis.

4. EXPERIMENTAL TEST

We consider an EPR experimental arrangement with glass plate polarizers in which the plates of each polarizer are parallel. The polarizers could consist either of a single glass plate or of a stack of plates. In the latter case the photon trajectories will be as shown in Fig. 4. Two small

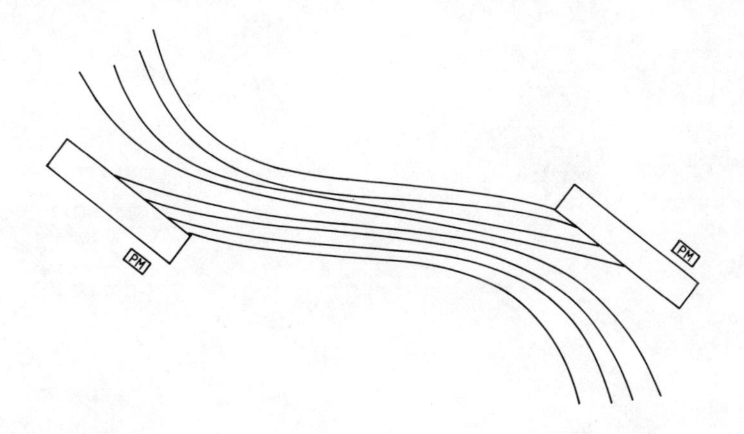

Fig. 4. Trajectories with two polarizers

apperture photomultipliers connected to a coincidence monitoring circuit are placed as shown at the edges of the polarizers. It can be seen that the flow lines require that the correlation should be -1 in this position. The photomultipliers are now moved towards the centres of the polarizers, where according to both our theory and quantum mechanics the correlation should be +1. As the photomultipliers are moved on towards the other edge of the polarizers, the correlation should fall again to -1. Thus the correlation should change from +1 to -1 and back again to +1 as the photomultipliers are moved across the face of the polarizers. This experiment can also be considered to test whether the photons reflected from a single glass plate come from one side of the incident beam only, as the Maxwellian analysis predicts.

5. REFERENCES

1. M. Born, Science 122, 675 (1955).
2. R.D. Prosser, Int. J. Theor. Phys. 15 181 (1976)
3. To be published.

ON NONLOCAL INFLUENCES

C.W. Rietdijk

Pinellaan 7, 2081 EH Santpoort-Zuid
The Netherlands

ABSTRACT

First, we give a graphical proof that quantum mechanics implies nonlocal influences. It is independent of the assumption of determinism and realism. Within this scope the confusion about realism is discussed. Nonlocality can be explained by means of retroactive effects. Further, we treat a thought experiment in which assuming retroactivity appears to be the only way to avoid a paradox. Retroactive influences only operate where causality does not: within quantal uncertainty margins. Thus they may restore determinism. It is explained why a + beam of photons and a × one are equivalent if we cannot know separate polarizations, whereas they are not if we can know the polarizations. Subsequently, we discuss far-reaching consequences of the nonlocal coherences in micro-processes. Thus, nature has to be conceived as a four-dimensional structure of events rather than objects, in which the action quantum h is the "atom of events" and a new action metric solves the nonlocality paradoxes. The Schrödinger-wave picture is translated into terms of action. Spinors and spherical rotation probably have some direct relation to both the "atom of occurrence" and spin.

1. GRAPHICAL PROOF THAT NONLOCAL INFLUENCES EXIST

In an EPR situation with correlated spin particles it can be proved that the measuring direction of the Stern-Gerlach apparatus on the right-hand side has an influence on some measuring results on the left-hand side, and/or *vice versa*.

In Fig. 1 it is understood that $\vec{a}, \vec{b}$ and $\vec{c}$ are situated in a

G. Tarozzi and A. van der Merwe (eds.), Open Questions in Quantum Physics, 129–151.

plane perpendicular to AB, as are $\vec{p}$ and $\vec{q}$. $\vec{b}$ and $\vec{p}$ are mutually parallel, as are $\vec{c}$ and $\vec{q}$. In all five directions we can measure either $+1/2\hbar$ or $-1/2\hbar$, whereas one will always find - in the $\vec{b}$ direction if + is found in the $\vec{p}$ direction. The same thing holds for the $\vec{c}$ and the $\vec{q}$ directions.

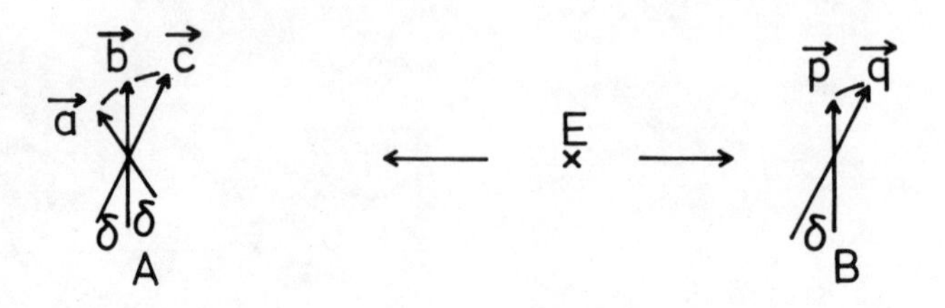

Fig. 1. EPR measurement situations with Stern-Gerlach apparatuses A and B.

In the first place, this implies that both at A and at B the + or - outcome has to be predetermined if one assumes locality. For if one measures the spin, say, in the $\vec{c}$ direction at A, the only thing one has to do in order to be in a position to predict the outcome is, first to measure the spin in the $\vec{q}$ direction at B. (We take AE>BE.) If this produces +, the prediction of a - result at A is 100 percent certain if we assume quantum mechanics to be correct. But if such outcome can be predicted, it has to be predetermined. In the same way, one could have predicted the outcome of any measurement at B (say, in the $\vec{p}$ direction) by first measuring the spin of the correlated particle in the direction parallel with $\vec{p}$ at a location C, so that CE<EB.

Moreover, on the assumption that no nonlocal influences exist, one can conclude that nothing at A can be different if the A measurement were not at all preceded by a B measurement. That is, the A outcome must have been predetermined in the latter case, too, while a similar thing can be said about the B measurement.

In our following argument we assume realism only in the sense that we can logically argue about some *situation* in the A and B regions. "Situation" is meant here in the most general sense. For example, if the A and B situations only existed in our heads, say, as projections or dream pictures, then the conclusions of our reasoning would relate to such dream pictures and the laws ("regularities") they conform to according to our experience. We do *not* assume realism in the sense that what we call objects of observation (which are often called the "outer world", e.g., the A and B situations) should be independent of the observer. We only assume this so far as we preliminarily assume locality, i.e., that the A and B situations are independent of *each other*. This includes that the A situation will preliminarily be assumed to be independent of B *observations*, too, and *vice versa*. This is the locality assumption.

Finally, we assume that the known logical and physical laws remain dependable in all situations in which we experienced them to be dependable up to now. This assumption allows us to argue "counterfactually," that is, to draw conclusions about how the outcome would have been of an experiment that is not actually performed, but of which experience and/or logic teach us how it would proceed.

From these starting points, we can argue as follows about the EPR experiment of Fig. 1:

(a) If we apply the direction $\vec{b}$ at A and $\vec{p}$ at B, we get complete anti-correspondence, say,

A result: + + + - + - - + - - + - + + - -,
B result: - - - + - + + - + + - + - - + +.

(b) If we "counterfactually" had applied the direction $\vec{c}$ at A and $\vec{p}$ at B for the same beam of spin particles, quantum mechanics implies that a fraction of $\sin^2 1/2\delta$ would not have anti-corresponded. We would have got, say,

A result: + + + - + - $\overset{\vee}{+}$ + - - $\overset{\vee}{-}$ - + + - -,
B result: - - - + - + + - + + - + - - + +.

Note that the B result would have been the same as in case (a) because of our assumption of locality: The change $\vec{b} \to \vec{c}$ at A has no influence at B. The *deterministic situation* at B would have resulted, therefore, in the same measurement series if we actually replaced $\vec{b}$ by $\vec{c}$ at A.

If $\sin^2 1/2\delta$ is 1/8, say, we could expect an A series as given above after an "A result", in which the ∨-marked elements, say, differ from the corresponding ones in the case (a).

(c) Combine the measuring directions $\vec{a}$ and $\vec{p}$. Again, the B series would have been the same, whereas the ∧-marked elements, say, of the A series would have been different from the corresponding ones of the A series in case (a).

A result: $\overset{\wedge}{-}$ + + - + - - + - $\overset{\wedge}{+}$ + - + + - -,
B result: - - - + - + + - + + - + - - + +.

It is clear that the A series in the $\vec{a}$ and the $\vec{c}$ case, respectively, would roughly differ by a fraction $f \leqslant 2\sin^2 1/2\delta = 1/4$ of their elements if $\vec{p}$ were applied at B each time. The $\leqslant$ sign applies instead of = because a ∨ and a ∧ mark might have been over the same element of the A series.

(d) Combine the directions $\vec{c}$ and $\vec{q}$. Then nothing would have been different from the situation (b) as regards the results at A, because the change $\vec{p} \to \vec{q}$ at B has no influence on the deterministic situation at A according to our locality assumption.

(e) Finally, combine the directions $\vec{a}$ and $\vec{q}$, which have a mutual angle of 2δ. Then, quantum mechanics requires that the fraction of non-anticorrespondences between elements of the a and q measurement series, respectively, is roughly

$$\sin^2(1/2\cdot 2\delta) = 4\sin^2 1/2\delta \cos^2 1/2\delta = 4\cdot 1/8(1 - 1/8) = 7/16$$

This implies that if we apply $\vec{q}$ at B, the $\vec{a}$ and the $\vec{c}$ series would have differed by a fraction $f' \approx 7/16 > 1/4$, which gives a contradiction with (c) above. Our suppositions, that substituting $\vec{q}$ for $\vec{p}$ has no influence on measurement results at A and that substitutions at A cannot influence results at B, therefore appear to violate quantum-mechanical predictions for the relevant experiments.

We observed above that, for our proof to be rigorous, it is only necessary that we can argue about "situations" at A and B; the nature of such a situation is not relevant to the validity of the argument. That is, if such situations and the "outer world" in general, e.g., would be dream pictures, such argument would establish that, if we apply the quantum predictions and the logic of our dream to the pictures of our dream relating to the A and B environments, a contradiction appears if we assume the pictures about A measurement results to be independent of what we experience in our dream as choices about measurement directions at B, and *vice versa*.

In such a case, a "nonlocal influence" from A to B would not be realistic in the traditional sense. But neither would electric attraction be or, say, Ohm's law: All of them would only coordinate dream pictures, making them more coherent. Therefore, nonlocality and nonlocal influences in any case do not constitute exceptions among physical relations and influences in general: Their reality is of the same kind.

Both as regards reality (i.e., situations) and counterfactual arguments, we only need to know from experience (and probably from innate logical intuition) to what physical and logical laws the situations that are considered will conform. The "proper" nature of such situations (dream pictures and traditional reality) is not relevant here. It is only the sense data and the regularities (laws) that we experience such data to be conforming to which matter. Our only assumption is that it is allowed to extrapolate such laws from our experience to similar situations -- which is an assumption of science in general.

Of course, if we are no longer interested in the "true" nature of situations about which we argue (or, rather, we cannot know it for certain), realism in our sense -- which means considering "situations" as parts or aspects of a really existing world -- becomes a truism. For, in any case, I myself am really existing, and so are the sense data I am experiencing. They are, at the very least, real as mental processes. If we apply the laws of our experience to such situations, the locality contradiction mentioned appears if we so often consistently apply these laws that we feel entitled to assume that they are also valid in nonverified cases that are identical to those in which verification actually took place ("counterfactual" situations). That is, we assume that we *would* have been dead *if* we jumped from the Empire State Building.

The general conclusion is that if we want to coordinate our experiences (sense data and logical intuition or intelligence) in an optimally consistent and coherent way, applying natural laws consistently to all (possibly "counterfactual") situations in which we experienced them to be valid -- without, moreover, bothering about the "proper" nature of such situations (parts of aspects of a really existing world) -- then nonlocal influences are necessary in order to avoid a contradiction.

We can generally say: Thought experiments about counterfactual situations are nothing but applications of known natural or logical laws to situations ("beables"). This only depends on our experience from which we derive such laws, and it is legitimate independently of whether such situations, e.g., are codetermined by us as observers; it is also independent of the proper nature of the situations. Both such possible codetermination and proper nature *have been counted in such laws.*

If, on the contrary, we do not require such consistency and coherence of the world as meant above, or if we are not interested in explanations at all (satisfying ourselves with descriptions), then we can do without nonlocal influences. However, then we do not need other "influences", either. We only *describe* the path of the earth around the sun, without being interested in any explanation in terms of gravitational "influences". Quantum physicists rejecting explanations in cases where models producing the Aha-Erlebnis *are actually available* (such as in our nonlocality case) are not consistent if they only do so in the quantum domain and not in other ones.

Of course, our "influences" may be illusions; but so may be all sense data and the causes or forces our minds use to economically coordinate them. The proper nature of the influences may be beyond our comprehension, as may be the proper nature of the other parts and aspects of reality which are the "beables" to which we apply natural and logical laws. The only thing we can say is: Situations, natural and logical laws, and gravitational as well as nonlocal influences all appear to be necessary *in a similar way* for optimally economizing such kind of comprehension of the world of which we appear to be capable in any case (up to now).

Bell proved that *if we add* parameters to quantum mechanics in order to determine the outcome of individual measurements, quantum mechanics implies that the setting of one instrument can influence the reading of another instrument, however remote.[1] The above proof goes one step further: Such nonlocal influence is implied by quantum mechanics *in any case*, unless we either deliberately content ourselves with doing without explanations of regularities and coherences -- which amounts to, e.g., refusing to accept electric forces -- or refuse to apply physical and logical laws consistently in conformity with experience.[2]

2. EXPLANATION OF NONLOCALITY BY A RETROACTIVE EFFECT

We now consider an EPR experiment with orthogonally polarized correlated photons. This is comparable with the experiment of Fig. 1 as regards the nonlocal correlation effects. Fig. 2 gives a four-dimensional picture of the course of matters in Minkowski space. A' and B' are the measurement events with respect to the photons γ_1 (world line E'A') and γ_2 (world line E'B'). The measurements are performed by means of two polarizers with mutually orthogonal axes (crossed polarizers).

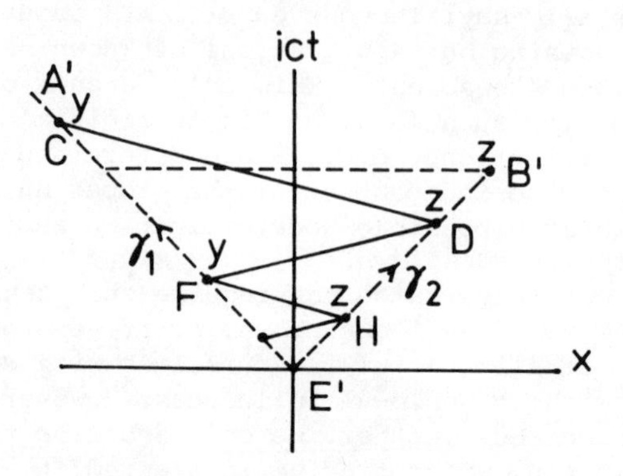

Fig. 2. Four-dimensional picture of an EPR situation with photons.

If then measurement B' finds a z polarization of γ_2, we know that a few moments later, at C, γ_1 with certainty would show a y polarization according to quantum mechanics. Because the orthogonality of the polarizations of γ_1 and γ_2 holds true in other inertial systems than our rest system, too, we can now conclude from γ_1's y polarization at C that γ_2 has a z polarization at point-event D if C and D are simultaneous in some inertial system. Subsequently, we find in the same way that γ_1 has a y polarization at F, that γ_2 has a z polarization at H and so on. That is, the polarizations of γ_1 and γ_2 must have been determined from their common emission at event E'. Because we could have chosen another couple (y', z') of orthogonal polarizer axes instead of (y, z), we see that the choice of such couple retroactively determines the polarization directions of γ_1 and γ_2 already at event E'.

Now the relation of this to the nonlocal influencing we found in Sec. 1 is that if B' can influence E' retroactively, whereas E' can influence A' (in a normal, causal way), then B' can influence A' "nonlocally". In the same way A' can influence B'.

The above explanation of nonlocality also makes it clear that it does not violate relativity.[3] For, relativity only prohibits energy or decodable information from being transmitted over space-like distances. Such transfer does not take place here because (a) only time-like distances are covered by "something" (i.e., causal and retroactive influences), (b) no decodable information can be transmitted over space-like distances in this way, whereas (c) it has not been found that the causal and retroactive influences mentioned under (a) involve any transfer of energy from A' to B', or *vice versa*.

Note, finally, that in the above proof of retroactivity and, therefore, of the existence of nonlocal effects, assumptions about determinism, realism, and the value of counterfactual arguments are not required, so that it only assumes quantum mechanics.

3. THE CONFUSION ABOUT REALISM

Quantum mechanics has led many to the conviction that there is no objective outer world, no objective reality in the sense that the outer world exists independently of the observer. Such conception is often carried through so far that one

(a) denies that comprehensible models of microprocesses can be constructed, so that one has to give up the idea of really explaining them;

(b) even denies that it makes sense to do more than arguing formalistically (instead of realistically) about sense data (positivism).

In any case, it is understandable that finding that the outer world is influenced by, "depends on", the observer (observation) suggested to many that such world in some way or other is *produced* by the observer or, at least, does not allow us to argue rationally about it as an object or situation that would also be there if the observer were not.

The demonstration that retroactive (and, therefore, nonlocal) influences exist, however, in such "technical" a way can explain the influences of observers on what they observe, that no more philosophical "devaluation" of reality and its rational coherence can be based on it. That is, it is clear from the discussion of Fig. 2 that, if the observational act B' can influence emission-event E' retroactively (and so can influence A' nonlocally, too) in an EPR experiment, it is by no means far-fetched to assume that similar influences of observational (and, more generally, "absorptional") events on "emission" events can appear more generally; that is, in other cases than the EPR one, too. Then, the process E' in the outer world is actually influenced by the observer producing a measurement-event B' with respect to E'. This does not however make E' or the outer world in general less objective, less real or less susceptible of rational argument (which susceptibility might lead to comprehensible models). We have to realize that our observational acts (and we ourselves) and the rest of the world are two-sidedly interrelated, i.e., the first are part and parcel of a realistic (really existing) nature. Therefore, dependable and understandable models of processes will in principle have to include effects of observational events, too.

Otherwise, *unrecognized* retroactive (nonlocal) influences from observational events may "miraculously" disturb the coherence of the "outer world", incorrectly suggesting that it has a "subjective", nonrealistic character and that it is incomprehensible in principle.

Actually, nonlocality can make the "new way of thinking" of the Copenhagen interpretation superfluous, and, in particular, the above conclusions (a) and (b). It can restore reality and the primacy of models and the Aha-Erlebnis: Though the outer world indeed co-depends on the observer, it has the same kind of reality, coherence, and rationality as a living cell under the microscope, which may also be influenced (e.g., destroyed) by our observational act. Humans and their observational acts are part and parcel of nature and of the outer world of other observers. They cannot be considered to be something extra- or supernatural, but at least they fit into the all-embracing model of reality the world itself is.

4. AN IMPORTANT THOUGHT EXPERIMENT IN CONNECTION WITH THE ROLE OF NONLOCALITY (RETROACTIVITY)

Consider Fig. 3. The photons γ_1 and γ_2 are correlated, having orthogonal polarization planes. A is a polarizer with vertical axis, B is a quarter-wave plate. We take AE<BE. A quarter-wave plate in the position of B transmits y- and z-polarized photons unchanged as to polarization. Linearly polarized photons with polarization directions making angles of $\pm 1/4\pi$ with the +y and +z directions are transmitted by B as either left-circularly or right-circularly polarized.

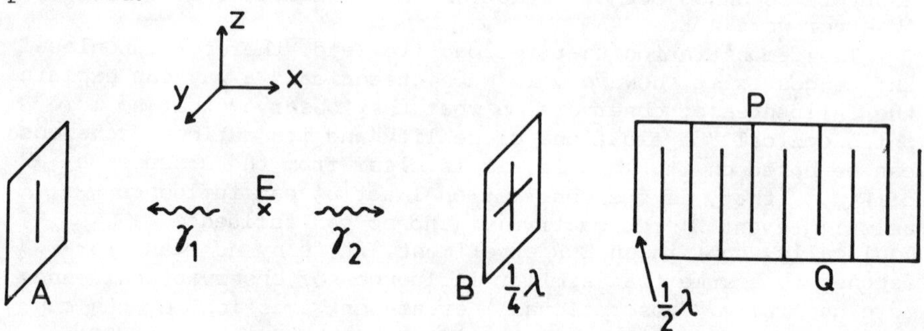

Fig. 3. EPR situation with photons. A is a polarizer with vertical axis, B a quarter-wave plate, and P and Q are systems of mutually connected half-wave plates.

P and Q are two systems of n mutually connected half-wave plates with their slow and fast axes in the y and z directions, respectively. A circularly polarized photon changes its helicity in passing each of the plates of P and Q. A left-circularly polarized photon sees its helicity change to $+\hbar$ from $-\hbar$ in passing each of the plates of P, and it sees it change to $-\hbar$ from $+\hbar$ in passing each of the plates of Q. For right-circularly polarized photons it is just the reverse.

Now we consider two experiments with our apparatus.

(a) First, we apply A in the drawn position with its axis in the z direction. If γ_1 passes A, quantum mechanics implies that γ_2 passes B, P, and Q as linearly polarized in the y direction. If γ_1 is not transmitted, γ_2 will pass B, P, and Q as linearly polarized in the z direction.

(b) Second, we apply A with its axis in the y' direction, which makes angles of $1/4\pi$ with both the y- and the z-axis. (See Fig. 4.) After a transmission of γ_1 quantum mechanics predicts that γ_2 will hit B with its polarization direction parallel to the z' axis, so that it will leave B as, say, right-circularly polarized.

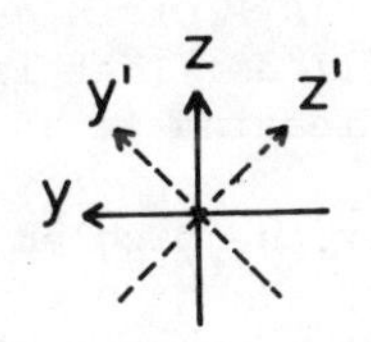

Fig. 4. The y-, z-, y'-, and z'-axes playing a part in the experiment of Fig. 3.

Subsequently, γ_2 is transmitted by P and Q, changing its helicity in each half-wave plate; to $-\hbar$ from $+\hbar$ in each of the n plates of P, and to $+\hbar$ from $-\hbar$ in each of the n plates of Q. That is, γ_2 transfers an angular momentum of $+2n\hbar$ to P and one of $-2n\hbar$ to Q. This implies that γ_2 reveals its angular momentum to P and Q because $2n\hbar >> \hbar$ is measurable in principle.

If γ_1 is not transmitted by A, γ_2 will hit B as linearly polarized with its polarization plane parallel to y', whereas it will leave B as left-circularly polarized. In a similar way as right-circularly polarized photons it will reveal its helicity to P and Q.

If we repeat the experiment with N photon couples, the γ_2 halves of which will hit P and Q as a series of randomly mixed left- and right-circularly polarized photons, the calculus of probability teaches us that the absolute value of the total angular momenta transfered to P and Q, respectively, will be $\approx 2\sqrt{N}\, n\hbar >> 2n\hbar >> \hbar$, so that the measurability is increased.

Now a paradox appears: If we take A's axis parallel to y', then P and Q in principle reveal the helicity of the separate photons γ_2; and if the left observer takes A's axis parallel to y, an observer on the right side *cannot* find the helicity of the photons γ_2 from the behaviour of P and Q. For, if he could, the photons on the right would not have remained *linearly* polarized, as they should according to quantum mechanics. Note here that photons whose helicity is known cannot be linearly polarized.

But then we could transmit signals with a velocity greater than that of light. For, by alternating A's axis, say, according to the Morse code, the left observer could cause the γ_2 photons to behave measurably differently in P and Q according to the same code.

One might suggest the idea of solving the paradox by assuming that also the γ_2 photons which leave B as *linearly* polarized (y- or z-polarized) reveal their helicities to P and Q (becoming circularly polarized in the process), as do the photons hitting B as y'- or z'-polarized. However, then again a paradox arises: Photons γ_2 which can be *predicted by an observer at A to certainly pass a polarizer at B with, say, a z-directed axis* -- and which always would actually appear to do so in experiments if quantum mechanics is correct -- *would nevertheless pass the system (B,P,Q) in quite another way than both y- and z-polarized photons will do.* That is, they would disclose their helicities to (P, Q), leaving it as photons with separately known helicities.

In order to find a solution for the problem of superluminal signals without getting a new paradox, we again consider Fig. 2. We saw in connection with it that in certain experiments $+$, $\times$, and similar "unpolarized" beams are actually retroactively attuned to the subsequent measurement events in a polarizer. Well, then the solution of our problem can be as follows.

(a) If γ_2 photons in our experiment are caught by a polarizer at B with its axis parallel or orthogonal to A's axis, the retroactive "attuning" discussed with Fig. 2 proceeds as usual, it predirecting the polarizations at E' in coherence with the situations at A and B.

(b) If, however, the γ_2 photons are caught by our (B, P, Q) system, the retroactive influence originating with such a completely different measurement event will be different, too. For example, it may stop working alltogether, leaving the γ_2 beam "completely unpolarized", with random polarization directions, $\ast$, *quite independently of the question whether A has its axis in the y, y', or any other direction.*

Thus, the existence of retroactivity would save both quantum mechanics (no measurable difference between $+$ and $\times$ beams in this case, too) and special relativity (no superluminal signals), whereas all predictions of quantum mechanics would remain valid.

Those accepting quantum mechanics but refusing to accept retroactive influences will at least need another kind of nonlocal influence here in order to avoid paradoxes.

We now apply retarding devices in our experiment that do not change polarization states, so that the A observer can let the B observer know in time whether a y-, z-, y'-, or z'-polarized photon is approaching his apparatus. One new phenomenon is then that the B observer is now in a position to actually separate the y and z photons of a + beam and, similarly, the y' and z' photons of a × beam. This will appear to help us in drawing important conclusions from our experiment, which can give more insight into the nature and function of retroactive influences.

In the first place, after the introduction of the retarding device, it would not even be possible to signal with a velocity greater than c if + and × beams would then start to behave differently. For, say, the Morse codes transmitted after such introduction would have a velocity $\leqslant$ c.

We predict that the + and × beams will indeed behave differently now. That is, if a retarding device is used, they change their behaviour by "suddenly" becoming mutually distinguishable. For if they did not, a paradox would again appear. This can be seen as follows. The B observer can separate the | and the — photons from a + beam and afterwards make the | and the — beams pass separately through the (B, P, Q) system. Of course, they will then behave as normal z- and y-polarized beams, continuing to do so in measurements after their transmission by (B, P, Q), so that the γ_2 photons will not reveal their helicities in (B, P, Q).

In the same way the B observer can separate the / and the \ photons from a × beam before they reach B and then make the / and the \ beams separately pass the (B, P, Q) system. Then the / beam will behave as a normal / beam, i.e., it will become a normal right-circularly polarized beam after having passed B and remain so after its passage through P and Q.

Now it is clear that also if the / and the \ photons are not separated, but pass (B, P, Q) while the B observer knows their separate polarizations, they will behave in the same way as when they are grouped into two polarized beams, disclosing their helicities to P and Q. A similar argument holds for the | and — photons which are known as such by B: They do not disclose their helicities to P and Q. So, the fact remains that + and/or × beams *behave physically differently* according as either only the A observer or both the A and the B observer know the polarization state of each separate γ_2. That is, at least some of the y, z, y', and/or z' photons *separately behave differently under such two alternative conditions*. For, if all four kinds behaved in the same way in both situations, one could discriminate between + and × beams in the first case, too, as one can in the second case. (And we saw that this would imply the possibility of superluminal signals.)

Now the question arises: How can the mere knowledge of the B observer about polarization states -- which knowledge is *additional* to that of the A observer, the latter knowing each separate polarization in any case if quantum mechanics is correct -- ever influence the behaviour of the γ_2's, as it seems to do? In any case, the difference made by the retarding device will have to change the behaviour of the γ_2 beams in order that a paradox be avoided.

We propose the following explanation.

If retroactive influences exist (as we proved), they will not cancel or frustrate the operation of causal influences; in any case, we have no indications of such effect. Then it is obvious that retroactivity only operates within the small margins of freedom which the quantum uncertainties, such as Heisenberg's Δx, Δp, etc., allow it and within which causality apparently does not settle matters. Generally, we can conceive of the retroactive influence as a -- nonlocal! -- hidden variable (co-)determining the outcome of experiments exclusively within the uncertainty margins characteristic for quantum mechanics.[4] Now, one of such margins relates to the "actual" polarization directions of the separate photons or spin particles in an unpolarized beam. So, the retroactive influence has the latitude to frame an "unpolarized" beam into, say, a + , × or ✳ beam in which the distribution of, e.g., the | and the — polarized photons is random. Or it may transform an "unknown" + beam into a $\hat{\times}$ one (i.e., one in which we cannot tell the / and the \ photons apart). But it can only do so if the polarizations of the "unpolarized" photons (say, of our γ_2 beam) separately are really undetermined *so far as causality is involved*, for retroactive influences cannot interfere with causal ones. They can only "fill the gaps" that the latter leave as regards a complete determination of a process. And now there is no uncertainty margin left at all by causal influences in our experiment if the B observer can know the polarization direction of each separate γ_2. For, the fact that he can know it with certainty means that *he can derive it with certainty from causal processes which then obviously do not leave any latitude for retroactive influences to be operative without contradictions would appear*. One can also say: The reason why B observers cannot know whether an approaching photon is polarized | , — , / , or \ is exactly the same as the reason why no causal influences can reach the B region in time to completely determine the polarization states, viz., the circumstance that both the relevant information and causal influences would have to have a velocity exceeding c in order to reach the B event from the A measurement.

We see here the very reason why a *physical* difference exists between, say, a + beam, of which one can *on the spot* tell the | and the — photons apart, and a + beam, of which we cannot. Only in the latter case does a subsequent measurement-event have a margin, or latitude, for retroactive influencing.

The above hypothesis -- i.e., that retroactive influences constitute a hidden variable that (co-)determines situations, concrete values of physical entities for which causality leaves an uncertainty margin -- not only can explain the behaviour of the γ_2's in our thought experiment, but it can also give a general explanation of the circumstance that + and × beams are experimentally mutually indistinguishable if and only if the experimenter cannot know the polarization of each separate photon. (We recall here that it appeared from our thought experiment that the possibility of knowing the separate polarization directions even influences the behaviour of individual photons.) For, only in the latter case can a retroactive influence "mould" an unpolarized (say, ✳) beam into a +, ×, or other shape depending on what circumstances -- i.e., the experiment to be performed -- require.

In short: If and only if a B observer can know in time (from A) what the separate polarizations are, *causal influences can reach B from A in time, too* (both the information and causal influences having velocities $\leqslant c$), they not leaving any latitude for any retroactive attuning of the γ_2 beam to a subsequent experiment, e.g., with either a polarizer or the (B, P, Q) apparatus. As regards A's quantum predictions about the behaviour of the γ_2 photons, e.g., about the orthogonality of the γ_2's with respect to the γ_1's in crossed polarizers experiments, we can say that such predictions (knowledge) are based upon his experience with respect to the relevant situations, *in which experience the retroactive influence inherent in quantum mechanics has been counted*. But then such experience about crossed-polarizer situations may not be indicative of what will happen to the γ_2's if the B situation is different, i.e., if the apparatus consists of the (B, P, Q) system. For, in such case, *a different retroactive influence* will be operative on the γ_2's, so that they may behave differently now. We saw that this is necessary indeed in order that a faster-than-light paradox be avoided in our thought experiment. In some other situations there may be no role at all for the retroactive influence, e.g., if we apply a retarding device.

We can conclude that the at first paradoxical circumstance, that the γ_2 photons behave differently according as either only an A observer or both A and B observers can have knowledge of all separate polarizations, fits completely into the function that retroactive influences probably perform as nonlocal hidden variables.

5. SOME CONSEQUENCES OF NONLOCALITY FOR PHYSICAL THEORY

We already discussed that nonlocal influences in principle can restore both realism and the possibility of constructing understandable models of micro-processes. In Sec. 4 we also saw that nonlocality, i.e., retroactive influences, can be conceived as hidden variables. Though we did not prove that such hidden variable actually determines the outcome of individual measurements -- let alone that it allows observers to predict such an outcome -- we can indirectly infer determinism from its existence. For if retroactive influences from the future are operative, such a future has to be real, must exist in one way or other; that is, it has to be determined.

However, such pre-existence of the future may have a more far-reaching consequence, too. For if retroactivity exists, and the future is so real that it can have influence on what happens now (as, e.g., holds for the emission-event E' in Fig. 2 that is influenced by A' and B'), we have to view the four-dimensional character of the world very seriously. We would have one more example of the fact that new insights in physics not seldom result from taking the mathematical formalism more seriously. In our case, the formalisms to be treated very seriously are:

(a) the quantum formula

$$\psi = \frac{1}{\sqrt{2}} [\psi_y(1)\psi_z(2) - \psi_z(1)\psi_y(2)]$$

for correlated systems and its nonlocality consequences, and

(b) Minkowski space and its metrical relations.

We refer to Refs. 3 and 4 for a more extensive treatment of what is summarized in this section.

Now the basis of our theory is that, the universe being truly four-dimensional, *events*, or occurrences, rather than objects, are the actual substance or building-blocks of the world, the real contents of the four-dimensional universe. We summarize some implications of this conception as follows:

1. If events rather than objects constitute reality, this has an important consequence for the *metric*. For as it is natural in configurations of objects to measure distances by means of objects, i.e., measuring rods, so in four-dimensional configurations of events, or occurrences, *it is equally obvious to measure the latters' distances by means of quantities of occurrence, that is, by asking how much "occurring", or action, it implies to get to event B from event A, or to transform event A into event B.*

2. If, e.g., a momentum carrier P approaches a grating G, we no longer ask: "What is the distance in meters between the *object* P in the alternative positions P_1, P_2, or P_3", but: "What is the difference, measured in occurrences, i.e., *the action difference for the process*, between the following alternative elements of occurrence:

(a) P approaches G via p_1,
and
(b) P approaches G via p_2."

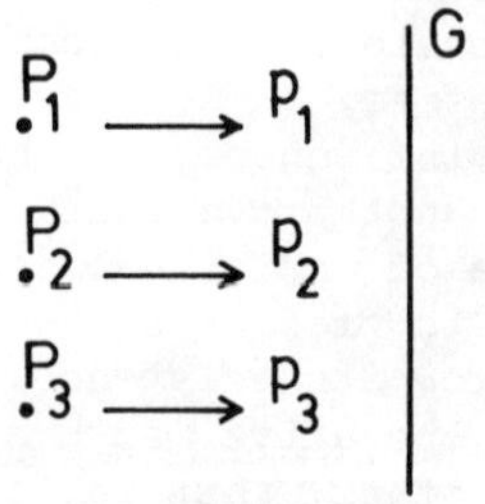

Fig. 5. Momentum carrier P approaching a grating G.

The answer -- i.e., the difference in terms of the action $Et - \vec{p}\cdot\vec{x}$ for the process -- is: exactly zero. This means that for nature the alternative events (a) and (b) are contiguous, not even really different at all as far as the process in question is concerned. Small wonder that it "emits" the parts p_1, p_2, p_3, ... of the "waves" representing the "particle" all "at the same time"! It is our scheme of reality in which the four-dimensional action process -- i.e., the momentum carrier P approaching G -- is in part distortingly resolved into terms of a particle having momentum and traveling from one location to another along a definite trajectory. Quantum mechanics already undermined the reliability of such scheme, but now we can see why it is partly unrealistic: It constrains physically relevant influences and distances to our now-hyperplane of the four-dimensional world, in which hyperplane they cannot be integrated into a consistent, understandable model.

3. Consider an action picture of an EPR experiment in which A and B represent diametrical measurement events (see Fig. 6).

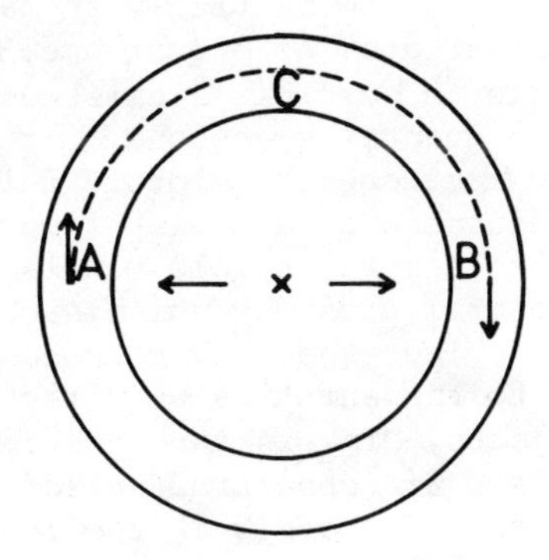

Fig. 6. The equi-action plane (sphere) ACB in an EPR situation. A and B represent measurement events.

Then the action difference between the events A and B, which defines the physically relevant distance between them, is zero because ACB is an equi-action plane (sphere). So we get a case similar to that in 2. above, apparently mutually distant elements of an integral process becoming integrated again. Therefore, A and B appear as physically contiguous events *in the internal action metric* of the process that the moving EPR systems constitute. Via ACB they can influence each other directly. They are nonlocally connected if we argue from the standpoint of our "coordinate scheme of objects", not reckoning with an action-based metric. The connection between the nonlocalities discussed with Figs. 2 and 6, respectively, is made clear in Refs. 3 and 4.

4. The essence of the above is that action differences, or action distances, are physically more relevant than Minkowskian ones (in micro-processes; see Refs. 3 and 4), in a similar way as

the latter are physically more relevant than Euclidean ones. If, then, we put such distances first and foremost within micro-processes the nonlocality paradoxes of quantum mechanics disappear, this aiding us in making consistent and coherent models of such processes.

In short: Action is primary in nature, also as regards metrics and models.

5. In the same vein, not only distances but everything else in physics is derived, or constructed, from action, action quanta, and their mutual relations in the structure or lattice S formed by all action quanta in the world; in fact, S constitutes the four-dimensional physical universe. Consequently, all nonlocal phenomena which can be explained by such action model of the world are relativistic. For action is relativistically invariant and all physical concepts and entities derived from it fit into the theory of relativity.

6. The quantum of action h quite naturally appears in this conception as the (indivisible) atom of occurrence, i.e., as the elementary event. For example, a particle's existence in time then consists of $\nu = mc^2/h$ action quanta per second.

7. In Sec. 3 we already mentioned "nonlocal" influences from the observer on observed phenomena as a cause of our not succeeding in constructing realistic and comprehensible models of the outer (micro-)world. As a more general cause by which such models remained elusive, we can now state the fact that our measurements, concepts, and formulas mostly relate to the entities we directly experience: objects, forces between objects, Minkowskian (instead of action) distances, etc. However, *these are unnatural building-blocks for the construction of understandable models in the real four-dimensional world*, which consists of processes, events, quantities of action, action distances and other concepts and entities constructed from action (quanta). Action distances and other action concepts are more realistic elements for constructing such under-standable, nonparadoxical models. The traditional concepts are comparable with Euclidean space.

Accordingly, our theory primarily contains a substitution of the latter concepts by action concepts, and the gist of such change can be defined as follows: What we do in positing that the action $Et - \vec{p}.\vec{x}$ is primary and that, e.g., E, t, $\vec{p}$, and $\vec{x}$ are derived quantities, goes one step farther than what special relativity did, i.e., *we integrated space-time and energy-momentum jointly into action*, as relativity already integrated, among other things, space and time into space-time *corresponding to a new (Minkowski) metric*. Our new metric is the action one which, in turn, leaves the Min-kowski metric as a "theoretical scheme" (like the earlier Euclidean metric) that produces metrical discrepancies if we "interpolate" it into micro situations where separate quanta and their internal action distances become important. Precisely such discrepancies underlie the nonlocality paradoxes.

The Minkowski scheme can be conceived as a macro-ordering scheme reflecting symmetries of the action-quantal lattice S referred to above. A more detailed study of some space-time characteristics of action quanta allows us to construct models that make it understandable both how retroactive effects come about and how they contribute to the explanation of the $|\psi|^2$ probability rule, e.g., in the interference of waves (see in particular Ref. 3 on this).

6. THE NEXT STEP: TRANSLATION OF THE QUANTUM FORMALISM INTO ACTION QUANTITIES AND CONCEPTS

1. After the in-principle substitution of a world and metric of objects by a world and metric of action (or occurrences) it is important to see how we can concretely *translate the quantum formalism into, and/or conceive of it in terms of, action quantities and concepts*. The current formalism is essentially a description of an action world and its physically relevant action metric in terms of our ordering scheme, in which space-time and energy-momentum have not yet been integrated into action. Such non-integration produces the nonlocality paradoxes in an analogous way as the non-integration of space and time produced paradoxes like the constant velocity of light. For, the latter non-integration caused our Euclidean space and time distances to show discrepancies with respect to the physically more relevant relativistic ones, whereas the former non-integration, in turn, causes Minkowskian distances to show discrepancies with respect to the physically more relevant action ones, which gives the impression of nonlocality in our scheme of reality in which Minkowski metric and three-dimensional objects (energy-momentum) have not been integrated into action. In other words: The translation of action phenomena and the action metric into our world of objects and the Minkowski ordering scheme implies metrical distortions in the microsphere that we experience as nonlocality.

2. A well-known example of a translation similar to the one meant at the beginning of this section -- i.e., from a world and metric of objects into ones of action -- is the translation of the classical equations of motion into the principle of least action, from which they can also be derived. Up to now, such a translation was found to be very convenient mathematically, but we did not take it very seriously from the physical viewpoint.

Now our point is that similar translations will be necessary for the quantum equations of motion, too, and that the latter will become understandable in the process. That is, our hypothesis is that, after such translation and/or conceptional changes in the direction of action concepts, the quantum equations of motion will appear to relate to realistic four-dimensional models of processes which are structures of action quanta playing the part of elementary four-dimensional building blocks of occurrences, or events. That is, such equations would no longer mainly be a mere formalism producing correct predictions. (Note further that events are not the same things as point events: They are instead occurrences with space-time dimensions in which energy-momentum is also involved.)

3. As regards the Schrödinger equation, the translation meant in 2. above is already possible. Consider, in the first place, Fig. 7, representing a four-dimensional picture of the wave pattern of one Fourier component of a Schrödinger-wave packet $\psi(x, t)$.

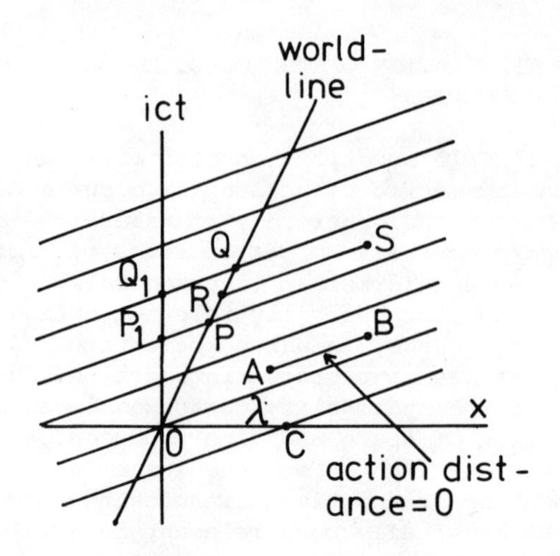

Fig. 7. Four-dimensional picture of the wave pattern of one Fourier component of a Schrödinger-wave packet.

This is explained more extensively in Refs. 3 and 4. The picture represents waves like $OC = \lambda = h/p$ in our now-hyperplane Ox and, at the same time, $P_1Q_1 = ic\,\Delta t$, in which $\Delta t = 1/\nu = h/E$. That is, the essential formulas $\lambda = h/p$ and $E = h\nu$ have been included in the representation. At the same time, world-line sections like PQ represent one "proper" quantum of action, of which there are $\nu_0 = m_oc^2/h$ per second in the existence in time of a particle; ν relates to our rest system, ν_0 to that of the particle. However, the slices forming a four-dimensional picture of the Schrödinger-wave pattern (in the case of a monochromatic wave packet) at the same time are "stretched" representations of action quanta in Minkowski space. For, the action of the process constituted by a uniformly moving particle is the same, e.g., in R and S, and also in A and B. Reckoned from the action zero at 0, it is (5/2)h at both R and S and 1/2h at both A and B. This means that R and S, in the same way as A and B, are mutually physically contiguous (or "nonlocally connected") in the action metric: their mutual physical distance is zero. One can also formulate it by saying that the quantum process taking place on -- actually, the quantum process constituting -- the section PQ of the particle's world line makes itself physically felt in the whole slice to which the point events P, Q,

R, and S belong. In the Minkowski space of Fig. 7 action quanta are stretched to the slices pictured, and that because of the above-mentioned discrepancy between the action metric, in which, e.g., the distance between R and S is zero, and Minkowski metric, in which such distance is finite. The slices, that is, four-dimensional pictures of matter waves, are at the same time (simplified) pictures of action quanta that are deformed in the Minkowski scheme. It is explained in Refs. 3 and 4 how the picture of quanta is modified if we also reckon with the Heisenberg margin Δp as regards the momentum, which gives rise to the various Fourier components of a wave packet.

In action metric the quanta are compact instead of stretched, the margins of "uncertainty" Δx and Δp being reduced to infinitesimal values. Δx corresponds here to the maximum length of AB in Fig. 7 in a wave packet with finite Δp, whereas Δp -- that is, the various Fourier components of a wave packet -- corresponds to a similar metrical stretching process as Δx does. Note here that $\vec{p}$ and $\vec{x}$ enter symmetrically in the action $Et - \vec{p}.\vec{x}$. The conditions of the relevant process allow mutually coherent margins Δx and Δp for x and p, respectively, within which the relevant action remains the same. (Compare again, as a simple example, A and B of Fig. 7 within the margin Δx; note that we are considering Δx from the standpoint of the rest system of the particle which, in Fig. 7, has a velocity unequal to zero.)

Thus the Schrödinger equation, which leads to a specific four-dimensional wave pattern described by the wave function $\psi(x,t)$ in question, has something very direct to do with the series of stretched action quanta (slices), which a process (e.g., a freely moving particle) actually is. That is, such a wave pattern corresponding to a wave function is nothing but a structure (e.g., a series) of action quanta as they are stretched in Minkowski space. *This is the translation of the quantum equation of motion, which the Schrödinger equation is, into a realistic model of an action-quantal structure* (which we call a process, or occurrence). The Schrödinger matter waves appear to be realistic quantal entities, i.e., stretched images of action quanta in Minkowski space.

Finally, hidden variables appear in this picture as nonlocal (among other things, retroactive) influences of integrated four-dimensional processes on their "separate" sub-events (emission, reflections, absorption, etc.).

4. Under point 3. we discussed the direct relation between action quanta and Schrödinger-wave functions defining wave patterns. Now spinors are the wave functions of the Dirac theory. Therefore, we may surmise that, just as Schrödinger-wave functions and the corresponding waves (slices) in a *rough* way correspond to action quanta, spinors can be seen as corresponding to (series of) action quanta in a more *precise* way, inasmuch as they incorporate spin and guarantee full relativistic invariance. Remember here that the slices are no more than rough pictures of quanta; they, e.g., reveal nothing about the proper nature of the actual process that a quantum of action is.

In this connection it is important to realize that spin probably has some very direct relation to action quanta. Thus, the spin rotation as a process may be a direct manifestation of the elementary process constituted by one quantum h. For example, one spin "rotation" corresponds to an action of $2\pi \times 1/2\hbar = 1/2h$, whereas *two* successive "rotations", not one, restore the spin particle's initial state, they at the same time completing one quantum of action. (Compare here: "According to quantum theory, a neutron or another particle with spin does not return to its initial state when its orientation is rotated through 360 degrees. Instead it takes two full turns, a 720-degree rotation, to restore the state of the particle to its initial condition."[5])

At the same time it is remarkable -- and this may be very important -- that the period of the spin for an elementary particle is the same as the period, the lifetime, $\Delta t = 1/\nu = h/mc^2$ of one action quantal section of the particle's world line if we start from the following preliminary realistic model of the spin-generating process in a particle.

Consider a spin particle to be a rotating dumbbell of two circularly rotating photons for which the Compton wavelength $\hbar/mc$ is the radius of the path and twice the length $2\pi\,\hbar/mc = h/mc$ of the circumference of the circle is the wavelength of the photons (see Fig. 8).

(a) Then the energy of the two photons combined is

$$2h\nu' = 2\hbar\, c/\lambda = 2h\, c/2h/_{mc} = mc^2,$$

equal to the energy of the particle.

(b) The angular momentum (spin) s of the dumbbell can be found from the momenta $h\nu'/c$ of both photons:

$$s = 2\, h\nu'/c \cdot \hbar/mc = 2\hbar\, h\nu'/mc^2.$$

Because

$$\nu' = c/\lambda = c/2h/_{mc} = 1/2\ mc^2/h,$$

one gets

$$s = \hbar.$$

We see that a factor 1/2 is lacking because the spin is $1/2\hbar$.

In (a) we see that $2h\nu' = mc^2$, in which ν' is the frequency of the photons. At the same time, $h\nu = mc^2$ if ν is the number of action quanta that "elapse" per second of the particle's existence. The period of rotation of the two photons is $2\pi\ \hbar/mc/c = h/mc^2 = 1/\nu$. *So we conclude that the rotation period of the photons corresponds exactly to the period of the quanta.* This makes it very probable that the spin periodic process and the quantal periodic process in the existence in time of a particle are mutually closely connected: Spin is a manifestation to us of the elementary

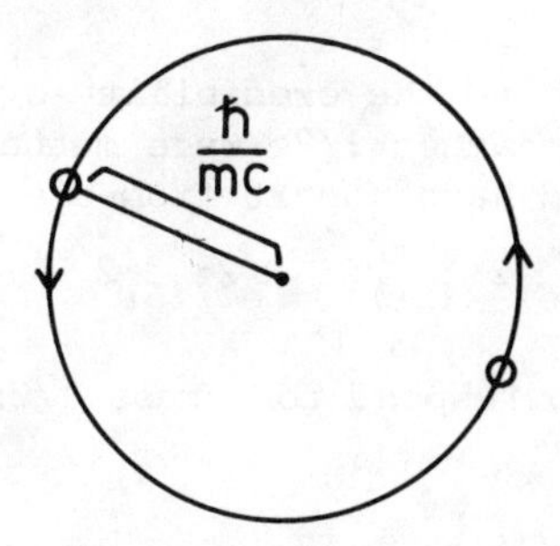

Fig. 8. The dumbbell model of an elementary particle.

event an action quantum is. Moreover, the relevant connection contributes to establishing the conviction that both spin and action quanta represent very realistic processes.

(c) The above "simplistic" model, involving at the same time elementary particles, spin processes, and action quanta, makes the magnetic moment more understandable, too: The magnetic moment μ of a circular electric current due to a circularly rotating electric charge q is $\mu = qvr/2c$, where r denotes the radius and v the velocity. If we assume that our two photons each correspond to $q = 1/2e$, and again take $r = \hbar/mc$ and $v = c$, substitution gives

$$\mu = e\hbar/2mc,$$

that is, for an electron with electric charge e, one gets the Bohr magneton.

(d) One problem remains: Does the energy $h\nu = mc^2$ of the particle only consist of the sum of the energies of the two photons; is there no binding energy? It might be there is none. For if we consider the radius $r = \hbar/mc$ of the internal movement to be parallel with the hyperplanes limiting the action slices, the action distance, for the process, of the photons P_1 and P_2 is

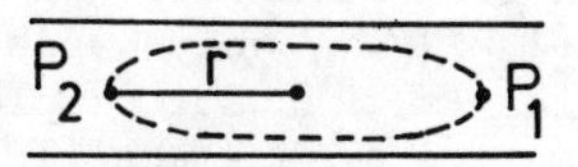

Fig. 9. The action distance between the elementary events represented by P_1 and P_2, situated in a quantum slice, is zero.

actually zero, which might -- we speculate -- correspond to a value zero of the relevant potential energy. Then, also the rest mass (energy) of elementary particles would be explained by our model.

(e) Another feature adding to the credibility of such a model is the following. If the two charges 1/2e were mutually attracted by an electric force, we would have (apart from the sign):

$$F = 1/2\ mc^2/r = (1/2e)^2/(2r)^2 = e^2/16r^2$$

(remember that both photons correspond to a mass $1/2m = 1/2\ h\nu/c^2$). Therefore,

$$1/2mc^2 = e^2/16r = e^2/16\ mc/\hbar$$

or

$$e^2/\hbar c = 8.$$

Actually, the fine structure constant $e^2/\hbar c \approx 1/137$. So, e^2 is much too small to produce the attractive force. If, however, we assumed that *nuclear* forces with a coupling constant G satisfying $G^2/\hbar c \approx 17$ caused the attraction between the two masses corresponding to the photons, we would get a 17·137 times greater force than the electric one, whereas we actually need a figure that is 8·137 times larger. So, only a factor of roughly 2 is lacking if we try to explain the centripetal force that our simplistic model needs by means of forces of the nuclear type.

(f) Finally, the Compton wavelength $r = \hbar/mc$ is well-known in connection with the *Zitterbewegung* and the "smearing out" in the Foldy-Wouthuysen transformation. That is, $\hbar/mc$ has something to do with realistic dimensions of elementary particles.

Of course, our model is simplistic indeed, but some characteristics of it are too remarkable to be purely accidental. In particular, the connection between the quantal period $\Delta t = 1/\nu$ and the rotation period of the "photons", i.e., of the spin, is striking. In the second place it is striking that after *two* rotations of the "photons" one $\lambda = c/\nu'$ has been completed, so that actually two rotations are necessary to restore the initial state, in agreement with the above quotation from Bernstein and Phillips. We can also find the factor 2 from $\nu' = 1/2\nu$, which appears from (b).

The most probable situation is that our simple, provisional model is a classical approximation of a so-called "spherical rotation" (see below). The rather remarkable circumstance that so "primitive" a particle and spin model can explain so much may give us hope that the proper spherical rotation can solve problems such as of the two lacking factors 1/2 and 2 of (b) and (e) above, respectively.

5. At the beginning of item 4, we already suspected that a direct connection exists between spinors and action quanta, which then at the same time would imply the connection between spinors and spin because of the further discussion in 4.

Spinors appear to be the most general entities for constructing

Lorentz-covariant equations. It might be, then, that such "most general entities" have something direct to do with, or even directly correspond to, the most general contents, or constituents, of the universe, which is elementary events: action quanta. Thus, spinors may represent action quanta in a more true-to-nature and complete way than Schrödinger-wave functions and/or wave packets do. In this sense, they would then generalize and "articulate" the (systems of) slices of Schrödinger wave packets as rough (and deformed) pictures or models of action quanta in Minkowski space. The spinor waves, too, would be realistic entities, i.e., "stretched" action quanta.

Accordingly, geometric models of spinors might be very important because they may teach us something about the functioning of the action-quantal process, about what quanta are as a physical phenomenon, and about the way in which the process produces spin in its acting. Probably, both action quanta as elementary processes and spin have something to do with spherical rotation, as have spinors. This direction of thinking could further contribute to the construction of realistic and understandable models of four-dimensional microprocesses, and in particular to the construction of such a model of the elementary action-quantal process. We refer to Refs. 5-8 for a discussion of geometrical models of spinors and the spherical rotation mentioned above.

REFERENCES

1. J.S. Bell: "On the Einstein Podolsky Rosen Paradox", *Physics* 1, 195 (1964).
2. Other proofs of nonlocality, also claiming to be independent of the assumption of determinism and realism, have been given, e.g., by H.P. Stapp: see *Found. Phys.* 9, 1 (1979); 10, 767 (1980); 12, 363 (1982); *Nuovo Cimento* 40B, 191 (1977); *Phys. Rev. Lett.* 49, 1470 (1982).
3. C.W. Rietdijk: "A Microrealistic Explanation of Fundamental Quantum Phenomena", *Found. Phys.* 10, 403 (1980).
4. C.W. Rietdijk: "On the Four-Dimensional Character of Microphysical Phenomena", in S. Diner, D. Fargue, G. Lochak, and F. Selleri (eds.), *The Wave-Particle Dualism*, D. Reidel Publ. Co., Dordrecht, Holland (1983).
5. H.J. Bernstein and A.V. Phillips: "Fiber Bundles and Quantum Theory", *Sci. Am.* July 1981, p. 96.
6. H. Hellsten, *A Visually Geometric Foundation of Spinor and Twistor Algebra*, Thesis, University of Stockholm (1980).
7. R. Kjellander: "A Geometrical Definition of Spinors from 'Orientations' in Three-Dimensional Space Leading to a Linear Spinor Visualisation", *J. Phys. A.: Math. Gen.* 14, 1863 (1981).
8. E.P. Battey-Pratt and T.J. Racey: "Geometric Model for Fundamental Particles", *Int. J. Theor. Phys.* 19, 437 (1980).

EINSTEIN LOCALITY FOR INDIVIDUAL SYSTEMS AND FOR STATISTICAL ENSEMBLES

F. Selleri

Dipartimento di Fisica, Università di Bari
INFN, Sezione di Bari, Italy

ABSTRACT

Einstein locality assumes the existence of a separate reality. Formulated deterministically it shows a sharp incompatibility at the empirical level with the predictions of quantum mechanics. This incompatibility is expressed: firstly, by the validity of Bell's inequality, which is here rigorously deduced with the logical elimination of the completeness idea; secondly, by an observable discrepancy with quantum mechanics for the coincident $\bar{K}\bar{K}$ - detection rate in the $J^{PC} = 1^{--}$ state of two K°'s. A probabilistic generalization of Einstein locality is presented: It leads to a new and more general inequality often violated by quantum mechanics.

1. REALITY AND SEPARABILITY

"If, without in any way disturbing a system, we can predict with certainty (i.e., with probability equal to unity) the value of a physical quantity, then there exists an element of physical reality corresponding to this physical quantity."

This famous "criterion of physical reality" proposed by Einstein, Podolsky, and Rosen in 1935[1] is a very weak criterion of reality, very carefully formulated and extremely general. It appeared to them (and it still appears today) "far from exhausting all possible ways of recognizing a physical reality." This crite-

G. Tarozzi and A. van der Merwe (eds.), Open Questions in Quantum Physics, 153–170.

rion can have many applications to the macroscopic domain where it appears as trivially true: For instance we can say that if we can predict with certainty that the length of a table is two meters (up to some error of, say, half a centimeter) then there exists an element of physical reality corresponding to the length of the table.

Some further specifications of the EPR reality criterion are the following:

(1) The element of physical reality is thought of as existing already before the act of measurement is concretely carried out. From a philosophical point of view this means that a step into ontology has been taken. Every strict positivist would have much to say against this hypothesis, which seems indeed very natural to many people if our planet has to be considered as existing even before man appeared on it (and this amounts to 99,9% of its history!) and the galaxies as existing also when no telescope is observing them.

(2) The element of physical reality is viewed as being associated with the measured object and *not* with the measuring apparatus: The latter has been checked to work properly, so that it could lead to different results of the measured physical quantity.

(3) The postulated element of reality is regarded as the *cause* of the exactly predictable outcome of the act of measurement: Realism and causality are therefore strictly tied together in the EPR reality criterion. The element of reality is also not considered necessarily coincident with the result of the measurement. Therefore, if we take for instance an electron with wave function

$$\Psi(x) = \Psi_0 \exp\left\{i\, \vec{p}\cdot\vec{x}\, /\, \hbar\right\},$$

we can predict with certainty that an apparatus measuring the electron's momentum will find the result $\vec{p}$. Then there exists an element of reality associated with the electron's momentum, and this statement is emphatically not identical with the idea that the electron *has* a well defined momentum before we measure it.

The idea of *separability* is the reasonable assumption that two arbitrary objects interact very little when their mutual distance is very large.

One can wonder what is a "very large" distance. Experience tells us that there is no universal definition for it, but that such a distance, different for each case, exists for all known interactions. A more precise definition of the assumption of separability is therefore the following one: *Given two physical systems, a distance between them exists beyond which their mutual interactions become negligibly small.*

The hypothesis of separability can be applied to quantum systems. These systems are known only through their wave functions, which can fill large regions of space; there are however also situations in which the spatial localization is very good: Take for instance two particles α and ß with wave functions $\Psi_\alpha(\vec{x}_1)$ and $\Psi_\beta(\vec{x}_2)$ respectively.

Suppose that $\Psi_\alpha(\vec{x}_1)$ is of Gaussian form with modulus appreciably different from zero in a region centered around x_{10} and with width r_1. Similarly let $\Psi_\beta(\vec{x}_2)$ be of Gaussian form around x_{20} with width r_2.

We will consider the two systems separated in the previously defined sense if the distance $|\vec{x}_{20} - \vec{x}_{10}|$ is much larger than either r_1 or r_2.

Separability will also be assumed to exclude any kind of propagation that takes place into the past.

In the following sections it will be shown that reality and separability are incompatible with quantum mechanics at the empirical level. This realization, due to the work of many people, has always led me to feel that it is correct to state [2]:

> ... Einstein turned away from Machian positivism, partly because he realized with a shock some of its consequences; consequences which the next generation of brilliant physicists, among them Bohr, Pauli and Heisenberg, not only discovered but enthusiastically embraced: They became *subjectivists*. But Einstein's withdraval came too late.
>
> Physics had become a stronghold of subjectivist philosophy, and it has remained so ever since.

To this I can only add that subjectivism seems to be tied to the very mathematical apparatus of the theory and to the rules giving empirical meaning to its symbols: In fact, the problem is that the *empirical predictions* of quantum mechanics are incompatible with the existence of a separable reality.

Some authors seem to believe that Aspect's recent experiment[3] settles the question against reality and separability. The situation is however much more complicated, and only future experiments will be able to solve the EPR pararadox, as will be discussed in the last section.

2. EINSTEIN LOCALITY AND THE SINGLET STATE

Consider the "singlet" state vector η_S of two spin-$\frac{1}{2}$ particles α and ß,

$$\eta_S = \frac{1}{\sqrt{2}} (u_+v_- - u_-v_+) , \tag{1}$$

where $u_\pm$ $[v_\pm]$ are eigenstates of σ_3 $[\tau_3]$, the third Pauli matrix for particle α [ß], with eigenvalues ± 1. This state will be the basis of the EPR paradox developed in the following.[1]

Let a molecule M be given capable of decaying into two spin -$\frac{1}{2}$ atoms α and ß. There are several concrete examples in which one knows that the spin state vector for α+ß must be η_S as given by (1). We limit our attention to one of these examples and suppose furthermore that the space part of the wave function is separated in the sense established in the first section.

Consider a large ensemble E composed of N decays M → α+ß of the molecules M into α and ß. The singlet state, which by hypothesis describes all these (α, ß) pairs, implies that if the observer O_α measures σ_3 and finds +1 (-1), then a subsequent measurement of τ_3 performed by O_β will give -1 (+1) with certainty.

The *EPR paradox* can now be formulated as follows:

- 1. On the systems α of an ensemble $E_1 \subset E$, O_α measures σ_3 and finds the results $A_1, A_2, \ldots, A_\ell$; any of these can only be +1 or -1, while ℓ is the number of αß pairs in E_1.

- 2. The results of *subsequent* measurements of τ_3 performed by O_β on the ℓ systems ß and E_1 can then be predicted *with certainty* to be $-A_1, -A_2, \ldots, -A_\ell$. This prediction can be checked on an ensemble $E_2 \subset E_1$. We assume it to be correct.

- 3. *Reality*. The certainty arrived at in the foregoing step allows us to use the EPR reality criterion and to attribute to ß an element of reality t_3 which determines the result of the future measurement of τ_3 for all ß $\in E_2$.

- 4. Assuming that t_3 is not generated retroactively in time by a future measurement on ß, we deduce that t_3 belongs to all ß $\in E_1$.

- 5. *Separability*. If α and ß are separated by a very large distance, we assume that the element of reality t_3 of ß cannot have been created by the measurement performed on α. Therefore it exists also if no measurement on α is carried out, and t_3 can be attributed to all ß $\in$ E.

- 6 *Completeness*. Assuming that quantum mechanics is complete, the element of reality t_3 must find a counterpart in the quantum theoretical description of ß. The only state vector describing as certain, and equal to -1 (or +1), the outcome of a measurement of t_3 on ß is v_- (or v_+).

- 7. Quantum mechanics predicts that the third component of the *total* spin of α+ß must be zero in all cases. The only state vectors for α+ß describing ß as either v_- or v_+ and giving zero for the third component of the total spin of the two atoms are u_+v_- or u_-v_+, respectively. This conclusion contradicts in an observable manner the description provided by η_S, as will be shown in a moment.

We thus arrive at an absurd conclusion. This means that among the previous points there are some which contain elements foreign to and incompatible with quantum mechanics. Einstein thought that the blame had to be put on the assumed completeness of quantum mechanics (point 6) and that an incomplete theory would be free of contradictions.

Modern research has however shown that this solution does not exist and that the incompatibility between reality, separability, and quantum mechanics exists *at the empirical level* also for an incomplete quantum mechanics. This will be discussed next by showing that Bell's inequality[5] is a necessary consequence both of the previous reasoning, in which quantum mechanics is assumed to be complete, and of a parallel reasoning in which the theory is taken to be incomplete from the start.

Consider again a very large number (N) of decays M → α+ß and suppose that the observer O_α measures on α the dichotomic observable $A(a)$, while in a distant region of space a second observer O_β measures on ß another dichotomous observable $B(b)$. The postulated dichotomy of these observables means that they can assume only two possible values, which for simplicity will be taken to be ±1. Notice that the observables A and B depend on the arguments a and b, respectively, which are supposed to be experimental parameters fixed in any given experiment but possibly variable in different experiments. Possible *examples* of such dichotomous observables are those corresponding to the spin matrices $\vec{\sigma}\cdot\hat{a}$ and $\vec{\tau}\cdot\hat{b}$, where the experimental parameters are the unit vectors $\hat{a}$ and $\hat{b}$.

When such measurements are made on all N pairs of the ensemble, O_α will register a set of experimental results $\{A_1, A_2, \ldots, A_N\}$

and O_β a similar set $\{B_1, B_2, \ldots, B_N\}$, all relative to fixed values of the parameters a and b.

The correlation function of these measurements is defined to be the *average product* of the results obtained by O_α and O_β from the same decays. It is given by

$$P(ab) = \frac{1}{N} \sum_{i=1}^{N} A_i B_i \ . \tag{2}$$

Since every product $A_i B_i$ is either +1 or −1 it follows that

$$-1 \leqslant P(ab) \leqslant +1 \ . \tag{3}$$

Now define the quantity

$$\Delta \equiv |P(ab) - P(ab')| + |P(a'b) + P(a'b')| \ .$$

This quantity is of great interest because we will see next that the requirements of reality and separability attribute to Δ a maximum value of 2. The inequality $\Delta \leqslant 2$ usually goes under the name of *Bell's inequality* and will now be shown to be a consequence of EPR's theorem.

Recall that the theorem leads from the singlet state to a mixture of factorable states: Consider then a general mixture of *factorable* states for the particles $\alpha\beta$, and call n_1, n_2, n_3, n_4 the number of cases with state vector u_+v_+, u_+v_-, u_-v_+, u_-v_-, respectively. One must have $n_1 + n_2 + n_3 + n_4 = N$, where N is the total number of pairs in the mixture.

A direct calculation of the correlation function P(ab) for this physical situation gives

$$P(ab) = \frac{1}{N}(n_1 - n_2 - n_3 + n_4)\, a_3 b_3 \ ,$$

where a_3 and b_3 are the third components of the unit vectors $\hat{a}$ and $\hat{b}$.

We can obtain without difficulty

$$\Delta \leqslant \frac{1}{N} |n_1 - n_2 - n_3 + n_4| \left\{ |a_3| \cdot |b_3 - b_3'| + |a_3'| \cdot |b_3 + b_3'| \right\}$$
$$\leqslant |b_3 - b_3'| + |b_3 + b_3'| \leqslant 2 \ ,$$

since $|a_3| \leqslant 1$, $|a_3'| \leqslant 1$, $|b_3| \leqslant 1$, $|b_3'| \leqslant 1$, $|n_1 - n_2 - n_3 + n_4| \leqslant N$.

Actually, the validity of Bell's inequality is a general property of all the conceivable mixtures of factorable states (therefore not only of factorable *spin* states).

We will carefully repeat the EPR argument, showing that it leads to a correlation function which contradicts in an observable manner the quantum mechanical one, even if one drops the assumption of completeness.

The starting point is the rotational invariance of the singlet state vector,

$$\eta_S = \frac{1}{\sqrt{2}} \left\{ u^n_+ v^n_- - u^n_- v^n_+ \right\}, \tag{4}$$

which follows from the fact that the form of η_S is always the same however the vector $\hat{n}$ may be chosen. Here $u^n_\pm$ and $v^n_\pm$ are eigenstates of $\vec{\sigma}\cdot\hat{n}$ and $\vec{\tau}\cdot\hat{n}$, respectively.

Consider a large ensemble E composed of N decays $M \to \alpha+\beta$ described by η_S. This state implies that if O_α measures $\vec{\sigma}\cdot\hat{n}$ and finds +1(-1), then a subsequent measurement of $\vec{\tau}\cdot\hat{n}$, performed by O_β, will give -1(+1) with certaintv.

The EPR theorem can be formulated step by step as follows:

- 1. On the systems α of an ensemble $E_1 \subset E$, O_α measures $\vec{\sigma}\cdot\hat{n}$ and finds the results $A_1, A_2, \ldots, A_\ell$.

- 2. The results of subsequent measurements of $\vec{\tau}\cdot\hat{n}$ performed by O_β on the ℓ systems ß of E_1 can then be predicted with certainty to be $-A_1, -A_2, \ldots, -A_\ell$. This prediction can be checked on an ensemble $E_2 \subset E_1$.

- 3. *Reality.* The certainty arrived at in the foregoing step allows us to use the EPR reality criterion and to attribute to ß an element of reality t_n which determines the result of the future measurement of $\vec{\tau}\cdot\hat{n}$ for all $\beta \in E$.

- 4. Assuming that t_n is not generated retroactively in time by a future measurement on ß, we conclude that t_n belongs to all $\beta \in E_1$.

- 5. *Separability.* The singlet state remains the same at arbitrary distances and is time-independent. Consequently, if α and ß are separated by a very large distance, we assume that t_n cannot have been created by the measurement performed on α. Therefore, t_n can be attributed to ß for all $\beta \in E$.

- 6 *Lack of completeness.* Assuming that quantum mechanics is *not* complete, we do not have to describe the element of reality with a state vector.

-7. The reasoning carried out in the first *five* steps has led to the conclusion that t_n belongs to ß even if no measurement on either α or ß is carried out. In this way the attribution of an element of reality has been completely disentangled from all acts of measurement. Given the symmetry of the problem in α and ß and given the invariant nature of η_S with respect to rotations, this allows us to introduce an arbitrary number of elements of reality for α and ß.

In particular, one can attribute to α two elements of reality s and s', corresponding to the observable $\vec{\sigma}\cdot\hat{a}$ and $\vec{\sigma}\cdot\hat{a}'$, respectively. Simultaneously, one can attribute to ß two elements of reality t and t', corresponding to the observables $\vec{\tau}\cdot\hat{b}$ and $\vec{\tau}\cdot\hat{b}'$, respectively. This can be done for all pairs (α, ß) of E.

- 8. The symbolic notation s = +1 (or −1) means that s is such that if a measurement of $\vec{\sigma}\cdot\hat{a}$ is performed on α the result +1 (or −1) is found with certainty. Similar meanings are attributed to the notations s' = ±1, t = ± 1, and t' = ± 1.

The statistical ensemble E of N pairs α, ß can then be subdivided into 2^4 subensembles having fixed values of all the four elements of reality. Calling n (s, s', t, t') the number of pairs with fixed s, s', t, t', one must obviously have

$$\sum_{ss'tt'} n\,(s\,s'\,tt') = N \quad . \tag{5}$$

- 9. The previous conclusion can be used to calculate the correlation functions P(ab), P(ab'), P(a'b), and P(a'b'). For instance,

$$P(ab) = \frac{1}{N} \sum_{ss'tt'} n(s\,s'\,t\,t')\cdot s\cdot t \quad .$$

Similar expression can be written down without difficulty for the other three correlation functions, and the inequalities

$$\begin{cases} \left|P(ab) - P(ab')\right| \leqslant \dfrac{1}{N} \displaystyle\sum_{ss'tt'} n(s\,s'\,t\,t')\cdot\left|t - t'\right| , \\ \left|P(a'b) + P(a'b')\right| \leqslant \dfrac{1}{N} \displaystyle\sum_{ss'tt'} n(s\,s'\,t\,t')\cdot\left|t + t'\right| \end{cases}$$

can easily be shown to hold. Adding these, one readily gets

$$\left|P(ab) - P(ab')\right| + \left|P(a'b) + P(a'b')\right| \leqslant 2, \tag{6}$$

which is Bell's inequality.

Obviously, this new form of the paradox eliminates logically the problem of completeness (or lack of completeness), so that the solution must be necessarily looked for in one of the following alternatives:

- 1. One must deny the validity of the EPR criterion of physical reality.[6]

- 2. The existence of signals propagating from the future into the past has to be admitted.[7]

- 3. Events are not separable and superluminal connections between distant events exist.[8]

- 4. Not all the predictions of quantum mechanics for correlated systems are correct.[(2)]

3. EINSTEIN LOCALITY AND THE $K^o\bar{K}^o$ SYSTEM

The quantum mechanical state vector for $J^{PC} = 1^{--}$ state decaying into $K^o\bar{K}^o$ has the form

$$|\Psi\rangle = \frac{1}{\sqrt{2}}\left[|K^o\rangle_a\,|\bar{K}^o\rangle_b - |\bar{K}^o\rangle_a\,|K^o\rangle_b\right]$$
$$= \frac{1}{\sqrt{2}}\left[|K_S\rangle_a\,|K_L\rangle_b - |K_L\rangle_a\,|K_S\rangle_b\right], \tag{7}$$

where a (left) and b (right) indicate the directions of motion of the kaons and $|K_S\rangle$ and $|K_L\rangle$ are states for short-and long-lived kaons, respectively (we neglect CP nonconservation in K^o decay). The time evolution of the latter states is given by

$$|K_S(t')\rangle = |K_S\rangle \exp(-\alpha_S t'),\quad |K_L(t')\rangle = |K_L\rangle \exp(-\alpha_L t'), \tag{8}$$

where t' is the <u>proper</u> time of the particle and

$$\alpha_S = \frac{\gamma_S}{2} + im_S\ ,\qquad \alpha_L = \frac{\gamma_L}{2} + im_L\ , \tag{9}$$

γ_S , γ_L denoting the total decay rates and m_S , m_L the rest masses.

The time evolution of the state (7) is assumed to be given by the product of the time evolutions,[9] so that

$$|\Psi(t_a t_b)\rangle = \frac{1}{\sqrt{2}}\Big\{|K_S\rangle_a\,|K_L\rangle_b \cdot \exp(-\alpha_S t_a - \alpha_L t_b)$$
$$- |K_L\rangle_a\,|K_S\rangle_b \cdot \exp(-\alpha_L t_a - \alpha_S t_b)\Big\}, \tag{10}$$

where t_a and t_b are the proper times of the particles moving to the left and to the right, respectively. The difference between the two exponentials in (10) generates a $\bar{K}^\circ\ \bar{K}^\circ$ component.[10]

The probability $w(t_a t_b)$ of a double $\bar{K}^\circ$ observation at times t_a and t_b is

$$w(t_a t_b) = \frac{1}{8}\left\{\exp(-\gamma_S t_a - \gamma_L t_b) + \exp(-\gamma_L t_a - \gamma_S t_b) -2\exp\left[-\gamma(t_a+t_b)/2\right]\cos\Delta(t_a-t_b)\right\}, \qquad (11)$$

where $\gamma = \gamma_S + \gamma_L$ and $\Delta = m_L - m_S$ is the $K_L - K_S$ mass difference.

Since Einstein locality is known to be in general incompatible with quantum mechanics at the empirical level, the problem arises of deciding which predictions of the theory to accept in any reasoning. The situation is here strongly reminiscent of the one encountered by Bohr and Heisenberg when they first departed from classical physics in building quantum theory. Their answer was contained in the correspondence principle, which opted to accept general structural properties of the old theory rather than specific predictions.[11]

We shall accordingly treat as acceptable the following quantum mechanical predictions:

- (P_1) Kaon decays are individual random processes occurring at constant rates. Populations decline exponentially.
- (P_2) All probabilities have left-right symmetry: For all purposes, a K_S going to the left and a K_L going to the right are physically equivalent to a right-bound K_S and a left-bound K_L.
- (P_3) All probabilities have particle-antiparticle symmetry.
- (P_4) Assuming exact time reversal, the rate of every transition equals that of the opposite transition.
- (P_5) If the "left" ("right") kaon has been observed to be a K_S, then the right (left) kaon can be predicted with certainty to be observed as a K_L at any future time.
- (P_6) If, instead, the left (right) kaon has been observed to be a K° at the proper time t_o, then the right (left) kaon can be predicted with certainty to be observed as a $\bar{K}^\circ$ at the *same* proper time t_o.

Einstein locality allows us to attribute elements of reality to $K^\circ\bar{K}^\circ$ pairs described by the state vector (10). The reasoning is the following:

-- If we measure CP on the a (left) - kaon of a pair described by (10), we find either +1 (K_S) or -1 (K_L). This allows us to predict *with certainty* that a future CP measurement on the b (right) - kaon will give the result -1 (K_L) or +1 (K_S), respectively. Using the EPR reality criterion, we can attribute an element of reality λ_1 to the right kaon.

-- Because of separability, the element of reality λ_1 is not created by the measurement made on the other kaon. Therefore it exists (even if unknown in the *individual* case) also if no measurement is made on the other kaon.

-- Because of the assumed lack of retroactive causality, the element of reality λ_1 is not created by a future measurement on the same kaon to which it belongs. Thus it exists also if no such a measurement is performed.

The situation is fully symmetrical between left and right. Therefore the previous reasoning allows us to associate a λ_1 element of reality both with the a (left) - kaon and with the b (right) - kaon.

Conclusion: *Every kaon of every pair described by the state vector (10) has an associated element of reality* λ_1 *which determines a well-defined CP value if and when a CP measurement is carried out.*

A second element of reality, connected with the strangeness S, can be introduced from Einstein locality.

The reasoning is now as follows:

-- If we measure S on the a (left) - kaon of a pair described by (10), we find either +1(K°) or -1($\bar{K}^\circ$). This allows us to predict with certainty that an S measurement at *equal proper time* on the b (right) - kaon will give the result -1($\bar{K}^\circ$) or +1(K°), respectively. Using the EPR reality criterion we can attribute an element of reality λ_2 to the right-kaon.

-- The element of reality λ_2 cannot have been created by the measurement made on the other kaon (separability). It therefore exists at least instantaneously also if no measurement is made on the other kaon.

-- The element of reality λ_2 cannot be created by a measurement on the same kaon to which it belongs (the instrument works properly). Therefore it exists also if no such measurement is performed.

Conclusion: *Every kaon of every pair described by the state vector (10) has an ssociated element of reality* λ_2 *which determines a well-defined S value if an S measurement is carried out.*

The first element of reality λ_1 describes the objective property of a kaon to decay either as a CP = +1 K_S or as a CP = −1 K_L.

The second element of reality λ_2 describes the objective property of a kaon to interact either as a S = +1 K° or as a S = −1 $\bar{K}^\circ$.

The important difference between λ_1 and λ_2 is that, while λ_1 describes a time independent property, λ_2 describes an instantaneous property: If a kaon of a pair is known at time t_0 to have the "property K°," one can expect it at a later time to have acquired the "property $\bar{K}^\circ$." Since at every instant the value of λ_2 is well defined, at every instant the kaon has either the "property K°" or the "property $\bar{K}^\circ$." Sudden jumps between the two situations are therefore possible.

Notice that Einstein locality has taken us outside quantum mechanics: No state vector is in fact known which can describe a given kaon as having *simultaneously* well-defined CP and S. This is of course a standard situation, since the attribution of elements of reality ("hidden variables") is known to lead generally to the existence of dispersion-free ensembles, which find no description within the existing quantum theory.

Having so introduced λ_1 and λ_2, it is rather straightforward to show the Einstein locality leads to [12]

$$w(t_a t_b) \leqslant \frac{1}{8} e^{-\gamma_S t_a} \left[1 + e^{-\gamma_S t}\right] \tag{12}$$

in the approximation $\gamma_L \ll \gamma_S$ and with $t = t_b - t_a$.

In the same approximation, (11) gives

$$w(t_a t_b) = \frac{1}{8} e^{-\gamma_S t_a} \left[1 + e^{-\gamma_S t} - 2 e^{-\gamma_S t/2} \cos \Delta t\right]. \tag{13}$$

The right-hand sides of (12) and (13) multiplied by $8 e^{\gamma_S t_a}$ are called $\gamma_E(t)$ and $\gamma(t)$, respectively, and are plotted in Fig.1. As one can see, (13) violates (12) by about 12% for $\gamma_S t \simeq 4 - 5$.

4. PROBABILISTIC SEPARABLE REALITY

There is a weak point in the EPR reality criterion in the predictability *with certainty* of the considered physical quantity: How can one ever be absolutely certain of anything in a concrete experimental situation?

The predictability *with certainty* appeared in 1935 to be a consequence of quantum theory for certain types of state vectors; the situation has however changed today, after the Wigner-Araki-Yanase theorem,[13] which proved the necessity of imperfect measurements, even in principle, if quantum theory has to be considered correct for all physical interactions, including measurements.

Insisting on the predictability with certainty has therefore no justification and leads furthermore to a narrow deterministic scheme which has unappealing features also for people who hold fast to a causal outlook.

We will therefore, in the following, abandon determinism and proceed to a probabilistic formulation of Einstein locality.[14]

Let a set α be given of physical objects of the same type (e.g., electrons). We call the objects $\alpha_1, \alpha_2, \ldots, \alpha_N$ and write

$$\alpha = \{\alpha_1, \alpha_2, \ldots, \alpha_N\}.$$

Suppose a measuring instrument $I(R)$ is given with which it is possible to measure a physical quantity R on the systems composing α. Suppose furthermore that the possible outcomes of the measurements of R are $r_1, r_2, \ldots, r_n$ and that N is so much larger than n that even the least probable of the results $r_1, r_2, \ldots, r_n$ is to be found by $I(R)$ a very large number of times.

If it is possible to detect the existence of a subset α' *of* α

$$\alpha' = \{\alpha_{i_1}, \alpha_{i_2}, \ldots, \alpha_{i_L}\},$$

without disturbing the objects composing α' *and* α *; if it is also possible to predict correctly that future measurements of R on* α' *will give the results* $r_1, r_2, \ldots, r_n$ *with respective probabilities* $p_1, p_2, \ldots, p_n$ *; then it will be said that a physical property*

$$\pi = \pi(R, p_1, p_2, \ldots, p_n)$$

belongs to α'.

The previous statement will be called the *generalized reality criterion* (grc). It consists essentially of two parts:

(i) *Existence of the subset* α'. This means that there is a method, not consisting of direct measurement or interaction, to ascertain the existence of a subset α' of α with interesting properties. In some cases one may be able to identify exactly which systems of α compose α'.

(ii) *Prediction of probabilities for* α'. This is supposed to be possible in a way peculiar to α' (and therefore in general with probabilities different from those which can be found in the full set α) without influencing the functioning of the measuring instrument I(R).

In general, the probabilities which become concrete when the observable R is actually measured, could *also* depend on the instrument I(R) that is used. This happens in many real experiments where the precision of the analyzers, the efficiency of the counters, and so on, enter in determination of the observed probabilities. Therefore, one can in general anticipate that these probabilities have, so to say, a double nature since they reflect the physical reality of the ensemble α' *and* of the instrument I(R). What is clearly needed for the future use of the reality criterion is that the concretization of the probabilities be *at least partly* due to the measured systems themselves, so that it may be said that π is the physical property of α' that contributes to the concretization of the probabilities $p_1, p_2, \ldots, p_n$ *when R is put in interaction with I(R).*

Let a second set ß of physical objects $ß_1, ß_2, \ldots, ß_N$ be given:

$$ß = \{ß_1, ß_2, \ldots, ß_N\},$$

and suppose that the particles α_i and $ß_i$ have been produced simultaneously by the decay of an unstable system M_i ($i = 1, 2, \ldots, N$). Suppose that at some time the quantum mechanical wave packets describing α_i and $ß_i$ are separated by a very large distance: This will be considered as a sufficient condition for the validity of the *generalized separability principle* (gsp), which is formulated as follows: *Measurements performed on the ß-systems do not generate the physical properties belonging to any subset* α' *of* α.

Of course the gsp is not assumed to provide an absolutely perfect separation of the two sets. Our point is simply that at very large distance the mutual interactions of two objects (or two sets of objects) are *small*.

5. A NEW INEQUALITY FOR THE SINGLET CASE

Given two spin $-\frac{1}{2}$ particles α_1 and β_1 described by a spatial wave function $\psi(\vec{x}_1, \vec{y}_1)$ (where $\vec{x}_1$ describes α_1 and $\vec{y}_1$ describes β_1) such that the probability clouds of the two particles are separated by a very large distance, let the spin part of the total wave function be given by the singlet state (1). This state is usually taken to imply that if a measurement of σ_3 performed by an observer O_α on α_1 gives the value +1, then a measurement of τ_3 performed by a second observer O_β on β_1 will give -1 *with certainty* (and *vice versa*).

As stated above, there are however good reasons for believing that certainty can never be attained. We will therefore assume that the practical meaning of η_S is the following: If the observer O_α measures σ_3 on α_1 and finds +1, then the probabilities that the observer O_β finds +1 and -1 by measuring τ_3 on β_1 are respectively given by

$$q_1 = \varepsilon \,, \qquad q_2 = 1 - \varepsilon \,.$$

The ideal case $q_1 = 0$, $q_2 = 1$ (certainty of finding -1) is obtained in the limit $\varepsilon \to 0$. The parameter ε will be assumed to be rotationally invariant and symmetrical in the following sense: It has the same numerical value however the third axis may be chosen and, furthermore, it keeps the same value if the roles of the two observers are interchanged.

Let a set of E pairs of correlated systems α_i and β_i be given:

$$E = \{\alpha_1 + \beta_1 \,;\, \alpha_2 + \beta_2 \,;\, \ldots \,;\, \alpha_N + \beta_N\} \,.$$

Notice that the set of the systems $\alpha_i(\beta_i)$ has been called $\alpha(\beta)$ in the previous sections. Thus E is, in a way, the physical union of α and β. Suppose that E is described quantum mechanically by the singlet state. It follows that:

(i) If O_β measures $\vec{\tau} \cdot \hat{a}$ on the systems β_i, one can consider a splitting of E into two subensembles $E_+(a)$ and $E_-(a)$. The former (latter) subensemble is composed of all the cases in which O_β has registerd the result -1(+1). For the former (latter) subensemble the probabilities that O_α measures $\vec{\sigma} \cdot \hat{a}$ and finds +1 and -1 are respectively given by $p_1 = 1 - \varepsilon$ and $p_2 = \varepsilon$ ($p_1' = \varepsilon$, $p_2' = 1 - \varepsilon$).
Assuming the generalized reality criterion, we attribute physical properties to $E_\pm(a)$:

$$\pi(\vec{\sigma}\cdot\hat{a}\,;\,1-\varepsilon\,,\,\varepsilon) \longrightarrow E_+(a),$$
$$\pi(\vec{\sigma}\cdot\hat{a}\,;\,\varepsilon\,,\,1-\varepsilon) \longrightarrow E_-(a). \tag{14}$$

Separability allows us to conclude that a meaningful splitting of E in two subensembles $E_+(b)$ and $E_-(b)$ exists, to which the previous attribution of physical properties is possible, *also if no measurement is made on the ß systems.*

(ii) If O_α measures $\vec{\sigma}\cdot\hat{b}$ on the systems α_i, there results a splitting of $E_+(b)$ and $E_-(b)$, the former (latter) containing the cases in which the result -1(+1) has been obtained. For $E_+(b)$ $[E_-(b)]$ the probabilities that O_β measures $\vec{\tau}\cdot\hat{b}$ and finds +1 and -1 are respectively given by $q_1 = 1-\varepsilon$ and $q_2 = \varepsilon$ $(q_1' = \varepsilon\,,\, q_2' = 1-\varepsilon)$. Assuming the generalized reality criterion, we attribute physical properties to $E_\pm(b)$:

$$\pi(\vec{\tau}\cdot\hat{b}\,;\,1-\varepsilon\,,\,\varepsilon) \longrightarrow E_+(b),$$
$$\pi(\vec{\tau}\cdot\hat{b}\,;\,\varepsilon\,,\,1-\varepsilon) \longrightarrow E_-(b). \tag{15}$$

Separability allows us to conclude here that a meaningful splitting of E in two subensembles $E_+(b)$ and $E_-(b)$ exists, to which the previous attribution of physical properties is possible, *also if no measurement is made on the ß systems.*

Thus, a simultaneous splitting of E into $E_+(a)$, $E_-(a)$, with the properties (14), *and* into $E_+(b)$, $E_-(b)$, with the properties (15), is possible. There are interesting consequences that can be deduced from this. In Ref. 14 we showed that the following inequality must hold:

$$|P(\hat{a}\hat{b}) - P(\hat{a}\hat{b}')| + |P(\hat{a}'\hat{b}) + P(\hat{a}'\hat{b}')| + 8|P(\hat{a}\hat{a})| \leqslant 10 \tag{16}$$

The comparison of (16) with quantum mechanics must be made in every concrete situation, taking into account the actual structure of the experimental apparata: There is, in fact, no *unique* quantum mechanical prediction following from the singlet state in the case of *real* apparata. Reference can be made, however, to the Wigner-Araki-Yanase theory, which takes into account systematic errors in quantum measurements and which therefore constitutes some kind of general theoretical model.

It can be shown that in the case of the singlet state this theory leads to

$$P(\hat{a}\hat{b}) = -\,|P(\hat{a}\hat{a})|\;\;\hat{a}\cdot\hat{b}\;.$$

Inserting this in (16) and maximizing the left-hand side over the angles, one deduces

$$2\sqrt{2} + 8 \leqslant \frac{10}{|P(\hat{a}\hat{a})|} .$$

It is an easy matter to show that this inequality can be violated only if

$$|P(\hat{a}\hat{a})| > 0.923 .$$

One can expect, therefore, a great sensitivity of the inequality (16) on the "forward" value of correlation function. This is also obvious from the large numerical coefficient of $|P(aa)|$ in (16). One can add that a saturation of (16) with local probabilistic models is to be expected,[15] so that the real experimental problem is to show that nature can violate (or satisfies) the inequality (16). It is therefore not correct to consider the problem already solved from the atomic cascade experiments. This conclusion is greatly strengthened by the fact that natural *local* hidden-variable models have been shown to exist fitting well the experimental data.[16] Furthermore, the interpretation of such experiments (the bibliography can be found in [16]) is obscured by the fact that a sizable fraction of the photons emitted in the atomic cascade undergo resonant rescattering in the beam before detection.[17] A new, more accurate generation of experiments is therefore needed.

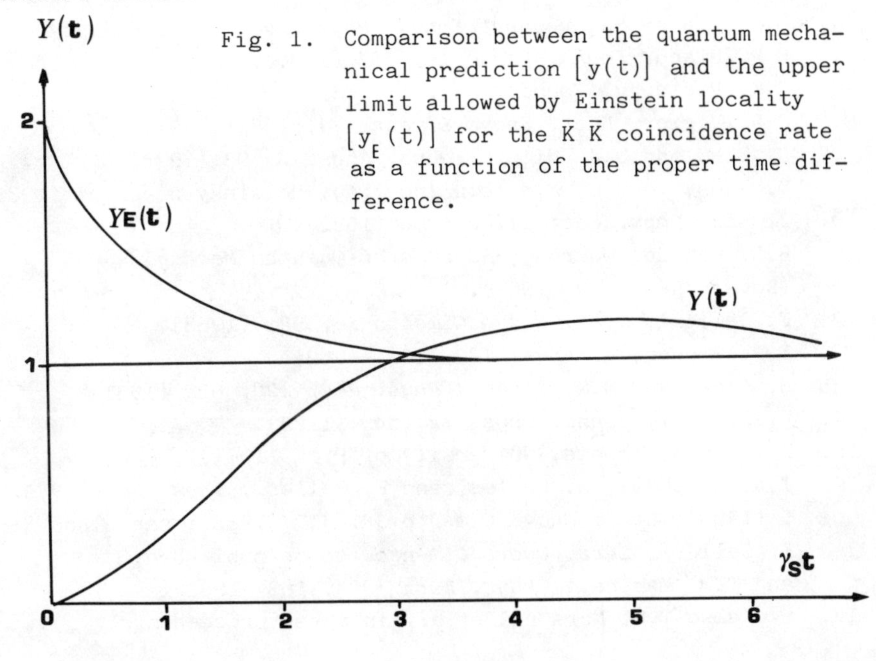

Fig. 1. Comparison between the quantum mechanical prediction $[y(t)]$ and the upper limit allowed by Einstein locality $[y_E(t)]$ for the $\bar{K}\,\bar{K}$ coincidence rate as a function of the proper time difference.

NOTES

(1) This presentation follows closely that of my forthcoming book.[4]

(2) This solution can be conceived especially because the problem of a separable reality played no role in the construction of the quantum paradigm. The EPR paradox was a later unexpected consequence of the theory.

REFERENCES

1. A. Einstein, B. Podolsky, and N. Rosen, *Phys. Rev. 47, 777* (1935).
2. K. Popper, *Unended Quest* (an intellectual autobiography) (Fontana/Collins, London, 1978), p. 152.
3. A. Aspect, J. Dalibard, and G. Roger, *Phys. Rev. Lett. 49,* 1804 (1982).
4. F. Selleri, *The Debate about Quantum Theory* (D. Reidel, Dordrecht, 1984), to be published; *Die Debatte um die Quantentheorie* (Vieweg, Braunschweig, 1983).
5. J.S. Bell, *Physics 1,* 195 (1965).
6. This was essentially Bohr's view; see N. Bohr, *Phys. Rev. 48,* 696 (1935).
7. O. Costa de Beauregard, *Found. Phys. 10,* 513 (1980). H.P. Stapp, *Found. Phys. 10,* 767 (1980). C.W. Rietdijk, *Found. Phys. 11,* 783 (1981).
8. J.P. Vigier, *Lett. Nuovo Cimento 24,* 258 (1979). D. Bohm and B.J. Hiley, *Found. Phys. 5,* 93 (1978).
9. M. Roos, *Test of Einstein locality,* Helsinky preprint (1980).
10. J. Six, *Phys. Lett. 114B,* 200 (1982).
11. B.L. van der Waerden, *Sources of Quantum Mechanics* (Dover, New York, 1967).
12. F. Selleri, *Lett. Nuovo Cimento 36,* 521 (1983).
13. E.P. Wigner, *Z. Phys. 133,* 101 (1952). H. Araki and M.M. Yanase, *Phys. Rev. 120,* 622 (1961).
14. F. Selleri, *Found. Phys. 12,* 645 (1982).
15. B. Galvan, Thesis, University of Pisa (1981).
16. T.W. Marshall, E. Santos, and F. Selleri, *Phys. Lett. 98A,* 5 (1983); *Lett. Nuovo Cimento 38,* 417 (1983); see also F. Selleri, *Lett. Nuovo Cimento,* to be published, and T.W. Marshall, *Phys. Lett. 99A,* 163 (1983).
17. See also T.W. Marshall *et al.* in these proceedings.

CAN NON-DETECTED PHOTONS SIMULATE NON-LOCAL EFFECTS IN TWO-PHOTON POLARIZATION CORRELATION EXPERIMENTS ?

J. Six
Laboratoire de l'Accélérateur Linéaire,
91405, Orsay, France

ABSTRACT

The recent experiments made on the analysis of the polarizations of the two photons emitted in a radiative decay of an excited calcium atom are reviewed. From the possible biases, only that coming from a conspiracy generated by non-detected photons in the apparatus cannot be completely excluded. However, this hypothesis is found unlikely to allow the possibility of some deterministic local theory. Non-local effects predicted by quantum mechanics are thus very difficult to reject.

It is the merit of Bell's inequalities, first introduced in 1965,[1] that they provide a possible test of quantum mechanics against any deterministic local theory.

The experiments devoted to this subject up to 1978 have been extensively reviewed.[2] From this review it appears that, if the Bell's inequalities seemed experimentally violated, as predicted by quantum mechanics, the results were not free from some ambiguities. Among the different types of experiments, the polarization correlation of the two photons emitted in a radiative cascade of some atoms may be the less ambiguous and the more conclusive one. Since that time, these correlations have been investigated with more completeness and greater precision in three experiments performed by the same group.[3,4,5] All these experiments indicated a clear violation of Bell's inequalities.

The aim of this paper is to investigate if, after the results of these experiments, we are obliged to abandon definitely the idea of a deterministic local theory.

G. Tarozzi and A. van der Merwe (eds.), Open Questions in Quantum Physics, 171–181.

1. SUMMARY OF THE EXPERIMENT[3,4,5]

The experiments referenced above have already been intensively described[3,7] ; we therefore give only a summary as an introduction to our discussion. These experiments investigated the two photons emitted in a $(J = 0) \rightarrow (J = 1) \rightarrow (J = 0)$ radiative cascade of an excited calcium atom. To conserve the angular momentum, the two photons have to be circularly polarized in opposite directions. In the first[3] and third experiments,[5] polarizers transformed this circular polarization into a polarization parallel to the polarizer variable axis $\vec{a}$ or $\vec{b}$ (Fig. 1). Photomultipliers could detect the transmitted photons.

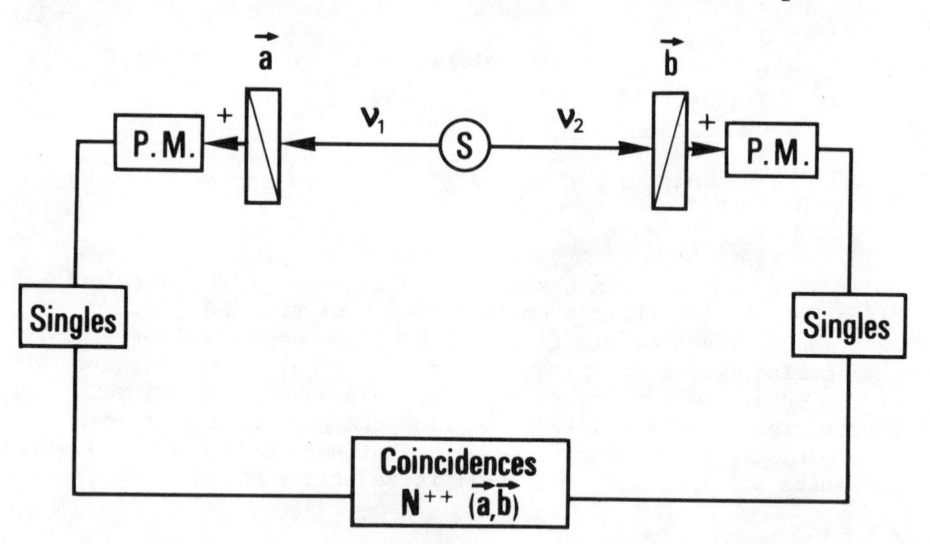

Fig. 1. Schematic diagram of the first experiment[3] where the photons ν_1 and ν_2 are analyzed by linear polarizers and may be detected by photomultipliers with single counting and coincidence electronic circuits.

Single rates $N_+(\vec{a})$, $N_+(\vec{b})$ and coincidence rates $N_{++}(\vec{a}, \vec{b})$ could be measured. The normalization of these experiments was made relative to the coincidence rates $N_{++}(\infty, \infty)$ when the polarizers alone were removed.

In the second experiment,[4] polarizers transmitted the parallel and perpendicular components of the circularly polarized photons, which were registered by photomultipliers (Fig. 2). Thus, the single counts and the coincidence rates $N_{++}(\vec{a}, \vec{b})$, $N_{+-}(\vec{a}, \vec{b})$, $N_{-+}(\vec{a}, \vec{b})$, $N_{--}(\vec{a}, \vec{b})$ were registered. The normalization was taken as the sum of these coincidences, $N_{++} + N_{+-} + N_{-+} + N_{--}$. In particular, for a given orientation $(\vec{a}, \vec{b})$, the correlation function was measured.

$$E = \frac{N_{++} + N_{--} - N_{+-} - N_{-+}}{N_{++} + N_{--} + N_{+-} + N_{-+}}$$

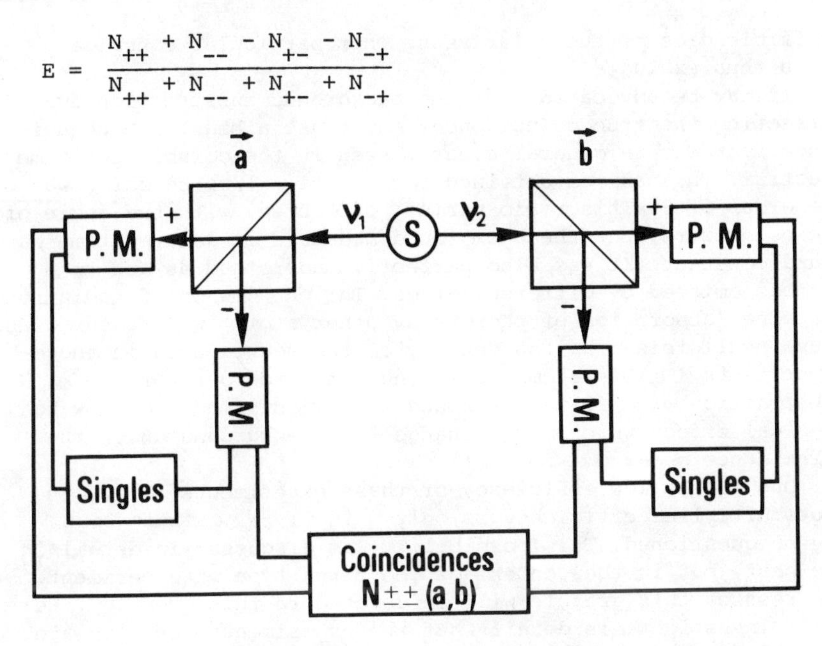

Fig. 2. Schematic diagram of the experiment[4] where the photons ν_1 and ν_2 are analyzed by two-channel polarizers, separating two orthogonal polarizations, and are detected by photomultipliers. The single countings and the coincidences N_{++}, N_{+-}, N_{-+}, N_{--} are registered by electronic circuits.

These experiments give a definite violation of Bell's inequalities and a complete agreement with quantum mechanics. Some possible biases are eliminated by these experiments :
1. The length between polarizers was L = 13 m (L/c = 43 ns) while the lifetime of the atomic cascade decay was ∿ 5 ns ; thus the exchange of some "information" between polarizers is ruled out.
2. A predisposition of the system to give a defined result, before the emission of the photons has been proposed.[1,8] This explanation is ruled out by the third experiment.[5] There, the photon was deflected rapidly towards one of two different polarizers. The switch time of the commutator was ∿ 10 ns, thus smaller than the total transit time L/c = 40 ns. Under these conditions Bell's inequalities were always violated. Despite the fact that the switch was activated periodically and not at random, the possibility of an artifact in the system seems excluded.
3. Some arguments[9] were advocated to accept the violation of Bell's inequality only if it is possible to have an experimental value $E(\vec{a}, \vec{b}) > 0.92$ for $(\vec{a}, \vec{b}) = 0$. In the second experiment this measured value is $E(o) = 0.954 \pm 0.024$, and it corresponds to the complete correlation $E(o) = 1$, corrected by the measured

inefficiencies of the polarizers. This particular conspiracy seems thus excluded.

4. It may be advocated that the background substraction for extracting the true coincidences may cause a bias. The coincidence peak was in general clearly seen in the coincidence time spectrum. The maximum obtained for the signal/noise ratio was of the order of 5. This ratio varies, of course, with the angle of the polarizers, but the background measured in delayed time was found constant. It was also perfectly understood as due to photons emitted by different atoms. The phenomenon of radiation trapping (absorption of photons by other atoms) was measured and found negligible[7] for the density of the source used in these experiments ($\sim 10^{11}$ atoms/cm^3). Thus the procedure of substraction of a flat background seems justified. We know no physical effect which might change this background under the coincidence peak.

5. Due to the low efficiency of these experiments (e.g., the photomultiplier efficiency is only $\sim$ 10 %), a possible bias may be questioned. This possibility was discussed in detail in the past, but in the context of the first type of experiment.[2] The rest of this article will be devoted to this question. We shall investigate in detail what the experimental results are and deduce from them some constraints which should be verified by this conspiracy hypothesis.

2. REVIEW OF THE BELL INEQUALITIES

2.1. Case of a Two-Outcome Experiment

Historically, Bell's inequalities were introduced in the scheme of Fig. 1 with two possible outcomes. We may call + and - the outcomes corresponding, respectively, to detected and non-detected photons. Supposing that the normalizations known, we can define the probabilities $p_+(\vec{a})$ or $p_+(\vec{b})$ for single counts of the first or second photomultiplier (Fig. 1) and $p_{++}(\vec{a}, \vec{b})$ for coincidences.

A first form of Bell's inequalities may be written :

$$-1 \leqslant \Delta \leqslant 0, \tag{1}$$

with $\Delta = p_{++}(\vec{a},\vec{b}) - p_{++}(\vec{a},\vec{b}') + p_{++}(\vec{a}', \vec{b}) + p_{++}(\vec{a}', \vec{b}') - p_+(\vec{a}') - p_+(\vec{b})$.

This first form of the inequalities was demonstrated to be valid for any deterministic local theory[10] or for any local hidden-parameter theory followed by some stochastic answer of the polarizer of the detector.[11] The latter validity is however restricted to an independence condition of stochastic behaviour for a given state λ, expressed as $p_{++}(\lambda,\vec{a},\vec{b}) = p_+(\lambda,\vec{a})\ p_+(\lambda,\vec{b})$.

A second form of Bell's inequalities may be written :

$$- 2 \leqslant S \leqslant + 2, \tag{2}$$

where $S = E(\vec{a},\vec{b}) - E(\vec{a},\vec{b}') + E(\vec{a}',\vec{b}) + E(\vec{a}',\vec{b}')$ and

$$E(\vec{a},\vec{b}) = p_{++}(\vec{a},\vec{b}) + p_{--}(\vec{a},\vec{b}) - p_{+-}(\vec{a},\vec{b}) - p_{-+}(\vec{a},\vec{b}).$$

Historically, this form of inequality was introduced firstly[1] by arguments concerning the correlation function $E(\vec{a},\vec{b}) = \int \rho(\lambda)\ A(\vec{a},\vec{\lambda})\ B(\vec{b},\vec{\lambda})\ d\lambda$. J.S. Bell[1] considered that the initial state, defined by hidden parameters λ, a state density $\rho(\lambda)$, and the locality condition, is given by the polarizer answer A (or B) assumed, independent of $\vec{b}$ (or $\vec{a}$). He took the dichotomic values A, B = ± 1. In these conditions, the function $E(\vec{a},\vec{b})$ has the form given above.

It is possible to show that inequalities (1) and (2) are equivalent. We may easily find the following equivalences, which allow us to derive (2) from (1), or <u>vice versa</u> :

$$E(\vec{a},\vec{b}) \equiv 1 + 4\ p_{++}(\vec{a},\vec{b}) - 2\ p_{+}(\vec{a}) - 2\ p_{+}(\vec{b})$$

and $\qquad S \equiv 2 + 4\ \Delta$

2.2. Case of a Three-Outcome Experiment

In the experimental configuration of Fig. 2, we have the possibility of three outcomes for a photon :

+ : parallel polarization detected in one photomultiplier ;
- : perpendicular polarization detected in the another photomultiplier ;
o : photon not detected.

We can define, as before, the various probabilities, and we have, for example, $p_{+}(\vec{a}) = p_{++}(\vec{a},\vec{b}) + p_{+-}(\vec{a},\vec{b}) + p_{+o}(\vec{a},\vec{b})$ and so on.

It is clear that the inequality (1) is always true. This inequality concerns only the outcome +, and we may split the old second outcome "-" into two outcomes "-" and "o", without changing anything in the demonstration.

However, if we take always the same definition as before for the quantities E and S, the inequalities (1) and (2) are not equivalent. We may indeed calculate the value of E. Calling, for simplicity, $p_{++}(\vec{a},\vec{b}) = p_{++}$, and so on, we find :

$$E(\vec{a},\vec{b}) \equiv 1 + 4\, p_{++} - 2\, p_{+}(\vec{a}) - 2\, p_{+}(\vec{b}) + (p_{+o} + p_{o+} - p_{o-} - p_{-o} - p_{oo})$$

and $\quad S \equiv 2 + 4\,\Delta +$ other terms.

Thus (1) and (2) are not necessarily equivalent. In the subsequent sections, we shall therefore consider only the inequality (1).

2.3. Experimental Symmetries

In the two-photon polarization experiments, the following relations were found to be true :

$p_{+}(\vec{a}) = p_{+}(\vec{a}') =$ constant, independent of the polarizer orientation ;

$p_{++}(\vec{a},\vec{b}) = p_{++}(\phi)$, where ϕ = angle $(\vec{a},\vec{b})$.

These relations were carefully tested in particular in the first experiment,[3] by comparing the single counts for different polarizer orientations and the coincidence counts for different configurations of polarizers, but with same ϕ. Of course, these tests were possible because the source was stabilized. The preceding relations were verified to a 1 % precision level.

With these symmetries, the Δ value in inequality (1) may be written :

$$\Delta = p_{++}(\alpha) - p_{++}(\alpha+\beta+\gamma) + p_{++}(\beta) + p_{++}(\gamma) - 2\,p_{+} \tag{3}$$

where the three relative angles α, β, γ correspond to the four orientations $\vec{a}$, $\vec{a}'$, $\vec{b}$, $\vec{b}'$.

2.4. Violation of Bell's Inequalities by Quantum Mechanics

For the experiments considered, the predictions of quantum mechanics are well known and simple (we assume here all efficiencies equal to unity) :

$$\begin{aligned} p_{++} &= p_{--} = \frac{1}{2}\cos^2\phi, \\ p_{+-} &= p_{-+} = \frac{1}{2}\sin^2\phi \end{aligned} \tag{4}$$

It is well known that for some orientation values, quantum mechanics violates the inequalities (1). It may be shown[2] that the maximum violation occurs when $\alpha = \beta = \pi/8$ or $= 3\pi/8$ in Eq. (3). This leads to the particular inequalities[2] which give a maximum violation :

$$\Delta_1 = 3\, p_{++}(\tfrac{\pi}{8}) - p_{++}(\tfrac{3\pi}{8}) - 2\, p_+ \leq 0, \quad (5a)$$

$$\Delta_2 = 3\, p_{++}(\tfrac{3\pi}{8}) - p_{++}(\tfrac{\pi}{8}) - 2\, p_+ \geq -1, \quad (5b)$$

$$\delta = \frac{1}{4}(\Delta_1 - \Delta_2 - 1) = p_{++}(\tfrac{\pi}{8}) - p_{++}(\tfrac{3\pi}{8}) - \frac{1}{4} \leq 0. \quad (5c)$$

The corresponding quantum mechanical predictions are + 0.35, - 1.2, and + 0.10 for Δ_1, Δ_2, and δ, respectively. It may be remarked that the last inequality, first introduced by Clauser,[12] contains only the probabilities of the coincidence rates. It is thus easier to test it by experiment.

3. EXPERIMENTAL RESULTS

We now consider the experimental results with the aim of establishing the laws for the theoretical probabilities. It has to be recalled that the detector efficiency is low, and thus the experimental normalization may be made only with coincidence rates. From the counting rates N_{+o}, N_{-o}, it is possible only to deduce the experimental efficiencies, but not to investigate the possibility to have non-detectable photons. We shall concentrate mostly on the more complete experiment,[4] which gives the measured quantities N_{++}, N_{+-}, N_{-+} and N_{--}.

3.1. Tests of Symmetries

Before making an experiment with four photomultipliers, it is necessary to align the system and to fix the high tension of the photomultiplier and the discriminator level associated with it. It was chosen to adjust these parameters to have equal countings N_+ and N_- for each polarizer. This means that it was not possible to test the equality between probabilities p_+ and p_-. But, taking into account the equality $p_{+-} = p_{-+}$ for reasons of symmetry, the relation $p_{++} = p_{--}$ was found to be verified to a 2 % precision level. On the other hand, as for p_+, the probability p_- was found independent of the polarizer axis.

3.2. Violation of Bell's Inequalities and Verification of Quantum Mechanics

Normalizing the experiment by the total number of coincidences $N_{++} + N_{--} + N_{+-} + N_{-+}$, the δ value of (5c) was found to be[7] $\delta_{exp} = (8.97 \pm 0.25)\ 10^{-2}$, in contradiction with Bell's inequality $\delta \leqslant 0$. The quantum mechanical prediction, corrected for experimental conditions,[7] is $\delta = (8.75 \pm 0.39)\ 10^{-2}$, in complete agreement with δ_{exp}. The experimental correlation $E(\vec{a},\vec{b})$ was measured by the quantity $N_{++} + N_{--} - N_{+-} - N_{-+}/N_{++} + N_{--} + N_{+-} + N_{-+}$. The ideal quantum mechanical prediction is $E(\vec{a},\vec{b}) = \cos 2\phi$ and the corrected value is $E(\vec{a},\vec{b}) = 0.954 \cos 2\phi$, taking into account the efficiencies of the polarizers.[4] The measured values are in complete agreement with this relation.

3.3. Stability of the Coincidences

Due to the stability of the source, it has been verified that we have, at a 1 % precision level, the following relation :

$$\Sigma N_{ij}(\phi) = N_{++}(\phi) + N_{--}(\phi) + N_{+-}(\phi) + N_{-+}(\phi) = \text{constant}, \tag{6a}$$

independent of the angle ϕ between polarizers. A measurement of the rate without polarizers allows us to derive :

$$\Sigma N_{ij}(\phi) \geqslant 0.86\ \Sigma N_{ij}\ \text{(without polarizer)}. \tag{6b}$$

The corresponding loss is attributed to the polarizer efficiency. This result (6b) allows to exclude some hidden variable models as we shall see below.

4. EXPERIMENTAL CONSTRAINTS ON A CONSPIRACY DUE TO NON-DETECTED PHOTONS

It is well known that in the type of experiments described here one has to manage various types of inefficiencies. These are the geometrical acceptance of the experimental system (~ 0.06), the polarizer efficiency (~ 0.95), and the photomultiplier efficiency (~ 0.10).

It is clear that the results are not affected if we venture the reasonable hypothesis that the various factors are independent of the state of the photon. This hypothesis means that the detected photons correspond to a non-biased sample. In that case the normalizations of these experiments are correct and the results non-biased.

However, this conclusion could be false if one supposes that a part of the inefficiency comes from a particular state of the photon. We can imagine, for example, some hidden variables which may cause a non-detection of the photon. We shall examine now if this type of hypothesis can be reasonably made.

4.1. Experimental Constraints for a Local Theory

If we leave open the possibility of some peculiar state for non-detected photons, we may derive from the experimental results, given above, the most general form of the theoretical probabilities, as follows :

$$\begin{aligned}
p_{++}(\phi) &= p_{--}(\phi) = \frac{1}{2} g \cos^2\phi, \\
p_{+-}(\phi) &= p_{-+}(\phi) = \frac{1}{2} g \sin^2\phi, \\
p_{+} &= \frac{1}{2} K_+, \\
p_{+o}(\phi) &= p_{o+}(\phi) = \frac{1}{2}(K_+ - g), \\
p_{-} &= \frac{1}{2} K_-, \\
p_{-o}(\phi) &= p_{o-}(\phi) = \frac{1}{2}(K_- - g), \\
p_{oo} &= 1 + g - (K_+ + K_-).
\end{aligned} \tag{7}$$

The two first relations are derived from the experimental relation $E(\phi) = \cos 2\phi$ and from the relation (6a), which imposes a g constant parameter. The other relations, with assumed symmetries between the two photons, take into account some possible loss of peculiar photons which would be non-detected. It is clear that these formulas are reducted to quantum mechanical formulas if we put $g = K_+ = K_- = 1$.

If a local theory is possible, it should obey Bell's inequalities,[5] which impose some restrictions on the parameters of Eq. (7). From (7),

$$p_{++}(\frac{\pi}{8}) = \frac{1}{4} g(1 + \frac{\sqrt{2}}{2}) \text{ and}$$

$$p_{++}(\frac{3\pi}{8}) = \frac{1}{4} g(1 - \frac{\sqrt{2}}{2}),$$ and from (5c) we derive the inequalities :

$$g\sqrt{2} \leqslant 1 \quad \text{or} \quad g \leqslant 0.71. \tag{8a}$$

Similarly, the combined inequalities (5a) and (5b) give :

$$1.2\ g \leqslant K_+ \leqslant (1 - 0.20\ g). \tag{8b}$$

Similarly we may derive the same inequality (8b) for K_- as well for K_+.

4.2. Is a Conspiracy of Non-Detected Photons Possible ?

The simpler form of conspiracy we can imagine would be due to some internal geometrical hidden variables that are sensitive to the polarizer orientations. Such models were proposed in the past.[11,13] If we restrict the conspiracy only to an effect between these internal variables and the polarizers, we cannot restore a local theory, as we should have $g \leqslant 0.71$ from (8a), and experimental results interpreted along these lines give $g \geqslant 0.86$, from (6b). All the class of models imagined with simple assumptions by Pearle,[13] with introduction of parameter $g(\phi)$ instead of the constant g and $K_+ = K_- = g(o)$, are completely excluded by experiment. Indeed, the data require $g(\phi) = g(o) = g$ = constant, and the inequality (8b) cannot be satisfied for the peculiar value $K_+ = g$.

Along that line of thought, it is thus necessary to imagine geometrical hidden variables sensitive not only to the direction of the polarizer but also for example to some part of the photomultiplier. There is no experimental evidence of an effect of the orientation on the detector efficiency, and thus this conspiracy seems highly improbable.

We cannot, however, exclude models where there could exist internal parameters giving two sorts of photons, detectable and non-detectable. It is possible in this case to conceive local models verifying the Bell's inequalities, but, to our knowledge, no simple model exist which might give the relations (7) with inequalities (8a) and (8b). On the other hand, an assumption of non-detectable photons would be a very strong assumption, which would change completely our actual physical scheme. It is probably less revolutionnary to admit non-local effects in physics.

5. CONCLUSION

From this study of recent experiments[3,4,5] on the polarization correlations between the two photons resulting from the radiative cascade decay of excited calcium atoms, we may conclude that non-local effects between separated particles, as predicted by quantum mechanics, are very highly suggested. Only one conspiracy, where the detector and some photon hidden variable interfere, cannot be completely excluded. However, we have shown that this hypothesis seems very unnatural.

A better hypothesis is to admit the existence of non-local effects. The question of understanding these non-local effects, by some interpretation or some refinement of quantum mechanics, remains open.

I am grateful to Drs. A. Aspect and P. Grangier for the quantitative precision level of the test measurements which are reported in this article. Thanks are also due to them for the classification of some experimental points and for numerous comments on this paper.

REFERENCES

1. J.S. Bell, Physics 1, 195 (1965).
2. J.F. Clauser and A. Shimony, Rep. Progr. Phys. 41, 1881 (1978).
3. A. Aspect, P. Grangier and G. Roger, Phys. Rev. Lett. 47, 460 (1981).
4. A. Aspect, P. Grangier and G. Roger, Phys. Rev. Lett. 49, 91 (1982).
5. A. Aspect, J. Dalibard and G. Roger, Phys. Rev. Lett. 49, 1804 (1982).
6. A. Aspect, Thesis Orsay, Université Paris-Sud, (Feb. 1983).
7. P. Grangier, Third Cycle Thesis (Paris VI, Feb. 1982).
8. D. Bohm and Y. Aharonov, Phys. Rev. 108, 1070 (1957).
   A. Aspect, Phys. Lett. 54A, 117 (1975) ; Phys. Rev. D14, 1944 (1976).
9. F. Selleri, Found. Phys. 12, 645 (1982).
10. E.P. Wigner, Am. J. Phys. 38, 1005 (1970).
11. J.F. Clauser and M.A. Horne, Phys. Rev. D10, 526 (1974).
12. J.F. Clauser, Phys. Rev. A6, 49 (1972).
13. P.M. Pearle, Phys. Rev. D2, 1478 (1970).

Part 2

The Stochastic Interpretation of Quantum Processes

QUANTUM THEORY OF MEASUREMENT WITHOUT WAVE PACKET COLLAPSE

Marcello Cini

Department of Physics
University of Rome, Italy.

ABSTRACT

A schematization of the measurement process in quantum mechanics is presented leading to the conclusion that the wave packet collapse should not be considered an additional postulate extraneous to the theory, but rather an approximate consequence of the standard Schrödinger time evolution. The equivalence - up to any conceivable level of accuracy - between the pure state vector of the total system "object + apparatus" and the statistical matrix representing the possible outcomes of the measurement, shows that the observer by no means determines its result, but merely obtains all the statistical information available from a description of the object together with the relevant part of the physical world interacting with it.

1. INTRODUCTION

The wave packet collapse postulate is an extra assumption one has to add to the closed system of rules forming the theory of quantum mechanics in order to give a physical meaning to its mathematical formalism.

It is well known that, if the state vector ψ of a system S is expanded in terms of the eigenvectors ϕ_n of an operator G representing an observable $\mathcal{G}$, the probability of finding any eigenvalue g_n as a result of a measurement of $\mathcal{G}$ by means of a suitable apparatus M_G is given by $|c_n|^2$, where c_n is the coefficient of the eigenvector in this expansion.

G. Tarozzi and A. van der Merwe (eds.), Open Questions in Quantum Physics, 185–197.

However, in order to justify this interpretation, one has to assume that, once a well determined result g_r has actually been obtained, the state vector is no longer the same ψ but has collapsed into ϕ_r, because the result of a measurement repeated immediately after the first one must be, with certainty, again g_r.

This postulate introduces therefore, in addition to the causal and reversible time evolution given by the Schrödinger theory, an acausal and irreversible source of sudden change of the state vector arising from the act of measurement.

It is useful to express this duality in terms of density matrices. Assuming the state vector of the system to be represented, at t = 0, by:

$$\psi = \sum_n c_n \phi_n \tag{1}$$

the corresponding density matrix W is defined as

$$W = \psi\psi^+ = \sum_{nn'} c_n c_{n'}^* \phi_n \phi_{n'}^+ \tag{2}$$

If the system is isolated, its time evolution is determined by the Schrödinger equation, and the density matrix W(t) will be given by

$$W(t) = \exp\left(-\frac{i}{\hbar} Ht\right) W \exp\left(\frac{i}{\hbar} Ht\right) . \tag{3}$$

However, when a measurement of the observable G is performed, the density matrix W(t) suddenly jumps into the form

$$\tilde{W} = \sum_n |c_n|^2 \phi_n \phi_n^+ \tag{4}$$

which represents a statistical mixture of the density matrices $\phi_n \phi_n^+$ corresponding to the measurement's possible outcomes. It is easy to see that $\tilde{W}$ can never be obtained by performing a unitary transformation, such as the one of Eq.(3), on the density matrix W of the pure state given by Eq.(2), because

$$\mathrm{Tr}(\tilde{W}^2) < \mathrm{Tr}(\tilde{W}) = 1 \tag{5}$$

while

$$\mathrm{Tr}(W^2) = \mathrm{Tr}(W) = 1 . \tag{6}$$

More directly, the difference between W and $\tilde{W}$ is shown by the different results one obtains in evaluating the mean value of an observable F that is incompatible with G, such that its commutator with G does not vanish:

$$[F, G] \neq 0 . \tag{7}$$

In fact one has

$$\bar{F} = \mathrm{Tr}\,(FW) = \sum_{nn'} c_n c^*_{n'} (\phi_{n'}, F\,\phi_n) \qquad (8)$$

while

$$\tilde{\bar{F}} = \mathrm{Tr}\,(F\tilde{W}) = \sum_n |c_n|^2 (\phi_n, F\phi_n) \neq \bar{F}\,. \qquad (9)$$

The jump

$$W(t) \to \tilde{W} \qquad (10)$$

can therefore never be described within the framework of quantum mechanics.

At this stage two different attitudes can be taken. The first is the one traditionally adopted, maybe without too much questioning, by the majority of the physicists. Namely that nothing can be done to understand this peculiar effect of the act of measurement: It is the observer who produces the wave packet collapse, and this effect can not be interpreted as a consequence of a physical interaction between the object and the measuring apparatus.

This amounts to accepting the von Neumann conclusion that only the observer's consciousness (not his material sense organs) is capable of performing an operation that eludes the laws of physics.

The second one, which in principle can be traced back to Bohr, but nevertheless is not usually considered to be a clearcut solution of the problem, consists in insisting that the actual interaction between the object and the measuring apparatus is the only source of information about the object's physical properties. This viewpoint refuses therefore to recognize any independent absolute meaning to the jump (10), which should be considered only as an idealized schematization of a physically understandable time evolution of the conventional Schrödinger type.

The purpose of the present paper is to show how this can be done, with the aid of a simple model of the interaction between the object and the measuring apparatus, thus providing an explicit example of approximate wave-packet collapse, which at the same time is in agreement with the established behaviour of macroscopic measuring instruments and with the rules of quantum mechanics [1].

2. CLASSICAL APPARATUS

Before going into the details of the proposed model, I will show briefly that the solution of the problem is trivial if one assumes outright that the measuring apparatus is a classical object. This goes of course too far, because one would like to derive the behaviour of classical bodies as a limiting case of the quantum laws, rather than postulating a dichotomy of the world into the two non-overlapping quantum and classical domains.

However this will be useful in order to point out the limit in which the concept of wave packet collapse, assumed to be only approximately valid in actual measurements, becomes exact.

In order to represent the interaction between the object S and the measuring instrument M, let us assume that the states of the latter are described by state vectors $|n>$ which are eigenstates of a macroscopic observable Γ with eigenvalues γ_n:

$$\Gamma |n> = \gamma_n |n> \tag{11}$$

Furthermore, we assume a one-to-one correspondence between the states of the microscopic object ϕ_n and the states $|n>$ of the apparatus. Then, if before the measurement the state vector Ω_o of the complete system S+M was described by

$$\Omega_o = \psi \otimes |0> \tag{12}$$

where $|0>$ is the initial state of the apparatus, then after the measurement the state vector Ω of the system S+M will be

$$\Omega = \sum_n c_n \phi_n \otimes |n> \tag{13}$$

At this stage the difference between the actual density matrix W

$$W = \Omega\Omega^+ = \sum_n \sum_{n'} c_n c^*_{n'} \phi_n \phi^+_{n'} \otimes |n><n'| \tag{14}$$

and the reduced density matrix $\tilde{W}$

$$\tilde{W} = \sum_n |c_n|^2 \phi_n \phi^+_n \otimes |n><n| \tag{15}$$

is still evident. However, if M is classical, its state will be completely specified by the values of all its observables that commute among themselves. Therefore, the eigenstates $|n>$ of Γ will be also simultaneous eigenstates of all the observables Δ^i (with eigenvalues δ^i_n) of M. This means that there will be no way to detect any difference between W and $\tilde{W}$, because there will be no off-diagonal contributions to the mean values of the observables Δ^i which might distinguish between them:

$$Tr(\Delta^i W) = Tr(\Delta^i \tilde{W}) = \sum_n |c_n|^2 \delta^i_n \phi_n \phi^+_n \tag{16}$$

For what concerns S, therefore, Eq.(16) shows that only the projection operators $\phi_n \phi_n^+$ of the measurement's possible outcomes contribute, each one with probability $|c_n|^2$. If the apparatus is classical, there is no need for the observer's consciousness. In this limit it has no power of "creating" reality, but merely obtains from an objective although probabilistic representation of reality, all the statistical information available about the microscopic object's physical properties.

The purpose of this paper can now be made clear. I want to investigate under what condition is it possible to say that a macroscopic apparatus can be considered, to a very good degree of approximation, a classical object. Or, in other words, I want to investigate to what extent the deviations from exact wave packet collapse, represented by the difference between the actual density matrix W (14) and the reduced one $\tilde{W}$ (15) may lead to detectable effects; and, should this be the case, what practical and theoretical consequences one would be forced to draw from it.

3. A MODEL OF COUNTER

In what follows I shall limit my discussion to the measurement of the spin component in a given direction σ_3 of a spin-one-half system S in a state

$$\chi = c_+ u_+ + c_- u_- \tag{17}$$

where $u_\pm$ are the usual σ_3 eigenstates "up" and "down".

A first possibility is to perform the measurement by means of a counter M made of a very large number N of particles, with whom the system S is brought to interact.

Each particle of the counter may be into one of two possible states ω_0 and ω_1, which will be denoted in analogy with real counters, as neutral and ionized respectively. All other degrees of freedom are neglected. This may seem, at first sight, to be an oversimplification. However, in a real counter, it is the presence of a large number $\underline{n}$ (of the order of N) of ionized particles which characterizes the discharged state, making it (macroscopically) different from its initial neutral state in which all its N particles are in their neutral states. Thus, our schematization which consists in suppressing the various coordinates that do not have a direct bearing on this dichotomy, does not alter substantially the main properties of the real atoms of which the real counters are made. The only thing it does is to largely overestimate their quantum mechanical coherence, by neglecting the phase randomization actually introduced by the neglected degrees of freedom. This approximation therefore goes just in the right direction, since we want to evaluate an upper limit for the quantum effects shown by the apparatus. The neglected complexities will only make the real effects much lower than the calculated ones.

The schematization proceeds with the choice of a suitable interaction between S and the particles of the counter. The mechanism involved is direct ionization of the latter by the former. Here also important complexities are neglected, such as all secondary ionization effects that arise from the mutual interactions of the counter's particles, thus again greatly enhancing the coherence of their dynamical evolution.

The final step consists in assuming that only one of the two independent states of S (say u_+) is capable of interacting with the counter, the other one (u_-) being incapable of triggering its discharge. This is what is meant by "polarized counter" namely a counter which selects between the two different states of polarization of the particle.

The term "polarized" stresses the difference with a counter used as a counting device, which simply detects the presence of a particle. Furthermore, since we want to treat the ideal case of a measurement of the first kind, the interaction of S should not change its state.

The interaction Hamiltonian will be therefore given by the expression

$$H = \frac{g}{\sqrt{N}} \frac{1}{2} (1 + \sigma_3) (a_0^* a_1 + a_0 a_1^*) \tag{18}$$

where, as usual, a_0^*, a_1^* are the creation operators in the states ω_0, ω_1, obeying boson commutation relations

$$\begin{aligned} \left[a_0, a_0^*\right] &= \left[a_1, a_1^*\right] = 1 \\ \left[a_0, a_1\right] &= \left[a_0^*, a_1^*\right] = 0 \end{aligned} \tag{19}$$

A given state of the counter will be defined by giving the number n of particles in the ionized state. This state will be therefore defined by

$$| n> = \frac{1}{\sqrt{n!}} \frac{1}{\sqrt{(N-n)!}} (a_0^*)^{N-n} (a_1^*)^n |> \tag{20}$$

We are ready now to work out the time evolution of the system (S+M), starting from an initial state

$$\Omega_0 = \chi \otimes | 0> \tag{21}$$

The solution of the Schrödinger equation

$$i\hbar \frac{\partial \Omega}{\partial t} = H\Omega \tag{22}$$

with initial condition (21) and Hamiltonian (18) is straightforward but lengthy. The details can be found in Ref.1. The result is

$$\Omega(t) = c_+ u_+ \left[\sum_{n=0}^{N} a_n(t) \mid n> \right] + c_- u_- \mid 0> \qquad (23)$$

with

$$a_n(t) = \frac{\sqrt{N!}}{\sqrt{n!}\ \sqrt{(N-n)!}} \left(\cos \frac{gt}{\hbar \sqrt{N}}\right)^{N-n} \left(\sin \frac{gt}{\hbar \sqrt{N}}\right)^{n} \qquad (24)$$

A first remark is trivial. Since the "down" state does not interact with the counter, at all times the state u_- is associated with the neutral counter's state $|0>$. A second remark is more interesting. The probability P_n of finding n ionized particles in the counter is simply the joint probability for n independent ionization events, each one with probability

$$p(t) = \sin^2 \alpha(t), \quad \alpha(t) = \frac{gt}{\hbar \sqrt{N}} \qquad (25)$$

the remaining N-n events corresponding to neutral particles. Thus

$$P_n(t) = \binom{N}{n} p^n (1-p)^{N-n} . \qquad (26)$$

This has an important consequence. For $n,N >> 1$ it is easily found that, for each value of the time t, there exists a number

$$\bar{n}(t) = N\ p(t) \qquad (27)$$

such that

$$P_{\bar{n}} = \binom{N}{\bar{n}} \left(\frac{\bar{n}}{N}\right)^{\bar{n}} \left(\frac{N-\bar{n}}{N}\right)^{N-\bar{n}} \simeq 1 \qquad (28)$$

the approximate Stirling formula having been used for the factorials. Since

$$\sum_n P_n(t) = 1 \qquad (29)$$

one has, in the stated approximation,

$$P_n(t) \simeq \delta_{n\ \bar{n}(t)} \qquad (30)$$

In other words, the superposition in (23) practically contains only two terms:

$$\Omega(t) = c_+ u_+ \mid \bar{n}(t)> + c_- u_- \mid 0> \qquad (31)$$

4. STERN-GERLACH MEASUREMENTS

Before discussing the consequences of (31) it is useful to examine another possibility of measuring the spin component of S. Rather than using a polarized counter, we might use a Stern-Gerlach device, namely we might first separate spatially two beams of particles, corresponding to the two spin states u_+, u_-, and then detect whether a given particle is localized in one or the other beam. The first stage (space separation) is, as we shall immediately see, mathematically quite similar to the measurement by means of a polarized counter already discussed. The second stage, namely the detection of the particle's presence or absence in a given beam, is conceptually similar to the problem of counting the number of ionized particles in a polarized counter, namely of detecting the difference between a neutral and a discharged counter.

Instead of using a conventional Stern-Gerlach beam splitting device, we will introduce a spin-orbit coupling between σ_3 and a given component L_1 of the particle's orbital angular momentum of the form

$$H = \frac{g}{\sqrt{2\ell}} \frac{1}{2} (1 + \sigma_3) \, 2L_1 \tag{32}$$

The reason for this choice will be seen in a moment. If the initial state is taken to be the eigenstate of L_3 with eigenvalue ℓ (equal to the total angular momentum quantum number ℓ),

$$\Omega_o = \chi \otimes Y_{\ell\ell}(\theta, \phi) \tag{33}$$

the solution of the Schrödinger equation can be written immediately if one takes into account the fact that the three components L_1, L_2, L_3 may be represented in terms of two independent sets of boson destruction and creation operators as follows

$$L_1 = \frac{1}{2} (a_0^* a_1 + a_0 a_1^*), \quad L_2 = \frac{i}{2} (a_0^* a_1 - a_0 a_1^*), \quad L_3 = \frac{1}{2} (a_0^* a_0 - a_1^* a_1) \tag{34}$$

These relations establish a formal correspondence between the measurement of the spin of a particle by means of a polarized counter and the measurement of the same spin by means of a Stern-Gerlach type setup. In fact, the Hamiltonians (18) and (32) become identical with the use of (34). Furthermore, the counter states (20) can be considered as eigenstates of L_3, since

$$L_3 \, | n \rangle = (\frac{N}{2} - n) \, | n \rangle . \tag{35}$$

One can therefore establish the correspondence

$$N = 2\ell \quad , \quad \frac{N}{2} - n = m \tag{36}$$

$$|n\rangle = Y_{\ell m}(\theta, \phi) \tag{37}$$

having adopted the usual notation for the eigenstates of L_3 in polar coordinates (spherical harmonics). From the solution (31) one derives immediately:

$$\Omega(t) = c_+ u_+ Y_{\ell \bar{m}(t)}(\theta, \phi) + c_- u_- Y_{\ell\ell}(\theta, \phi) \tag{38}$$

with

$$\bar{m}(t) = \frac{N}{2} - \bar{n}(t) \; . \tag{39}$$

The dynamical evolution (38) can be interpreted as follows. At $t = 0$ the angular momentum component L_3 has the value ℓ . Since $\ell \gg 1$, the other two components are zero, with uncertainties

$$\Delta L_2 \simeq \Delta L_1 \simeq \sqrt{\ell} \ll \ell \tag{40}$$

The larger ℓ is, the better can we say that the angular momentum lies in the direction of the 3-axis. When the spin orbit interaction is turned at $t = 0$, the two components u_+ and u_- are decoupled. The latter continues to be associated with $L_3 = \ell$, while the former becomes associated with $L_3 = \bar{m}(t)$ and $L_2 = \sqrt{\ell^2 - \bar{m}^2}$. ($L_1$, being the Hamiltonian, is a constant of the motion.) The angular momentum rotates in the L_2L_3 plane forming an angle $2\alpha(t)$ with 3-axis:

$$\alpha(t) = \frac{gt}{\hbar \sqrt{2\ell}} \tag{41}$$

It should be noted that this entails a space separation of the two states u_+, u_-, exactly as in the conventional Stern-Gerlach experiment. In fact, since the eigenfunction of L_3 corresponding to the eigenvalue ℓ is, in polar coordinates,

$$Y_{\ell\ell}(\theta, \phi) = A(\sin\theta)^{\ell} \tag{42}$$

one sees immediately that, for $\ell \gg 1$, the probability density is practically zero for any value θ except $\theta = \pi/2$. In other words, the orbital plane of the particle is perpendicular to L_3. Similarly, if we denote by $\bar{\theta}$ the polar latitude referred to the direction of $\vec{L}$ at time t, we will have

$$Y_{\ell \bar{m}(t)} \simeq A(\sin\bar{\theta})^{\ell} \tag{43}$$

For the reason discussed above this means that the orbital plane of the particles with spin up (u_+) will form with the orbital plane of the particles with spin down (u_-) an angle 2α. Therefore, the two beams are spatially separated and their detection in one plane or the other one is a means of measuring their spin.

The necessity of $\ell \gg 1$ is evident by now. A system with angular momentum of the order of $\hbar$ cannot be used as a measuring device, exactly as it is impossible to use a single ionized particle as a counter. Of course, a single ionized particle can trigger an amplification device, but this means that a macroscopic counter is needed to detect the single particle ionization produced in the primary interaction.

5. INTERFERENCE EFFECTS IN TWO SUCCESSIVE MEASUREMENTS

We arrive now to our most important question. How can one distinguish between the two density matrices

$$\tilde{W} = |c_+|^2 \, |\bar{n}><\bar{n}| \otimes u_+u_+^+ + |c_-|^2 \, |0><0| \otimes u_-u_-^+ \tag{44}$$

and

$$W = \tilde{W} + c_+c_-^* \, |n><0| \otimes u_+u_-^+ + c_+^*c_- \, |0><n| \otimes u_-u_+^+ \tag{45}$$

corresponding, respectively, to the wave packet collapse postulate and to the Schrödinger evolution?

The answer is easy. Let us perform a second measurement on S of a different spin component, say σ_2. We will need a second polarized counter M_2 which selects v_+ from v_-, where

$$\sigma_2 \, v_\pm = \pm \, v_\pm \tag{46}$$

$$v_\pm = \frac{1}{\sqrt{2}} (u_+ \pm u_-) \tag{47}$$

The total state vector Φ of the complete system ($S+M_1+M_2$) will be

$$\Phi = \frac{1}{\sqrt{2}}\left[c_+|\bar{n}>_1 + c_-|0>_1\right] v_+ \, |\bar{n}'>_2 + \frac{1}{\sqrt{2}}\left[c_+|\bar{n}>_1 - c_-|0>_1\right] v_- \, |0>_2 \tag{48}$$

Since we are not going to make any more measurements on S, we construct a density matrix W_{12} in the product Hilbert space of the two counters by evaluating the trace on the S variables of $\Phi\Phi^+$:

$$W_{12} = \mathrm{Tr}_S(\Phi\Phi^+) = \frac{1}{2}\left[c_+|\bar{n}>_1 + c_-|0>_1\right]\left[c_+^* \; {}_1<\bar{n}| + c_-^* \; {}_1<0|\right] |\bar{n}'>_2 \, {}_2<\bar{n}'|$$

$$+ \frac{1}{2}\left[c_+|\bar{n}>_1 - c_-|0>_1\right]\left[c_+^* \; {}_1<\bar{n}| - c_-^* \; {}_1<0|\right] |0>_2 \, {}_2<0| \tag{49}$$

On the other hand, had we assumed exact wave packet collapse after the first measurement, the corresponding density matrix $\tilde{W}_{12}$ would have been

$$\tilde{W}_{12} = \left(|c_+|^2 |\bar{n}\rangle_1 \, {}_1\langle\bar{n}| + |c_-|^2 |0\rangle_1 \, {}_1\langle 0| \right) \left(\frac{1}{2} |\bar{n}'\rangle_2 \, {}_2\langle\bar{n}'| + \frac{1}{2} |0\rangle_2 \, {}_2\langle 0| \right) \quad (50)$$

In Eq.(50) the interference terms

$$\pm \left(c_+ c_-^* \, |\bar{n}\rangle_1 \, {}_1\langle 0| + c_- c_+^* \, |0\rangle_1 \, {}_1\langle\bar{n}| \right) \quad (51)$$

appearing in (49) are missing. But these terms give a non vanishing contribution only in the mean value of observables Δ such that

$$\langle\bar{n}| \, \Delta \, |0\rangle \neq 0 \quad (52)$$

These would have the meaning of observables capable of transforming back a discharged state $|\bar{n}\rangle$ into the initial neutral state of the counter. Obviously, for real counters there are no such observables. Of course, in our case, one could write a mathematical expression containing a product of n factors a_1 and n factors a_0^* with the property (52). But there is no physical quantity corresponding to this operator. In fact, should anyone be capable of inventing a procedure for detecting the presence of interference terms in measurements with counters, the result would be quite sensational.

For the Stern-Gerlach setup the interference terms can be evaluated. They are at most

$$Y_{\ell\bar{m}}(\theta,\phi) \, Y_{\ell\ell}(\theta,\phi) < e^{-\ell\alpha^2(t)} \quad (53)$$

for $\theta = \pi/2 + \alpha(t)$, namely in the plane bisecting the two orbital planes.

These results show that there is no practical way to detect any difference between W and $\tilde{W}$. This amounts to saying that, in actual measurements with good measuring instruments, as far as the system S is concerned, everything happens as is the wave packet collapse had occurred. This, however, is not a consequence of the observer's consciousness, but, more simply, of the interaction between measured object and measuring apparatus, according to the laws of quantum mechanics.

6. CONCLUSIONS

A last point is worth mentioning. There is indeed a difference between a measurement made with a real polarized counter and one performed by means of a Stern-Gerlach setup. A real counter has an enormous number of degrees of freedom. Hence its discharge, unlike the one described by my model, is irreversible. I have already commented on this point at the beginning, pointing out that this makes the interference effects by far even more undetectable. There is no way to find that each ionized electron should go all the way back to regain its place in the neutral atom to which it belonged, and each atom should find the position it had before the discharge. To say that irreversibility makes the (approximate) wave-packet collapse irreversible is not therefore a tautology. It means that the effect of macroscopicity (N large) is going to remain forever when the number of degrees of freedom is also very large.

However, for a Stern-Gerlach device, a large value of ℓ (or a large separation Δz of the two beams in the conventional experiment) is still a manifestation of a single degree of freedom. This means that the time evolution of the system (S+M) may be reversible. The two beams, after having been widely spacially separated, may be brought back together. In this case the two beams will interfere again. Experiments of this type have been made both with photons [2] and with neutrons [3] and they definitely show that the interference does occur even when a single particle is present in the apparatus. But in this case, of course, one is no longer making a measurement, because the one-to-one correspondence between the (macroscopically) different states of the measuring setup and the states of the measured microscopic observable, is lost. In other words, the interference effects destroy the possibility of a reliable measurement.

If, on the other hand, a counter is placed along the path of one of the beams, the interference effects are destroyed, even if the two beams are brought together again. Formally this result follows from the stated equivalence between W_{12} and $\tilde{W}_{12}$ given by Eqs.(49) and (50), provided M_1 is regarded as a detecting counter placed along the path of the beam with spin up (u_+) and M_2 as a polarized counter placed to measure the spin where the two beams are re-united.

This result has some relevance in connection with the proposal for experiments designed to detect the path followed by a particle without destroying the interference pattern [4]. In the light of the preceding discussion, one can safely assert that the usually predicted impossibility of achieving this is not a consequence of the wave-packet collapse postulate, but rather follows directly from the standard Schrödinger time evolution of the system ($S+M_1+M_2$). It seems therefore that the chance that the proposed experiments would yield positive results is as unlikely as the failure of Schrödinger's equation.

REFERENCES

1. M. Cini, *Nuovo Cimento 73B*, 27 (1983)
2. A. Gozzini, private communication.
3. H. Rauch, communication at this conference.
4. A. Garuccio and F. Selleri, communications at this conference.

A CAUSAL FLUIDODYNAMICAL MODEL FOR THE RELATIVISTIC QUANTUM MECHANICS

N. CUFARO PETRONI

Istituto Nazionale di Fisica Nucleare
Dipartimento di Fisica dell'Università
70100, Bari, Italy

ABSTRACT

The physical basis for a fluidodynamical model of quantum mechanics and the subsequent possibility of a causal, non-local explanation of the EPR paradox are shortly discussed.

1. INTRODUCTION

This paper constitutes the physical basis for the deterministic explanation with action at a distance of the Einstein-Podolsky-Rosen paradox, which is discussed by Prof. Vigier in his paper.

The starting point of our work is the causal fluidodynamical interpretation of quantum mechanics given by Bohm.(1,2) So we effectively regard the wave function of an individual electron as a mathematical representation of an objectively real field. This field exerts a force on the particle by means of the well-known quantum potential.

However this interpretation was criticized by Takabayasi,(3) Pauli (4) and others. (5) We will list here only the problems that we consider connected with our research:

1. the assumption that $|\psi|^2$ coincides with the probability density is not appropriate in a theory aimed at giving a causal explanation of quantum mechanics;
2. the ψ field was so different from other objectively real fields that it does not provide a satisfactory model for quan-

G. Tarozzi and A. van der Merwe (eds.), Open Questions in Quantum Physics, 199–213.

tum phenomena; in a word: what is a Madelung fluid?
3. this interpretation did not include spin;
4. the quantum potential for the N-body equation is non-local, so that the quantum mechanical forces may be said to transmit uncontrollable disturbance instantaneously from a particle to another through the medium of the ψ field.

We think that this set of questions still constitutes an open field of research, and we therefore shall try to give an outline of an organic answer: a stochastic hydrodynamical model of the relativistic quantum equations in a material aether and in the presence of a relativistic action at a distance.

2. "IS THERE AN AETHER?"

First of all we expose our point of view on the material subquantum medium in which particles are imbedded. We adopt the idea, first presented by Dirac,(6) that in the light of quantum mechanics "the aether is no longer ruled out by relativity". In fact, the velocity of the aether in each point (which would destroy the Lorentz isotropy of the vacuum) is subject to the uncertainty relations. So, the velocity of the aether at a certain point will be distributed over various values, and we may set up a wave function which makes all value of the velocity equally probable. Of course (see Fig.1) this wave function must be constant on the hyperboloid

$$v_0^2 - v_1^2 - v_2^2 - v_3^2 = 1. \tag{1}$$

In other words, Dirac has bypassed all former relativistic objections to the aether's existence by introducing for it a chaotic subquantal behaviour.

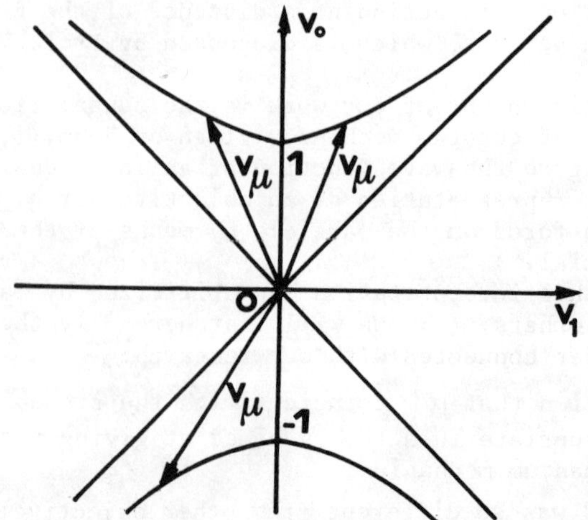

Fig.1. The velocity of the aether at a point is uniformly distributed on the hyperboloid(1).

A complete and satisfactory theory of Dirac's aether still doesn't exist. We list here only some questions and proposals about this problem.

1. To meet the question of the viscous drag in the aether we can build a superfluid model as Sudarshan et al. (7) did.

2. We can try (8) to connect the aether with the "negative energy sea", which still remains an essential basis for the second-quantization formalism. In this case, as shown in Fig. 2, the four-momenta of the aether will be distributed on the lower mass hyperboloid in an uniform way, so that

$$dN = K\sqrt{|ds^2|} \quad , \tag{2}$$

where K is a constant and dN is the number of states in a section ds of the hyperboloid.

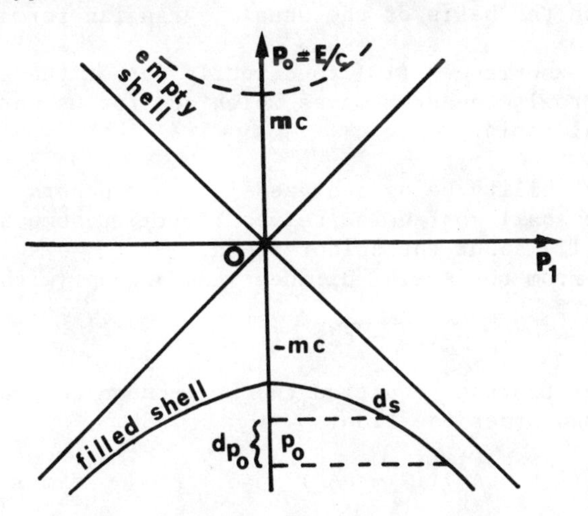

Fig. 2. Dirac's aether and negative energy sea

In this case the distribution in energy is not a constant:

$$\rho(p_0) = \frac{Kmc}{\sqrt{p_0^2 - m^2c^2}} \tag{3}$$

Of course, the number of the almost light-like four-momenta is predominant.

3. We can introduce the spin. (9)

4. We need a deeper analysis of the interactions between the aether and the particles.

3. HYDRODYNAMICAL ANALYSIS OF A SPINOR FIELD

In our model the quantum waves are perturbations on the aether produced by the presence of a particle. From that standpoint, the particles are "corpuscles" imbedded in the aether and in interaction with it by means of the quantum potential. Here $\hbar$ is a sort of coupling constant. In other words, an electron is a particle *IN* the wave, and ψ is a *REAL* wave that pilots particle-like concentrations of energy-momentum corresponding to electrons.

Coherently with this standpoint, we can now:

1. make a hydrodynamical analysis of quantum waves and attribute to them the usual quantities (like momentum, energy angular momentum) on the basis of the usual Lagrangian formalism;

2. design new experiences (10) to directly detect the existence of the de Broglie quantum waves, which are for us not only mathematical tools.

The second possibility being analyzed in other papers for this conference, we shall confine ourselves to quoting some hydrodynamical results (11) about the spinor field.

Starting from the scalar bilinear Lagrangian (with $\hbar = c = 1$)

$$L = m\bar{\psi}\psi - \overline{(i\partial\!\!\!/ - eA\!\!\!/)\psi}\,(i\partial\!\!\!/ - eA\!\!\!/)\psi \tag{4}$$

we get as Euler-Lagrange equation the well-known Feynman and Gell-Mann (12) second order equation

$$[(i\partial_\mu - eA_\mu)(i\partial^\mu - eA^\mu) - \tfrac{1}{2}\sigma_{\mu\nu} F^{\mu\nu}]\,\psi = m^2\psi \tag{5}$$

with

$$\sigma_{\mu\nu} = \frac{i}{2}[\gamma_\mu,\gamma_\nu] \text{ and } F^{\mu\nu} = \partial^\mu A^\nu - \partial^\nu A^\mu . \tag{6}$$

Of course, the eq. (5) is not equivalent to the ordinary Dirac equation, which in this formalism plays the role of a constraint on the solutions of (5) that select spinors with positive-definite conserved density.

The current is now

$$J^\mu = \frac{i}{2m}[\bar{\psi}\partial^\mu\psi - \partial^\mu\bar{\psi}\psi - i\partial_\nu(\bar{\psi}\sigma^{\mu\nu}\psi)] - \frac{e}{m}A\bar{\psi}\psi, \tag{7}$$

so that the free part coincides with the Gordon form of the usual spinor current and is equivalent to Dirac's usual expression only for solutions that obey to the Dirac constraint.

If now we posit

$$\psi = Qw, \tag{8}$$

where $Q^2 = |\bar{\psi}\psi|$ is a real positive function and w an unitary spinor with $\bar{w}w = \pm 1$, the imaginary part of (5) furnishes

$$\partial_\mu J^\mu = 0, \tag{9}$$

where the current is given by (7) and the real part corresponds to a generalized Hamilton-Jacobi equation

$$\bar{w}(i\partial_\mu - eA_\mu)w\ \bar{w}(i\partial^\mu - eA^\mu)w - \frac{\Box Q}{Q} - \frac{e}{2}\bar{w}F^{\mu\nu}w\ \bar{w}\sigma_{\mu\nu}w - m^2 = 0 \tag{10}$$

with quantum potential

$$U = \frac{\Box Q}{Q} . \tag{11}$$

From the same Lagrangian (4) we can also calculate all the usual physical quantities associated with our wave (energy-momentum tensor, angular momentum, spin, etc;). We follow in this calculation the general method given by Halbwachs (12) in his book on the spinning fluids.

We stop here this argument by recalling that along these lines we can:

1. elaborate a two-particle case and explore in a direct way, for example, and in term of forces and torques, what the word "correlation" means in the instance of correlated electrons;

2. employ the usual formalism of stochastic derivation of quantum equation.

4. CONNECTION WITH STOCHASTIC PROCESSES

We do not recall here the results of the derivation of the relativistic quantum equations starting from Nelson's equations. (13) We remark only that, from the relativistic second order Nelson's equation, we can get only second-order quantum equations: It is because we adopted the second-order Feymann and Gell-Mann equation for our spinor field in the preceding section.

In order to preserve the quantum equations in our model, we present here (8,14) another derivation suggested to us by a seminar of Prof.A.Avez (15) on the probabilistic interpretation of the hyperbolic partial differential equations. We are able at present to derive the relativistic quantum equation for spinless particles.

To simplify our demonstration, we limit ourselves to the case in which it is assumed that random walks occur on a square lattice in a two dimensional space-time (see Fig.3). To analyze our random walks we describe our two dimensional space -time with the coordinates x^0, x^1 and consider a limiting process where in each step it is supposed that our particle, starting from an arbitrary point $P_0(x^0, x^1)$, can make only jumps of fixed length and always at the velocity of light. Of course, this prescription completely determines the lattice of all the possible positions of the particle.

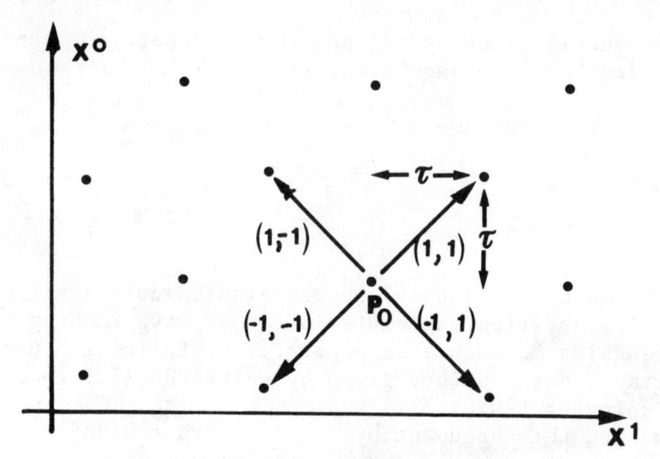

Fig. 3. Space-time lattice of dimension τ and starting point P_0 ; for each possible direction of the first jump we marked the corresponding value of the couple (t,s).

On this lattice the particle can follow an infinity of possible trajectories. In our calculation we will consider first a lattice with fixed dimensions. Indeed, for each jump we posit:

$$\Delta x^0 = t\tau \quad \text{and} \quad \Delta x^1 = s\tau, \quad (t,s = \pm 1), \tag{12}$$

so that for the velocity we always have

$$v = \Delta x^1/\Delta x^0 = s/t = \pm 1 \tag{13}$$

Here τ is the parameter that fixes the lattice dimensions; of course, in order to recover the quantum equations, we will consider later the limit $\tau \to 0$. Moreover, it is clear that we also consider the possibility of trajectories running backward in time: We will interpret them as trajectories of antiparticles running forward in time, following the usual Feynman's interpretation.(16)

In order to describe random walks on this lattice, let we consider the following Markov process on the set of the four possible directions of the velocity. We define two sets of stochastic variables $\{\varepsilon_j\}$, $\{\eta_j\}$, in such a way that the only possible values of each ε_j and η_j are ± 1, following this prescription:

$$\varepsilon_j = \begin{cases} +1 \\ -1 \end{cases} \text{if in the (j+1)-th jump the sign of velocity} \begin{cases} \text{doesn't ch.} \\ \text{changes} \end{cases}$$

$$\eta_j = \begin{cases} +1 \\ -1 \end{cases} \text{if in the (j+1)-th jump the direction of time} \begin{cases} \text{doesn't ch.} \\ \text{changes} \end{cases}$$

with respect to the preceding j-th jump. It means that the realization of the signs of ε_j, η_j determines one of the four possible directions of the (j+1)-th jump on the basis of the direction of the j-th jump, as can be seen in Fig.4.

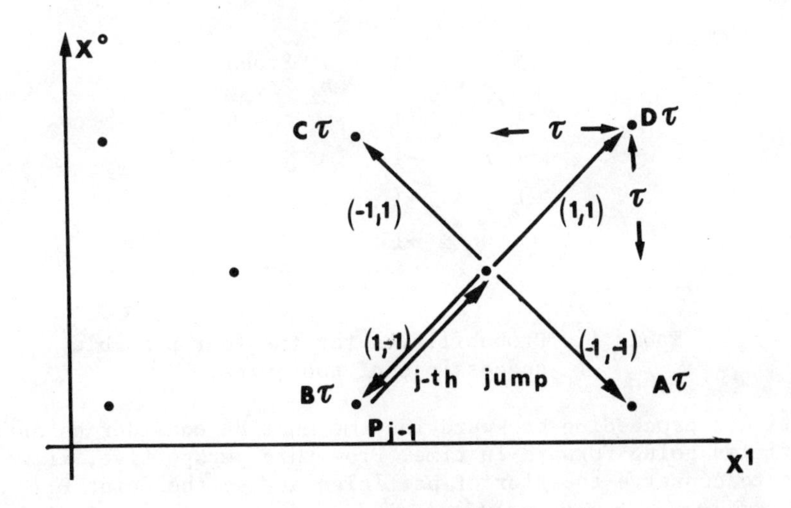

Fig.4. An example of the four possible successions of two jumps. For each possible (j+1)-th jump we marked the value of the couple (ε_j, η_j) and the corresponding probability.

Of course a sequence $\{\varepsilon_j, \eta_j\}$, with $j \in N$, of values of these stochastic variables completely determines one of the infinite possible trajectories, except for the first jump, because there is no "preceding" jump for it. Thus, starting from $P_0(x^0, x^1)$, in the first jump we can get one of the four possible points

$P_0(x^0+t\tau, x^1+s\tau)$ and, after N jumps, one of the points $P_N(x^0+tT_N, x^1+sD_N)$ where

$$T_N = \tau(1+\eta_1+\eta_1\eta_2+\ldots+\eta_1\eta_2\cdots\eta_{N-1}),$$

$$D_N = \tau(1+\varepsilon_1\eta_1+\varepsilon_1\varepsilon_2\eta_1\eta_2+\ldots+\varepsilon_1\varepsilon_2\cdots\varepsilon_{N-1}\eta_1\eta_2\cdots\eta_{N-1}). \quad (14)$$

We come now to the problem of the assignment of a statistical weight to each trajectory. in order to do this e introduce for each jump a probability for each realization of the signs of the corresponding couple ε_j, η_j. We listed these probabilities in Table 1. Moreover we, suppose that A,B,C,D are constant and positive over all of space-time.

Among these four constants we can also posit a relation that can be justified as a principle of mass-flux conservation. If we consider, e.g., as in Fig. 4, a particle arriving at P_j after its j-th jump, we must remember, for the (j+1)-th jump, that the

ε_j	η_j	Probability
-1	-1	$A\tau$
+1	-1	$B\tau$
-1	+1	$C\tau$
+1	+1	$D\tau$

Table 1. Probabilities for the four possible successions of two jumps.

particles proceeding backward in time must be consider as antiparticles going forward in time. From this perspective, if we want to conserve the flux of particles across the point P_j between the j-th and the (j+1)-th jumps, we must remark that:

(a) in the j-th jump we have only a particle going to the right;

(b) in the (j+1)-th jump we have a "fraction" $D\tau$ of particles and $B\tau$ of antiparticles going to the right and a "fraction" $C\tau$ of particles and $A\tau$ of antiparticles going to the left.

If we remember that particles and antiparticles have the same mass, the mass flux conservation across P_j finally gives

$$(-A + B - C + D)\tau = 1. \quad (15)$$

We now consider a function $f(x^0,x^1)$ defined over all of space-time and, generall speaking, having complex values and then we introduce the following set of functions:

$$F_N^{t,s}(x^0,x^1) = \langle f(P_N)\rangle \quad , \tag{16}$$

where $\langle\cdot\rangle$ indicates an average over all the possible points P_N that are reached along trajectories consisting of N jumps, starting from P_0, with a first jump made in the direction fixed by (t,s). Of course, because of the arbitrariness of the starting point P_0, the functions $F_N^{t,s}$ are defined over all of space-time.

We now make our average for the first jump, so that, from (15) and passing to the limit $N \to \infty$(for fixed τ), we have

$$\begin{aligned} F^{t,s}(x^0,x^1) &= F^{t,s}(x^0+t\tau,x^1+s\tau) \\ &A\tau\,[F^{-t,s}(x^0+t\tau,x^1+s\tau) + F^{t,s}(x^0+t\tau,x^1+s\tau)] \\ &B\tau\,[F^{-t,-s}(x^0+t\tau,x^1+s\tau) - F^{t,s}(x^0+t\tau,x^1+s\tau)] \\ &C\tau\,[F^{t,-s}(x^0+t\tau,x^1+s\tau) + F^{t,s}(x^0+t\tau,x^1+s\tau)] \;, \end{aligned} \tag{17}$$

where $F^{t,s}$ are our functions for $N \to \infty$.

In the limit $\tau \to 0$, when our lattice tends to recover all of space-time, we get the following set of four partial differential equations:

$$\begin{aligned} -\partial_0\, F^{t,s} = \frac{s}{t}\,\partial_1 F^{t,s} + \frac{A}{t}(F^{-t,s}+F^{t,s}) + \frac{B}{t}(F^{-t,-s}-F^{t,s}) + \\ + \frac{C}{t}\,(F^{t,-s}+F^{t,s}). \end{aligned} \tag{18}$$

If we define now the following four linear combinations of $F^{t,s}$:

$$\begin{aligned} \phi &= F^{++} + F^{--} + F^{+-} + F^{-+}, \\ \chi &= F^{++} + F^{--} - F^{+-} - F^{-+}, \\ \psi &= - F^{++} + F^{--} - F^{+-} + F^{-+}, \\ \omega &= - F^{++} + F^{--} + F^{+-} - F^{-+}, \end{aligned} \tag{19}$$

we can build a new set of equations equivalent to (18):

$$\begin{aligned} \partial_0\phi + \partial_1\chi &= 2(C - B)\psi \;, \\ \partial_0\chi + \partial_1\phi &= 2(A - B)\omega \;, \\ \partial_0\psi + \partial_1\omega &= 2(A + C)\phi \;, \\ \partial_0\omega + \partial_1\psi &= 0. \end{aligned} \tag{20}$$

By differentiation and successive linear combination one gets

$$\begin{aligned}\Box\phi &= 2(A-2B+C)\,\partial_0\psi - 4(A-B)(A+C)\,\phi\,,\\ \Box\chi &= 2(A-2B+C)\,\partial_0\omega\,,\\ \Box\psi &= -\,2(A+C)\,\partial_1\chi + 4(A+C)(C-B)\,\psi\,,\\ \Box\omega &= 2(A+C)\,\partial_0\chi - 4(A+C)(A-B)\,\omega\,,\end{aligned} \tag{21}$$

(where $\Box$ is a two-dimensional d'Alembert operator).
Setting:

$$B = \frac{A+C}{2}\,, \qquad 2(A^2-C^2) = m^2\,, \tag{22}$$

one is finally led to

$$\begin{aligned}(\Box + m^2)\phi &= 0,\\ \Box\chi &= 0,\\ (\Box + m^2)\psi &= -\,2(A+C)\,\partial_1\psi\,,\\ (\Box + m^2)\omega &= 2(A+C)\,\partial_0\chi\,.\end{aligned} \tag{23}$$

We interpret the first equation as a Klein-Gordon equation for a function ϕ which is the average of an arbitrary function f over all the final points that an reached along all possible trajectories realized in an infinite number of jumps; in this average we consider also the first jump by supposing that the four possibilities for the signs of t,s are equiprobable.

In the previous derivation we proved that each solution (ϕ,χ,ψ,ω) of (20) is a solution of (23), but it is possible to show that not all the solutions of (23) are solutions of (20). In other words, we proved the statement "the function ϕ defined as the average (19) always is a solution of a Klein-Gordon equation";now, what about the inverse statement, "all the solutions of a Klein-Gordon equation are interpretable as averages satisfying a system like (20)"? It is easy to show that, if ϕ is an arbitrary solution of the Klein-Gordon equation, we always can determine the functions χ,ψ,ω in such a way that (ϕ,χ,ψ,ω) is a solution of (20). In fact, if ϕ is an arbitrary solution of the Klein-Gordon equation, we choose χ as an arbitrary solution of $\Box\chi = 0$ and then postulate

$$\begin{aligned}\psi &= \frac{1}{C-A}\,[\partial_0\phi + \partial_1\chi]\,,\\ \omega &= \frac{1}{A-C}\,[\partial_0\chi + \partial_1\phi]\,.\end{aligned} \tag{24}$$

It is only a question of calculation to show that this (ϕ,χ,ψ,ω) is a solution both of (20) and (23).

5. TWO-PARTICLE SYSTEM AND NON-LOCAL INTERACTIONS

The stochastic derivations of relativistic quantum equations can be generalized (at least starting from Nelson's equation (13)) to the case of two particles. The method (17) is simply based on the introduction of an eight-dimensional configuration space, so that the pair position is defined by an eight-component vector.

The same results were also obtained by Namsrai (18) in a recent paper.

We now remark that in this case, as first observed by Bohm, the quantum potential will be non local. Moreover, this result is in agreement with

(a) Ghirardi's observation (19) that the usual stochastic models of quantum mechanics can not eliminate the quantum non-locality;
(b) the consequences of the hydrodynamical analysis (11) of the two particle quantum field which led to instantaneous interactions.

Here we claim that this non-locality is based on the real action of a real quantum potential in a real material aether and that this non-local interaction solves the EPR paradox. (20) We shall devote here only a few words to explaining how we see the problem of the causal anomalies.

In our opinion, in a completely deterministic world there is no place for what we mean by the word "signal". With "completely deterministic world" we mean that we can uniquely predict all the word lines of all the particles by solving a Cauchy (with or without action at a distance) and starting with initial conditions given on a space-like surface. In such a world we claim that, even if we have non-local potentials or action at a distance, we cannot have causal anomalies because we can not have superluminal signals.

In fact, let us suppose that we are in such a completely deterministic world where particles interact even by means of non-local potentials. In that world we can solve the Cauchy problem for relativistic equations and then we can uniquely determine the world lines of our interacting particles by means of non-local potentials. Let us now consider two such particles (see Fig.5a), and let us explore the possibility for using action at a distance (i.e., the non-local potential connecting our two particles) to send superluminal signals from (1) to (2). How can we do it? Of course by disturbing the world line of (1) at a given time: This disturbance will propagate faster than light, or even instantaneously (see Fig.5b). Moreover, how can

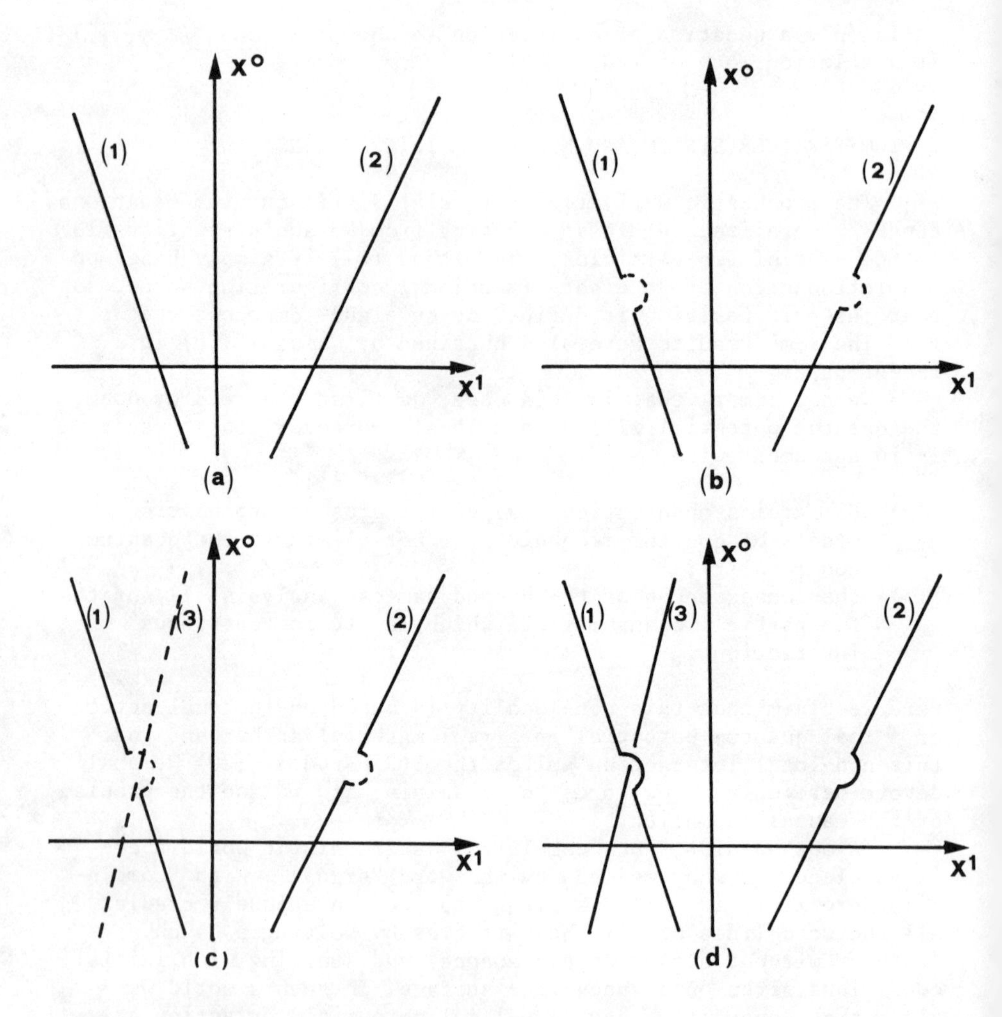

Fig. 5. What the word "signal" means in a completely deterministic world.

we disturb the particle (1)? Of course, by another particle (3) (see Fig.5c). But, if the world is uniquely determined, the world line of (3) was at its place in space-time before I decided to disturb (1), so that Fig. 5c can not be a real modification of Fig.5a. In a completely deterministic world there is no possible "modification": The world *IS* and we can not intervene from exterior in its tissue in order to modify it because we are *IN* the world. If the particle (3) exists, it produces no disturbance of (1) or (2): We have only to solve our Cauchy problem for three particles and we would have three world lines without disturbance, and thus without "signal" (see Fig. 5d).

In fact, it is clear that a "signal" always need a free will that is external to the physical world and that, at a given time, decides to modify a regular behaviour in order to send a message. Indeed, a regular behaviour never constitutes a signal. For example, we can always arrange a line of lamps so that they all light up in a very short time interval, independently of one another. In that case, if we look at them without knowing the arrangement, we could have the illusion of something traveling faster than light, but in reality it cannot constitute a signal at all.

From that standpoint, if, as we believe, the particles of the human brain are in the physical world and behaves like other particles, there are no "signals" at all. In other words, the world is an unique configuration of events in space-time, and our problem is not to find how to "produce" events but to find the general laws that express the disposition of the events. We can predict or remember, hope or regret, because we have a mind and a memory, and then we explore the space-time in a "historical" way and have the illusion of the "production" of the events by other events because we have a partial (in space and time) knowledge of the world. In a certain sense, our idea of causality seems to us to be the awareness of determinism together with the illusion of an event production , given by our lack of information. So, I cannot kill my grandfather before I was born; but I will say more: I cannot kill my grandchild, if that it is not already written in the space-time. In the words of a rationalist philosopher of the XVII century, "Men are wrong in considering themselves free, and this opinion consists only in the fact that they are conscious of their actions and unaware of the causes which determines them ".(21)

In conclusion, if our world is completely deterministic, we have no problems about signals. We shall not discuss here the philosophical problems of a deterministic standpoint, but we will limit ourselves to posing the main physical question: Is a relativistic and completely deterministic picture of the world compatible with action at a distance? concerning this problem, Currie, Jordan and Sudarshan (22) demonstrated in 1963 the well-known "non-interaction theorem". Briefly this theorem states that, if in a classical phase space one wishes to describe a system of particles in such a manner that (1) the dynamics is given by a Hamiltonian, (2) the theory is relativistically covariant, (3) the coordinate variables of the individual particles transform correctly under Lorentz transformations, then the only such system is a collection of free particles. Of course, under these condition it is completely impossible to build a covariant deterministic world with non-local potentials.

A recent advance in analyical mechanics is the result that a predictive mechanics is possible if some particular conditions of compatibility are satisfied by the non local potentials. (23) In this way the door to construct a completely deterministic and covariant world with action at a distance is reopened; and, from the standpoint of a deterministic interpretation of the quantum mechanics, the main question becomes: Does the quantum potential for two correlated particles satisfy the compatibility conditions of predictive mechanics? The answer is yes (20); the detail of this demonstration is discussed in the paper of Prof. Vigier.

REFERENCES

1. E.Madelung,*Z.Phys.* *40*, 332 (1926).
2. D.Bohm, *Phys.Rev.* *85*, 166,180 (1952);*Phys.Rev.* *89*,458(1953); *Prog.Theor.Phys.9*,273 (1953).
3. T.Takabayasi,*Prog. Theor.Phys.* *8*,143 (1952);*Prog.Theor.Phys.* *9*, 187 (1953).
4. W.Pauli,*Louis de Broglie,physicien et penseur* (Albin Michel, Paris,1953).
5. J.B.Keller,*Phys.Rev.* *89*, 1040 (1953).
6. P.A.M. Dirac,*Nature 168*, 906 (1951) ; *Nature 169*, 702(1952).
7. K.P.Sinha,C.Sivaram, and E.C.G. Sudarshan, *Found.Phys.6*,65, 717 (1976).
   K.P.Sinha and E.C.G.Sudarshan, *Found.Phys.8*,823 (1978).
8. N.Cufaro Petroni and J.P.Vigier, *Found.Phys.13*,253 (1983).
9. J.P.Vigier,*Lett.N.Cim.* *29*,467 (1980); *Astron.Nachr.303*,55 (1982).
10. F.Selleri,*Lett.N.Cim.* *1*,908 (1969).
    A.Garuccio, K.Popper and J.P.Viger, *Phys.Lett.* *86A*,397 (1981).
    J.et M.Andrade e Silva, *C.R.Acad.Sc.Paris 290*,50 (1980).
11. N.Cufaro Petroni,Ph.Gueret and J.P.Vigier, "Hydrodynamical Analysis of Spinor Fields" (Bari University preprint, 1983).
12. F.Halbwachs, *Theorie relativiste des fluides à spin* (Gauthier-Villars,Paris,1960).
13. E.Nelson, *Phys.Rev. 150*,1079 (1966).
    L.de la Peña Auerbach, E.Braun and L.S.Garcia Colin, *J.Math. Phys. 9*, 668 (1968).
    L.de la Peña Auerbach, *J.Math.Phys.* *10*, 1620 (1969).
    L.de la Peña Auerbach and A.M. Cetto, *Found.Phys.5*, 355(1975).
    W.Lehr and J.Park,*J.Math.Phys.18*, 1235 (1977).
    F.Guerra and P.Ruggiero, *Lett.N.Cim.* *23*, 529 (1978).
    J.P.Vigier, *Lett.N.Cim.* *24*, 265 (1979).
    N.Cufaro Petroni and J.P.Vigier, *Phys.Lett.* *73A*, 289 (1979); *Phys.Lett.* *81A*,12 (1981).
14. N.Cufaro Petroni and J.P.Vigier, "Random Motions at the Velocity of Light and Relativistic Quantum Mechanics", in press on *J.Phys. A*.
15. A.Avez, "Interpretation Probabiliste d'Equations aux Derivées Partielles Hyperboliques Normales" (preprint Paris).
16. R.P.Feynman, *Phys.Rev.76*, 749,769 (1949).
17. N.Cufaro Petroni and J.P.Vigier, *Lett.N.Cim.* *26*,149 (1979).

18. Kh.Namsrai, *Sov.J.Part.Nucl.* *12*, 449 (1981); *J.Phys.* *14A*,1307 (1981).
19. G.C.Ghirardi, C.Omero, A.Rimini and T.Weber, *Riv.N.Cim.* *1*,1 (1978).
20. N.Cufaro Petroni, Ph.Droz Vincent and J.P.Vigier,*Lett.N.Cim.* *31*, 415 (1981).
N.Cufaro Petroni and J.P.Vigier, *Phys.Lett.* *88A*, 272 (1982); *Phys.Lett.93A*, 383 (1983).
21. B.Spinoza, *Ethica*, Part 1, Appendix; Part 2, prop. XXXV(scolium). (UTET, Torino, 1972).
22. D.G.Currie,T.F.Jordan and E.C.G.Sudarshan, *Rev.Mod.Phys.* *35*, 350 (1963).
23. Ph. Droz Vincent, *Phys.Scripta* *2*, 129 (1970).
L.Bel, *Ann. Inst. H.Poincaré* *3*, 307 (1970).
Ph. Droz Vincent, *Ann. Inst. H.Poincaré* *27*, 407 (1977).
A.Komar, *Phys.Rev.* *18D*, 1881, 1887, 3617 (1978).
L.Bel,*Phys.Rev.* *18D*, 4770 (1978).
Ph. Droz Vincent, *Phys.Rev.* *19D*, 702 (1979);*Ann. Inst. H.Poincaré* *32*, 377 (1980).
R.P.Gaida, *Sov.J.Part.Nucl.* *13*, 179 (1982).

ON THE NON-LINEAR SCHRÖDINGER EQUATION

A.B. Datzeff

Faculty of Physics
University of Sofia
Sofia, Bulgaria

ABSTRACT

When a homogeneous electron beam passes through a small hole, a well-known diffraction picture is obtained according to quantum mechanics. This picture does not depend on the density of the electrons, since the Schrödinger equation is linear. Experiment contradicts however this deduction, because in reality the beam dilates. Accordingly, a generalized nonlinear Schrödinger equation is here proposed which gives both a diffraction picture and the dilatation phenomenon. Attempts to check these effects may turn into a crucial experiment.

The Schrödinger equation, which is a linear one, is confirmed by the experiment in an enormous number of cases. Nevertheless, its different nonlinear generalizations, especially in the one-dimensional case, were proposed during the last years, with a view to some concrete applications. Physical reasons for such generalizations are however not given. This is easy to understand as the physical bases of quantum mechanics (QM) are not clear and are still being debated. We have proposed a nonlinear generalized Schrödinger equation that is based on physical arguments and aspires to be a fundamental equation.

We have developed a new point of view concerning quantum mechanics in a number of publications. These have been generalized in the monograph *Mécanique quantique et réalité physique* (1).

G. Tarozzi and A. van der Merwe (eds.), Open Questions in Quantum Physics, 215–224.

Our basic hypothesis is the introduction of a material carrier for the electromagnetic field, which we have called the *subvac* (substance of the vacuum). This is assumed to have a discrete structure. In other publications (2) we have shown that such a material carrier of the field is quite compatible with the special theory of relativity. We treat a microparticle, say an electron, as something that can, in principle, be localized. Since it interacts with the elements of the subvac, which have a chaotic character, the electron's movement cannot be described by means of classical mechanics, and it is necessary to introduce a probability function $F(\underset{\sim}{r},t)$. In looking for a new description, we have obtained an equation similar to Schrödinger equation and we have been able to recover the entire mathematical formalism of quantum mechanics. We shall call this equation the *probability equation,* to distinguish it from the Schrödinger's wave equation with wave function Ψ.

In Ref.(2) we have shown how, in our opinion, the basic modifications should be introduced into quantum mechanics and especially in accordance with Schrödinger's equation as the next stage of development. A nonlinear generalization of the Schrödinger equation was obtained there for the case of a free particle and for the bound state of the particle in an atom, molecule, solid body, etc., when the density of the particles is quite large and where Schrödinger's equation ceases to be valid. Later we shall discuss in more detail one of these cases, viz., the passage of a beam of electrons through a hole in a screen. This example is sufficiently general to enable us to draw a conclusion of prime importance. It also leads to observable effects which cannot be explained by the present quantum mechanics and which may lead to crucial experiments.

Let us examine the case of a dense beam of electrons moving with uniform velocity $\underset{\sim}{v}$. Let it be incident perpendicular to a nontransparent unlimited plane screen A_1 with a circular aperture σ and be absorbed on a second unlimited plane screen B_1 parallel to A_1. The diffraction effects are well described by classical wave optics as well as by quantum mechanics. This picture obviously does not depend on the density of the incident electron beam, as Schrödinger's equation is linear. Indeed, if the latter equation has a solution Ψ, then the function $\Psi_1 = A\Psi$, in which A is an arbitrary constant, is also a solution. This should be true for arbitrarily large values of $|A|$. (The zeroes of the function Ψ_1 do not depend on $|A|$, and neither does the relative density of the electrons that have fallen on the screen B_1 .) Experiments show, however, that the diameter of the electron beam increases behind screen A_1 and gets larger as B_1 is moved further away from A_1 .

This dilatation encountered for example in electron tubes, is usually attributed to the Coulomb force of repulsion existing between the electrons in the beam. This solution of the problem is however unsatisfactory for the following reasons:

The explanation is entirely classical, no use being made of QM. Indeed, the explanation is based on the classical notion of space charge, which is not a quantum mechanical notion. With its help, engineers explain in some simple practical cases the beam dilatation (in electron tubes, for instance) with an approximation which could be sufficient in today's practice. But then all the diffraction phenomena, which, as is known, are well described by QM, are absent. In practice these are indeed insignificant, but theoretically they cannot be neglected unless we wish to forget the existence of QM. On the other hand, the insertion in the Schrödinger equation of such a space charge, suggested by experiment, although able to explain approximately the said dilatation of the beam, is evidently a violation of the Schrödinger equation. So, the solution of the problem cannot be phrased in these terms.

Although we shall not discuss in detail the question of the physical meaning of quantum mechanics with which we are often faced here, it is necessary to elucidate the following point.

The well-known Schrödinger equation $H\Psi = i\hbar(\partial\Psi/\partial t)$ for the wave function $\Psi(x,y,z,t)$ is usually called a one-particle equation because x,y,z are treated as coordinates of one particle (and similarly for the two- and many-particle equations). It is generally accepted that this equation well describes the probabilistic behaviour of one particle. But is this equation in reality associated with only one electron as its name would imply? Of course not. All physicists agree on the statistical meaning of quantum mechanics. In fact the problem refers to an ensemble of large number $N (N \to \infty)$ of noninteracting particles (electrons) that occupy the same physical state (in our case - with equal average velocity $\underset{\sim}{v}$ and equal average density $\varrho = |\Psi|^2$). Since the Schrödinger equation is a linear one, ϱ is proportional to $|A|^2$ (A the normalization constant), i.e., ϱ increases indefinitely along with $|A|$. We claim that this interpretation solely agrees with the basic formula for the density flux that follows from the Schrödinger equation and which is always confirmed by experiment.

Accordingly, one electron $P(x,y,z)$ formally represents the whole ensemble by means of the function Ψ . In this fact lies, in our opinion, the profound sense and force of the Schrödinger equation. If, however, a Coulomb interaction between the electrons is admitted, as it is in reality, then, even if the average electron density in all the space is arbitrarily low, the Coulomb potential energy of every electron becomes infinitely large simultaneously with $N \rightarrow \infty$. Then some N-particle Schrödinger equation for these N electrons will be senseless.

This confrontation of a particle and an ensemble is immediately seen, for example, in the case of a free movement ($U = 0$), where $\Psi = a \exp i(-\omega t + pr)$. Then, for any fixed value of the constant A determining with certainty one electron in a given volume V_0, there will exist an infinite number of electrons in the whole space ($N = \int |\Psi|^2 d\tau = \infty$). But this equality represents physical nonsense -- a good physical theory should not introduce nonexisting infinities. But this is a simple consequence from the principles of quantum mechanics, and it cannot be removed by any artificial reasoning.

The hydrogen problem is a typical two-particle problem where the two-particle Schrödinger equation is employed. Using relative coordinates and center of mass coordinates, one has for the movement of the center of mass G a plane wave connected with an infinite number of particles (hydrogen atoms). This solution, being not approximate but absolute, does not permit any forced alteration of the physical content of the problem as described by the laws of quantum mechanics. A quite similar conclusion follows for the many-particle Schrödinger equation (pertaining for example, to a molecule with n particles). In fact, there exists the exact theorem of the center of mass uniform movement and the conclusion that the wave function Ψ describes the behavior of N molecules ($N \rightarrow \infty$). In other words, the designation of the Schrödinger equation as a one-particle, two-particle, or multiparticle equation is only a *conditional name* given to underline the properties of a large ensemble of identical particles or identical systems of 2,3,...,N particles occupying the same state of motion.

We shall consider below the generalized equation for the electron beam in question. We shall discuss only this concrete case because, on the one hand, it is quite simple and leads to the theoretical deductions on which we can take a firm stand, and, on the other hand, it leads to a simple experimental check. At the same time the considerations are sufficiently general to permit the conclusion that, while in this case the Schrödinger equation contradicts the experiment, the generalized equation can explain it. Similarly, for the hydrogen gas contained in the volume V, the one-particle Schrödinger equation describes well the Balmer spectral lines, but it would be in contradiction with the experiment in the case of very dense gas (for which the spectral lines broaden). The latter case can be treated by the generalized equation we have proposed (1), and the result could be submitted to an experimental check.

Let us return to the problem of generalizing the Schrödinger equation. In the case of free movement of electrons between the screens A_1 and B_1, let $F(x,y,z)$ be the probability function for an electron $P(x,y,z)$ moving in an external potential field $U^o(x,y,z)$. The effect of all the other electrons on P can be taken in account by means of a supplementary term $\lambda U(x,y,z)$, which expresses the average Coulomb potential energy caused by the other electrons. So we admit that in this case of the free movement of the electrons the function F will satisfy the following integro-differential probability equation, which generalizes the Schrödinger one-particle equation, namely,

$$\Delta F - \frac{2m}{\hbar^2}(U^o + \lambda U)F = -\frac{2im}{\hbar}\frac{\partial F}{\partial t},$$

$$U(\underset{\sim}{r},t) = \int_V \frac{|F(r',t)|^2\, d\tau'}{|\underset{\sim}{r} - \underset{\sim}{r}'|}, \quad \lambda = l^2 N, \quad N = \int_V |F|^2_{\lambda=o}\, d\tau. \tag{1}$$

(In other words, Eq.(1) takes into account the interaction of one representative electron P with all the other electrons in the ensemble; this interaction is not present in the usual Schrödinger equation for the same case.) The integral in U and in N is taken over the volume V between the two screens, and it is always convergent.

Equation (1) for the stationary case has an integral of the type $F = f(x,y,z)\exp(-iEt/\hbar)$, whence we obtain the amplitude probability equation

$$\Delta f + \frac{2m}{\hbar^2}(E - U^o - \lambda U)f = 0, \quad U(\underset{\sim}{r}) = \int_V \frac{|f(r')|^2 d\tau'}{|\underset{\sim}{r} - \underset{\sim}{r}'|}, \quad N = \int_V |f|^2_{\lambda=o}\, d\tau. \tag{2}$$

In (1) and (2) the parameter $\lambda = e^2N$, where e is the electron charge and N is the average number of electrons between the screens A_1 and B_1 at any moment. N is given by the approximate formula $N = \varrho v \sigma_0 (l/v) = \varrho \sigma_0 l$, where σ_0 is the area of the aperture σ, ϱ the electron density, and l stands for the distance between the screens A_1, B_1.

Let the incident beam be characterized by the function $\bar{f} = A \exp(i\underset{\sim}{k} \cdot \underset{\sim}{r})$, normalized in a unit volume. The boundary conditions determine the solution of Eq.(2) between A_1 and B_1. These are in this case, if the origin O is in σ, the axis $OZ \perp A_1$, and one uses semipolar coordinates ϱ, φ, z:

$$f(P) = \bar{f}(P \in \sigma), \quad f(P) = 0(P \in (A_1 - \sigma)), \quad (z = 0),$$
$$f(P) = 0, \quad (0 < z < l, \ \varrho \to \infty). \tag{3}$$

(In reality the boundary condition (3) on the screen A_1 may be realized through some electron source placed on the aperture σ and ensuring such electron flow as this one given by the wave $\bar{f}$.) The values of f on the screen B_1 are not given beforehand, but can be obtained as a solution of the problem using the conditions (3). (Here we have a problem of emission.)

Equation (2) is a generalization of Schrödinger's time-independent equation, and in the case of small densities of the electron beam, or, rather, when the term λU can be neglected relative to $E - U^0$, becomes identical with it. As in the case of Schrödinger's equation, Eq.(2) seems to be a one-particle equation, because it uses the coordinates of one particle $P(x,y,z)$ as a representative of the ensemble. But in contrast to Schrödinger one-particle equation (which is obtained from Eq.(2) by setting $U = 0$), the statistical meaning of (2) can be seen directly, because it obviously contains the total number of electrons between A_1 and B_1.

By writing down the generalized equation (1) we are supposed to deal with particles having electrical charge, and this leads to the supplementary energy U (1). If $U = 0$, we obtain again the Schrödinger equation valid for free particles of any kind. If, in the case of any noncharged particles, we know some phenomenological law for the two-body interaction potential $V(r)$ as a function of the mutual separation (e.g., the Yukawa potential for neutrons, for instance), then the generalized equation (1) will hold good for these particles as well if $1/r$ is replaced by $V(r)$ under the integral for $U(1)$. But it is clear that the problem considered for a beam of particles is practically without interest for particles that interact through short-range forces. Then the Schrödinger equation would be a good approximation even in the case of relatively large densities.

At first sight Eq.(2) resembles the Hartree-Fock equation, used in the method of self-consistent field, because of the integral term U. This term, which has the form of a Coulomb-type potential for a continuous charge density $|f|^2$, is indeed similar to the corresponding potential terms U in the Hartree-Fock equations. However, the analogy between Eq.(2) and Hartree-Fock's equations (more exactly, the Hartree equations if the effect of the spin is neglected) is only apparent, and in fact the former completely differs from the latter for the following reasons:

The system of n Hartree equations was proposed as an approximation to the n-particle Schrödinger equation ($n>2$) in the case of bound movement (for example, for an n-electron atom of total energy $E<0$). But if $E>0$, which could be a real case, then the integrals of the type (2) become divergent, i.e., the Hartree equations would be meaningless. On the other hand, the Hartree equations are not written for a case of free electron movement ($E>0$), as the initial functions, necessary for writing the successive approximations, are non-normalizable ones. (The integrals of the type $U(2)$ are divergent.) In our case we have a free electron movement, and our non-linear equation (2) has nothing to do with a system of Hartree equations, which are even not defined here.

Another question which naturally occurs in connection with Eq.(2) is the following: Since the electrons in the beam between the screens A_1 and B_1 interact through the Coulomb potential, is it not possible to obtain the beam dilatation as a consequence of Schrödinger's equation for n particles? Or in this way may one arrive at a corrected one-particle generalized Schrödinger equation of the type, say, of Eq.(2)? It is true that no one has done this so far, but can it be done? We claim this is indeed impossible for the following reasons.

Let us admit that one can explain the beam dilatation starting from a $\bar{N}$-particle Schrödinger equation $H\Psi_{\bar{N}} = i\hbar(\partial\Psi_{\bar{N}}/\partial t)$, where $\bar{N}\gg1$ is a fixed number chosen for some reasons, and the Hamiltonian H contains the electron interactions $U_{\bar{N}}$. But using the Jacobi coordinates, the last equation decomposes, as is known, into two equations: (1) one-particle Schrödinger equation (without a potential term) for the center of mass G with a mass $M = \bar{N}m$ and a wave function Ψ_G (a non-normalizable plane wave); (2) an $(\bar{N} - 1)$-particle Schrödinger equation for the relative movement, with a wave function $\Psi = \Psi'_{\bar{N}-1}$. On the other hand, if the electron interactions are neglected, the last equation will decompose into $\bar{N} - 1$ independent equations, whose solutions are non-normalizable functions. (All these results are absolutely exact, although they lead to physical paradoxes.) At the same time one cannot find any new, non-linear equation using such functions and the apparatus of QM. Hence one cannot explain the beam dilatation in this manner, using any number $\bar{N}$ of electrons.

At the same time the following question remains open. What is the number $\bar{N}$ and how is it connected with the electron beam problem? In the last case the following numbers appear in a natural way: N_1 particles/cm^{2}sec, the density of the incident beam, and $N_2(t) = \alpha' t$ (α' is a constant), the total number of electrons incident on screen B_1 during time t. It is clear that neither N_1 nor N_2 could be taken instead of $\bar{N}$. Likewise, the average number N_0 of electrons between screens A_1 and B_1 is determined in a stationary case. But the electrons appear at the aperture σ and disappear at the screen B_1. Schrödinger's equation for one or more particles can be written in terms of their coordinates and assuming that they exist eternally, i.e., that particles are not created or destroyed. There is no Schrödinger equation for an undetermined number of particles being replaced by new ones. On the other hand, the number N_0, which can be expressed by means of $|\Psi|^2$ and which is connected with the average interaction between the electrons, could be inserted into the one-particle Schrödinger equation as a supplementary potential energy of one electron (for instance, by the way used here to write the proposed equation (1)). This is, however, a procedure which does not follow in a logical way from the principles of QM. Hence the equation (1) is not a consequence of these principles.

It is clear that by replacing the linear one-particle Schrödinger equation by another nonlinear one on the basis of any considerations used, one changes in principle the basic apparatus of quantum mechanics, no matter what the concrete form of such a nonlinear equation is. The nonlinear generalized equation (1) we proposed is exactly of such a nature. It is, however, not a consequence of the principles of quantum mechanics, but essentially a new postulate.

It is proposed that Schrödinger's equation be replaced by Eq.(1) or Eq.(2), as they are more general, both in the above examined case and in all similar cases of free particles (a multitude of apertures on A_1, differing in shape and geometric conditions), a case of scattering, etc. This is also true when screen B_1 is eliminated, but then the problem becomes nonstationary. One must then proceed from Eq.(1) and the boundary conditions (3) on screen A_1 for $t > t_0$. In principle, a solution can be found for every fixed $t > t_0$, because then the integral U (1) is convergent. Equation (1) can also be applied directly to a limited motion (say, N electrons enclosed in a space limited by surface S on which F = 0), but that case is of no interest at this moment.

An approximate solution of Eq.(2), leading to a satisfactory description of the beam dilatation in some real case, is given in Ref.(1). A complete analytic solution of equation (2) is given in Ref.(7).

DISCUSSION AND CONCLUSION

Equation (1) (or (2)) generalizes Schrödinger's equation and leads to results differing from it. These results can be grouped into two types of processes: macro- and microprocesses.

(1) *Macroprocesses*. A case in point is the above beam dilatation. A large number of similar macrotypes can be proposed by means of various modifications of the geometric or physical parameters in the above case. Thus, for example, if an electron beam passes through two (or more) neighboring apertures on a screen, then the beams that have passed through will move away from each other because of mutual repulsion. This result is, of course, known from experience. However, it follows from Eq.(2) and not from Schrödinger's equation. One must seek an exact quantitative description of this result by means of (2) and compare it with an exact experimental measurement, so as to subject Eq.(2) to a thorough check.

(2) *Microprocesses*. There are various diffraction processes with electrons. As has been indicated already, the diffraction phenomena will be obtained around the boundary of a cylindrical beam of electrons. These cannot be observed during ordinary experiments in electronics, which are macroscopic, as long as $\lambda \sim 10^{-8}$cm. However, they exist and in principle can be observed by means of sufficiently accurate experiments. It is

this type of experiment that accounts for the success of quantum mechanics. They are essentially diffraction experiments for the determination of de Broglie wavelength λ' (passage of electrons through a crystal, through a hole, etc.). For the situation of Eq.(2) without an external field ($U^o = 0$), when according to Schrödinger's equation one would have $\lambda' = h/mv$, λ' will in our case be a complicated function of position. An approximate evaluation of the change of λ' in some real case is given in Ref.(1).

If the experiment should confirm the consequences of the aforementioned two basic effects (beam dilatation and change of λ' as a result of diffraction) and also, in general, the new consequences of the probability equation (2), this will lead us to the conclusion that current quantum mechanics, based on the theory of linear operators, is only a first approximation, valid at relatively small particle densities ϱ ($\varrho < \varrho_o$, ϱ_o given), or relatively large velocities. This will also be an occasion for a more thorough discussion of the physical content of QM and also to consider in more detail the point of view that we are proposing. Thus the above experiments on the dilatation of the electron beam and on the modification of de Broglie's wavelength may become crucial experiments.

REFERENCES

1. A.B. Datzeff, *Mécanique quantique et réalité physique* (Sofia, 1969).
2. A.B. Datzeff, *Compt. Rend. 245,* 827, 891 (1952); *Cahier de Phys.,* No. 115, 99 (1960); *Ann. Univ.,* Sofia, Fac. Sci., *52,* 7 (1957/58).
3. A.B. Datzeff, *Nuovo Cimento 29,* B, No. 1 (1975)
4. T.R. Pierce, *Theory and Design of Electron Beams* (Van Nostrand, New York, 1954), 2nd. ed.
5. P.T. Kirstein, K.S. Gordon, and W.E. Eaters, *Space-Charge Flow* (New York, McGraw-Hill, 1967).
6. W. Claser, *Grundlagen der Optik* (Vienna, 1952).
7. A.B. Datzeff, *Phys. Lett. 29A,* No. 3 (1976); Preprint ICTP (78)5, Trieste; *Bulg. J. Phys. 6,* 3 (1979); *6,* 4 (1979).

RECENT PROGRESS IN DE BROGLIE NONLINEAR WAVE MECHANICS

Ph . Guéret

Institut de Mathématiques Pures et Appliquées
Université Pierre et Marie Curie
4 Place Jussieu, 75230 Paris Cedex 05

ABSTRACT

Within the de Broglie wave-particle duality theory, the relativistic covariance of the phase connection principle implies the existence of a nonclassical and nondispersive wave packet. This is a solution of relativistic wave equations with quantum potential nonlinearity. Physical implications follow from this mathematical model.

1. SOME REMARKS ON THE CLASSICAL DE BROGLIE WAVE PACKET

The idea of a wave associated with a particle occurred first to de Broglie (1) and was embodied in his well-known doctoral thesis of 1924.(2) His assumptions were that the particle has an internal vibration and moves in the space-time of special relativity, where at every moment it is localized.

A free particle at rest has mass m_o and energy $E = m_o c^2$. If this energy is taken to equal one quantum $h\nu_o$, the internal particle vibration frequency is defined by $\nu_o = m_o c^2/h$ and a wave function Ψ_o can be associated with this vibration so that

$$\Psi_o = a \exp(2\pi i \nu_o \tau), \qquad (1)$$

τ beeing the proper time of the particle. Ψ_o is uniform throughout space and does not serve in any way to locate the particle.

Now, if it is supposed that the same particle is moving freely with uniform velocity v in the +x direction, a Lorentz transformation gives

G. Tarozzi and A. van der Merwe (eds.), Open Questions in Quantum Physics, 225–236.

$$\tau = (t-\beta x/c)/\surd(1-\beta^2) \tag{2}$$

($\beta = v/c$) and, for an observer at rest, the wave function Ψ_o becomes

$$\Psi(x,t) = a \exp\left[2\pi i \frac{\nu_o}{\surd(1-\beta^2)}(t-\beta x/c)\right] \tag{3}$$

or, on taking $\nu_o/\surd(1-\beta^2) =$ and $c^2/v = V$

$$\Psi(x,t) = a \exp\left[2\pi i\nu(t-x/V)\right] \tag{4}$$

Note that, under the Lorentz transformation, the wave frequency becomes $\nu = \nu_o/\surd(1-\beta^2)$, while the particle frequency becomes $\nu = \nu_o\surd(1-\beta^2)$ according to the clock slowdown formula. However, *the internal vibration and the de Broglie wave remain in phase where the particle is located*, with de Broglie wave consequently acting a guide for the particle motion.

According to (4), for an observer at rest, the particle motion is associated with the propagation of a plane wave (B-wave) of frequency ν and phase velocity $V > c$. On defining, as usual, the wavelength by $\lambda = V/\nu$, one gets for the B-wave

$$\lambda = c^2h/Ev = h/p \tag{5}$$

E and $p = |\vec{p}|$ respectively representing the energy and relativistic momentum of the moving particle.

On the basis of classical reasoning in wave theory, (3) one can consider a wave packet (P-wave) made up of a superposition of plane waves with closely related frequencies. Using $k = \nu/V$, such a packet can be written as

$$\Phi(x,t) = \int_{\Delta k} \{a \exp\left[2\pi i(\nu t-kx)\right]\} \, dk \tag{6}$$

To apply this formula to matter waves, one puts $k = p/h$ and $E = h\nu$ obtaining

$$\Phi(x,t) = \int_{\Delta k} \{a \exp\left[\frac{i}{\hbar}(Et-px)\right]\} \, dp \tag{7}$$

Let p_o be a central value in the wave packet and assume that E varies slowly enough with p to justify the Taylor expansion

$$E = E_o + \frac{dE}{dp}(p-p_o) + 0\left[(p-p_o)^2\right] \tag{8}$$

Substitution of Eq.(8) into Eq.(7) yields

$$\Phi(x,t) \sim a \exp\left[\frac{i}{\hbar}(E_o t-p_o x)\right]\int_{\Delta p}\{\exp\left[\frac{i}{\hbar}(\frac{dE}{dp}t-x)(p-p_o)\right]\}dp \tag{9}$$

In Eq(9) the exponential term outside the integral represents a plane wave moving with constant velocity, and the integral behaves like a wave packet when

$$\frac{dE}{dp} t - x = 0 \tag{10}$$

One thus obtains for the group velocity U of the wave packet

$$U = \frac{dE}{dp} = \frac{x}{t} \tag{11}$$

and one notes that

-- *in the nonrelativistic case* (Schrödinger waves) : $E = p^2/2m$, $p = mv$, and thus

$$U = \frac{dE}{dp} = \frac{d}{dp}\left(\frac{p^2}{2m}\right) = v \tag{12}$$

-- *in the relativistic case* (de Broglie waves) : $E^2 = p^2c^2 + m_o^2c^4$, $p = m_o v/\sqrt{(1-\beta^2)}$, and

$$U = \frac{dE}{dp} = \frac{d}{dp}(p^2c^2 + m_o^2c^4)^{1/2} = \frac{pc^2}{E} = \frac{\sqrt{(1-\beta^2)}\ m_o vc^2}{\sqrt{(1-\beta^2)}\ m_o c^2} = v \tag{13}$$

In both cases the identification of the group velocity of a wave packet with particle velocity holds. This coincidence is a common source of confusion between the de Broglie wave (phase velocity $V = c^2/v$, frequency $\nu = E/h$) and the Schrödinger wave (phase velocity $V = v/2$, frequency $\nu = mv^2/2h$). Moreover, the Schrödinger which is a statistical featured wave without local concentration of energy, cannot be likened to a real physical wave. For this reason, de Broglie (4) expressed his double solution hypothesis : *To each solution* Ψ = a exp(iS) *of the propagation equation, there must correspond a solution* U_o = f exp(iS) *with the same phase* S, *but one whose amplitude* f *exhibits a singularity moving with the particle velocity.* In his reasoning, the U_o singularity corresponds to an extended wave phenomenon centered on a very small region, standing, strictly speaking, for the particle. The U_o wave would be the solution of a yet unknown nonlinear equation and the Ψ wave the solution of its linear approximation, at least outside the singularity.

If we take Ψ = a exp(iS), the linear Klein-Gordon equation

$$\Box \Psi - (m_o^2c^2/\hbar^2)\Psi = 0 \tag{14}$$

splits into

$$\begin{aligned} &(C) \quad \nabla(a^2\nabla S) = 0 \\ &(J) \quad (\nabla S)^2 + m_o^2c^2 = \hbar^2 \Box a/a \end{aligned} \tag{15}$$

(C) is a continuity equation and (J) a Jacobi equation.

On the right-hand side of Eq(15,J) one recognizes the relativistic generalization of the quantum potential introduced by de Broglie, which expresses the reaction of the wave deformed in the presence of obstacles to its propagation. In particular, the explicit calculation of this quantum potential in a two-slit situation (5) shows how to obtain interference without the need to abandon the notion of well-defined trajectories.

2. CONSTRUCTION OF A NONDISPERSIVE DE BROGLIE WAVE PACKET

It is known that the wave packet built in the previous paragraph actually spreads with time. However, on the basis of the *phase connection principle*, which states that at all points in which the particle is located the phase of its internal vibration coincides with the phase of the B-wave, and bearing in mind that this equality must be a covariant one, identical for all the observers, it is possible to construct a new nondispersive wave packet, as was shown recently by Mackinnon. (6) This is a convenient model of a U_o wave, which henceforth we shall call an S-wave.

To show this, let us assume that in a Minkowski diagram space-time is described by the axes (0; x, ct) of a stationary observer who sees a particle of rest mass m_o move with a constant velocity v in the negative x direction. Its associated frame (0; x_o, ct_o) is symmetrical with respect to the second bisectrix of the axes (0; x, ct), and - Ox_o makes an angle α with Ox such that $tg\alpha = v/c$ (see Fig. 1).

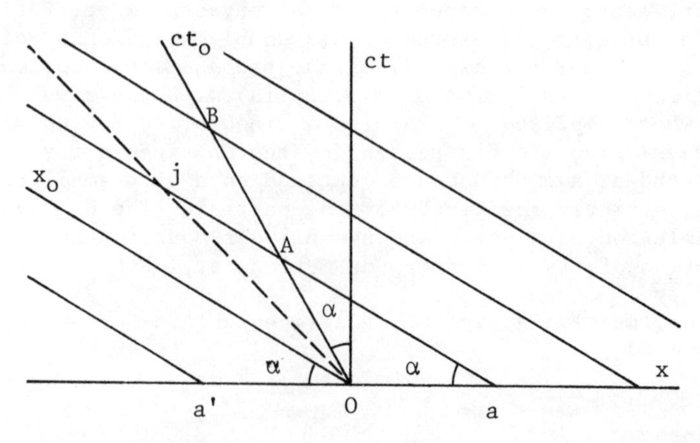

Fig. 1. Minkowski diagram pertaining to a particle moving in the -x direction with uniform velocity v. Its associated wavelength is given by Oa.

If the particle contains an intrinsic oscillation of frequency ν_o, a comoving observer will see it in the same state after equal time intervals (1/c)OA, (1/c)AB, etc., given by the rest period $T_0 = 1/\nu_0 = h/m_0c^2$.

The lines parallel to Ox_0 passing through A, B, etc. are "equal phase lines" for the observer moving with the particle, and the points •••, a', 0, a, ••• represent their intersections with the space axis of the observer at rest time 0, so that $0a=\lambda$ represents the de Broglie wavelength as seen by the observer at rest.

Let us now consider a second observer moving at uniform velocity v' in the +x direction. Its associated frame (0; x', ct') is symmetrical with respect to the first bisectrix of the axes (0; x, ct), and 0x' makes an angle δ with 0x such that $tg\delta = v'/c$ (see fig. 2). From the point of view of this second observer, the

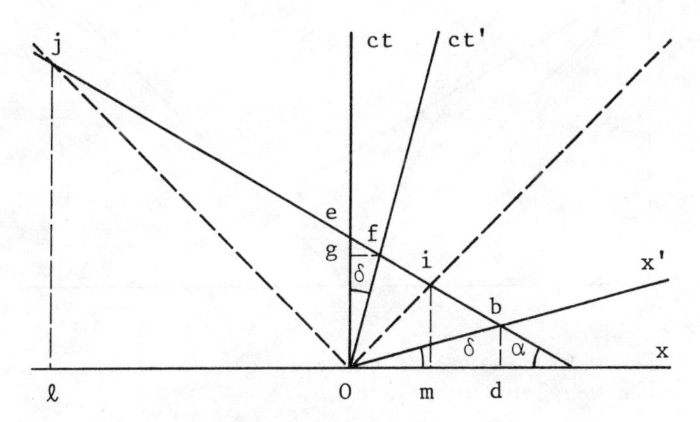

Fig. 2. Minkowski diagram representing the same particle as Fig.1 and the motion of an observer in the +x direction with uniform velocity v'. 0a still denotes the particle's wavelength in the rest frame. 0b furnishes the corresponding wavelength for the moving observer and 0d that as seen from the rest frame.

de Broglie wavelength is changed and appears to the observer at rest with the value $\lambda_1 = 0d$.

The velocity of the moving observer with respect to (0; x,ct) can vary from -c to +c, and determines the boundaries between which all wavelengths can vary in turn. The observer's frame which moves with velocity +c is mapped in the rest frame with the first bisectrix of (0; x,ct), and the observer's frame which moves at velocity -c is mapped in the rest frame with the second bisectrix of (0; x, ct). In this way the de Broglie wavelengths, measured respectively by 0i and 0j in the frames which have the limiting velocities +c and -c, appear respectively as 0m and 0ℓ for an observer at rest.

From the point of view of an observer at rest, there now

appears a wave packet associated with the particle. It is built from a superposition of B-waves with wavelengths greater than $O\ell$ moving in the $-x$ direction and of B-waves with wavelengths greater than Om moving in the $+x$ direction.

Let us now calculate the limiting values $O\ell$ and Om. In Fig. 3, one can write

$$Oj^2 = j\ell^2 + O\ell^2 = 2\ O\ell^2 \quad , \quad Oi^2 = jm^2 + Om^2 = 2\ Om^2 \qquad (16)$$

The triangles jAO and iAO have two equal sides, so that

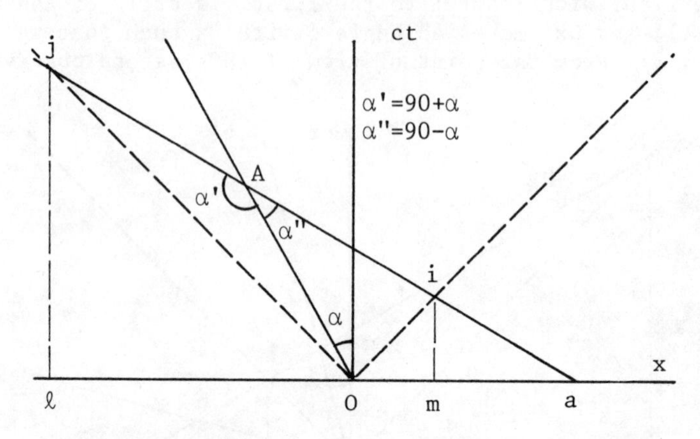

Fig. 3. Minkowski diagram of the triangles that yield the limiting wavelengths $O\ell$ and Om.

$jA = AO = Ai$. Moreover, we have

$$Oj^2 = 2\ AO^2[1 - \cos(90 + 2\alpha)]\ ,\ Oi^2 = 2\ AO^2[1 - \cos(90 - 2\alpha)]$$

Replacing the Oj and Oi by their values deduced from (16), we obtain

$$O\ell^2 = AO^2(1 + \sin 2\alpha) \quad , \quad Om^2 = AO^2(1 - \sin 2\alpha)$$

Since $\beta = v/c = \mathrm{tg}\alpha$ one gets $\sin 2\alpha = 2\beta/(1 + \beta^2)$, so that

$$O\ell^2 = AO^2(1 + \beta)^2/(1 + \beta^2) \quad , \quad Om = AO^2(1 - \beta)^2/(1 + \beta^2)$$

Noting that $AO/c = h/m_oc$, there results finally

$$\begin{aligned} O\ell &= (h/m_oc)[(1 + \beta)(1 + \beta)]^{1/2} \\ Om &= (h/m_oc)[(1 - \beta)(1 + \beta)]^{1/2} \end{aligned} \qquad (17)$$

We obtain in this way a wave packet (S-wave) for which the

wave vectors recovers the domain

$$\Delta k = 2\pi(O\ell^{-1} + Om^{-1}) = m_o c/\hbar\sqrt{(1-\beta^2)}$$

The central wave vector k_o from k can be written

$$k_o = [\,2\pi(Om^{-1} - O\ell^{-1})]\,/2 = m_o v/\hbar\sqrt{(1-\beta^2)}$$

If $m = m_o/\sqrt{(1-\beta^2)}$ is the apparent mass for the observer at rest, one can thus finally write $\Delta k = mc/\hbar$ and $k_o = mv/\hbar$.

Let us now show that the S-wave does not spread with time. Reverting to Fig. 2, we have seen that

$Oa = \lambda$ = de Broglie wavelength for the observer at rest;

$Ob = \lambda'$ = de Broglie wavelength for a moving observer corresponding to an apparent value $Od = \lambda_1$ for the observer at rest.

We obtain in the same manner :

$Oe = cT$ = de Broglie wave period for the observer at rest;

$Of = cT'$ = de Broglie wave period for a moving observer corresponding to an apparent value $Og = cT_1$ for the observer at rest.

We thus obtain through a simple geometric reasoning the relations:

$fg/Og = bd/Od = tg\delta = v'/c$, i.e. $fg/cT_1 = bd/(\lambda-\lambda_1) = v'/c$;

$gl/fg = bd/ad = tg\alpha = v/c = \beta$, i.e,
$c(T-T_1)/fg = bd/(\lambda-\lambda_1) = v/c = \beta$;

$Oe/Od = tg\alpha = v/c = \beta$, i.e. $cT/\lambda = v/c = \beta$

By eliminating fg and bd in this system, and putting $\omega(k) = 2\pi/T$, with $k = 2\pi/\lambda$, and $k_o = 2\pi/\lambda_o$, we get $\omega(k) - \omega(k_o) = v(k - k_o)$. Since v, k_o and $\omega(k_o)$ are constants, it follows that $(d/dk)[\omega(k)] = v$ for all ω and $(d^2/dk^2)[\omega(k)] = 0$. This is the necessary condition for ensuring that the S-wave preserves its shape for all time.

One thus sees that the S-wave preserves its shape under Lorentz transformations and appears as a relativistic effect.

If an inverse Fourier transformation is now utilized, we can express the S-wave shape (up to a multiplicative constant) in space-time in the form

$$F(x,t) = \frac{\sin \Delta k(x-vt)}{\Delta k(x-vt)} \quad \exp\{i[\,\omega(k_o)t-k_o x]\} \qquad (18)$$

One notices that this form of the S-wave is obtained without approximation, that it implies its nondispersive character in time, and that it is directly deduced from the basic assumptions of wave mechanics. The spatial distance between the two first zeros of the S-wave is $h/mc = (h/m_o c)\sqrt{(1-\beta^2)}$, so that it is subject to the usual relativistic contraction of length with velocity.

In the system tied to the particle, v = 0, and the S-wave amplitude is proportional to (sin kx)/kx, with $k = m_o c/h$, so that the spatial distance between the first two zeros of the S-wave

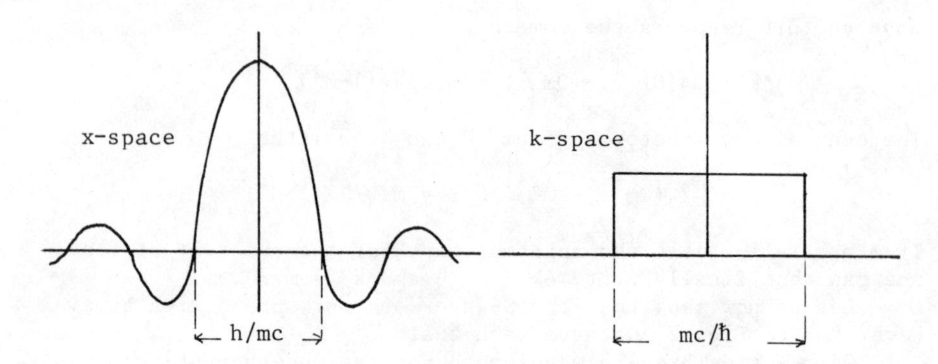

Fig. 4. The form of the nondispersive wave packet in the x-space and in the k-space.

is just equal to Compton's wavelength h/m_oc, i.e., 2.43×10^{-13}m for an electron and 1.32×10^{-15}m for a proton.

3. RELATIVISTIC WAVE EQUATIONS WITH QUANTUM POTENTIAL NONLINEARITY

Let us now deduce the form of the nonlinear equation which accepts F as a soliton-like solution. Writing F = GH, where G(x,t) and H(x,t) denote respectively $\exp\{i[\omega(k_o)t - k_ox]\}$ and $[\sin \Delta k(x - vt)]/\Delta k(x - vt)$, and calculating $F_{xx} - F_{tt}/c^2$, one checks that the imaginary part in the multiplier of G cancels out and what result is the modified Klein-Gordon equation

$$\Box F - \kappa^2 F = (1 - \beta^2) GH_{xx} \tag{19}$$

$\Box$ is the dalembertian operator in the two-dimensional space-time and $\kappa = m_oc/\hbar$. Moreover, one has $\theta(x,t) = \Delta k(x - vt)$, from which $H(x,t) = H[\theta(x,t)]$, and one finds $H_{xx} = (\Delta k)^2 H_{\theta\theta} = (mc/\hbar)^2 H_{\theta\theta}$, so that H_{xx} depends on θ but not on the precise form in which θ depends on x and t. Relation (19) can be written more generally in the form

$$\Box F - \kappa^2 F = \kappa^2 GH_{\theta\theta} \tag{20}$$

or, still more generally, as

$$\Box U - \kappa^2 U = \kappa^2 GH_{\theta\theta} \tag{21}$$

Here $U = U_o + \Phi$, in which

-- U_o is the S-wave defined by $U_o(x,t) = aF(x,t) = aH(x,t)e^{iS}$, a beeing a constant factor and $S = (Et - \vec{p}\vec{x})/\hbar$;

-- Φ is a P-wave of same phase as U_o : $\Phi(x,t) = R(x,t)e^{iS}$, which satisfies the Klein-Gordon condition $\Box \Phi = \kappa^2\Phi$.

One encounters in this way the de Broglie double-solution model of wave mechanics.(4)

Note that, in the nonrelativistic approximation, one has $\sin \Delta k(x - vt) / \Delta k(x - vt) \to 1$, so that the S-wave does not appear at that stage. As for the P-wave equation, Eq(21) becomes a stationary Schrödinger equation.(7)

In equation (19) one can express G and H_{xx} as a function of F. Indeed, since $H = (\text{sgn } H)|F|$ and $G = (HG)/H = F/(\text{sgn } H)|F|$ with sgn H standing for "sign of H," one has $GH_{xx} = (F/|F|)|F|_{xx}$. As a consequence, if one accepts the physical reality of the B-waves, a two-dimensional space-time wave equation of de Broglie's double solution theory is

$$\Box u = \kappa^2 (1 + \frac{\ell^2 |u|_{xx}}{|u|}) u \tag{22}$$

(ℓ^2 is a homogeneity constant). Equation (22) is a nonlinear Klein-Gordon equation in which U_o describes a soliton-like solution.

Now, let us examine some extensions of Eq.(22). First, let us consider a nondispersive wave packet of arbitrary shape, exactly as in the case of a nonlinear quantum potential Schrödinger equation,(8)

$$u(x,t) = R(\theta) e^{iS} \tag{23}$$

$\theta = \Delta k(x - vt)$. Then, in Eq.(22) the nonlinear term $(|u|_{xx}/|u|)u$ becomes $(\Delta k)^2 R_{\theta\theta} \exp(iS)$, with

$$(\Delta k)^2 R_{\theta\theta} \, e^{iS} = \frac{1}{\kappa^2} \frac{\Box |u|}{|u|} u \tag{24}$$

after trivial manipulations. Thus, Eq.(22) exhibits a nonlinear quantum potential term when it is written as

$$\Box u - \kappa^2 u = \frac{\Box \sqrt{\rho}}{\sqrt{\rho}} \tag{25}$$

where $\sqrt{\rho} = (\bar{u}u)^{1/2}$. This new equation satisfies the superposition principle, in conformity with the Guerra and Pusterla requirements,(9) and describes the propagation of two-dimensional space-time kinks and solitons of arbitrary shape. However, only the wave packet (18) is made of waves which will have the same phase at the same place where the particle is, in a covariant way.

In writing now $\Box = \partial_\mu \partial^\mu = \partial^2/\partial x^2 - \partial^2/c^2\partial t^2$, it can be shown that Eq.(25) stems from the Lagrangian

$$\mathcal{L} = \frac{1}{4} [\frac{u}{\bar{u}}(\partial_\mu \bar{u})(\partial^\mu \bar{u}) + \frac{\bar{u}}{u}(\partial_\mu u)(\partial^\mu u)] - \\ - \frac{1}{2}(\partial_\mu \bar{u})(\partial^\mu u) - \kappa^2 \bar{u}u \tag{26}$$

From this Lagrangian can be derived the same current as in the linear case, i.e.,

$$J_\mu = -(ie/\hbar c)[\bar{u}(\partial_\mu u) - u(\partial_\mu \bar{u})] \tag{27}$$

and the energy-momentum tensor

$$T_{\mu\nu} = \frac{1}{2}\Big[\frac{u}{\bar{u}}(\partial_\mu \bar{u})(\partial_\nu \bar{u}) + \frac{\bar{u}}{u}(\partial_\mu u)(\partial_\nu u) - - (\partial_\mu u)(\partial_\nu \bar{u}) - (\partial_\mu \bar{u})(\partial_\nu u)\Big] - \delta_{\mu\nu}\mathcal{L} \tag{28}$$

In four dimensional space-time, instead of (23), the non-dispersive wave packet associated with a particle of rest mass m_o, and traveling in the +x direction, are defined by

$$u(r,t) = R(r)e^{iS} \tag{29}$$

where $r = |\vec{r}| = \left[\frac{(x-vt)^2}{1-\beta^2} + y^2 + z^2\right]^{1/2}$. They are solutions of

$$\Box u - \kappa^2 u = \left(R_{rr} + 2\frac{R_r}{r}\right)e^{iS} \tag{30}$$

But a simple calculation shows immediately that, for u given by (29),

$$\frac{\Box\sqrt{\rho}}{\sqrt{\rho}}\, u = (\Box R)e^{iS} = \left(R_{rr} + 2\frac{R_r}{r}\right)e^{iS} \tag{31}$$

so that Eqs (30) and (25) are equivalent in four dimensions.

Among the solutions (29) there are solutions for which arise simultaneously :

$$\Box u = 0 \quad \text{and} \quad m_o^2 + \frac{\hbar^2}{c^2}\frac{\Box\sqrt{\rho}}{\sqrt{\rho}} = 0 \tag{32}$$

i.e. the Eulerian differential equation

$$R_{rr} + 2\frac{R_r}{r} + \kappa^2 R = 0 \tag{33}$$

which admits the general solution

$$u(r,t) = \left(A\frac{\sin \kappa r}{\kappa r} + B\frac{\cos \kappa r}{\kappa r}\right) e^{iS} \tag{34}$$

where A and B are constants. If one now chooses to reject the cosine term by setting B = 0, on grounds of the akward infinity at r = 0, one is left with

$$u(r,t) = A\frac{\sin \kappa r}{\kappa r}\, e^{iS} \tag{35}$$

This precisely defines the nondispersive wave packet constructed by Mackinnon (10) as a generalization of (18) to four-dimensional space-time. This solution is well-known to represent the superposition of two spherically symmetrical waves, one converging and the other diverging.(4) In electromagnetism it behaves as a phase-locked cavity similar to those analyzed by Jennison (11) and having the inertial properties of classical particles.

Now, let it be suppoed that a particle, of rest mass m_o and charge e, is traveling freely with uniform velocity v in the +x direction in a region where there exists an uniform electromagnetic potential A . In such a case, Eq.(25) becomes

$$(\partial_\mu + \frac{ie}{\hbar c} A_\mu)(\partial^\mu + \frac{ie}{\hbar c} A^\mu)u - \kappa^2 u = \frac{\Box \sqrt{\rho}}{\sqrt{\rho}} u \tag{36}$$

This equation can be derived from the Lagrangian

$$\mathfrak{L} = \frac{1}{4}[\frac{u}{\bar{u}}(\partial_\mu \bar{u})(\partial^\mu \bar{u}) + \frac{\bar{u}}{u}(\partial_\mu u)(\partial^\mu u)] - \frac{1}{2}(\partial_\mu \bar{u})(\partial^\mu u) + + \frac{ie}{\hbar c}[A^\mu \bar{u}(\partial_\mu u) - u(\partial_\mu \bar{u})] - \frac{e^2}{\hbar^2 c^2} A_\mu A^\mu \bar{u} u - \kappa^2 \bar{u} u \tag{37}$$

in conformity with the statements of Guerra and Pusterla.(9)

Four main implications result from the physical conjectures introduced to construct this mathematical model :

(1) As has been pointed out in the two-dimensional case, this model is consistent with the double solution interpretation of wave mechanics. On account of the superposition principle, to each solution (18) can be added, with the same phase S, a solution Φ of the linear Klein-Gordon equation, and it can be shown that the guiding principle is avalaible.(12)

(2) Following the stochastic interpretation of wave mechanics, the linear waves can be considered as Brown-Markov stochastic waves on a random covariant thermostat (Dirac aether).(13,14, 15) The continuity equation (15 C) is equivalent to a forward and backward Fokker-Planck equation, i.e., in the Guerra-Ruggiero notation

$$\frac{\partial \rho}{\partial \tau} + \nabla v_{\pm} \rho \mp D \Box \rho = 0 \tag{38}$$

The corresponding diffusion coefficient $D = \hbar/2m$ results (15) from the need to preserve phase coherence in quantum jumps at the velocity of light.

The nonlinear quantum potential term can be associated with a time reversible frictional force (important in the particle region, but negligible elsewhere), in a way generalizing to Dirac vacuum the Einstein-Smoluchowski treatment of Brownian motion.

(3) If the nondispersive wave packet is in a uniform electromagnetic potential, it is necessary to assume that the vector potential is always zero in the rest frame of the particle described by the packet.(16)

(4) The fact that an interference pattern is not dependent on the observer may be explained by de Broglie waves, but cannot

be explained by Schrödinger waves.(17) For this reason, if one accepts the physical reality of de Broglie waves, one must take into consideration relativistic waves, even in the nonrelativistic limit.

REFERENCES

1. L. de Broglie, "The Beginnings of Wave Mechanics" in *Wave Mechanics. The First Fifty Years,* W.C. Price et al., eds. (Butterworths, London, 1973).
2. L. de Broglie, *Ann. Phys.* (Paris) 3, 22 (1925).
3. E. Mac Kinnon, *Am. Jour. Phys.* 44, 1047 (1976).
4. L. de Broglie, *Nonlinear Wave Mechanics* (Elsevier, Amsterdam, 1960).
5. C. Philippidis, C. Dewdney, and B.J. Hiley, *Nuovo Cim.* 52 B,1 (1979).
6. L. Mackinnon, *Found. Phys.* 8, 157 (1978).
7. E.C. Kemble, *Fundamental Principles of Quantum Mechanics*, MacGraw-Hill, New York, 1937 .
8. R.W. Hasse, *Zeit. Phys.* B, 37, 83 (1980).
9. F. Guerra and M. Pusterla, *Lett. Nuovo Cim.* 34, N°12, 351 (1982).
10. L. Mackinnon, *Lett. Nuovo Cim.* 31, 37 (1981).
11. R.C. Jennison, *J. Phys.* A 11, 1525 (1978).
12. Ph. Guéret and J.P. Vigier, *Found. Phys.* 12, N°11 (1982).
13. W. Lehr and J. Park, *J. Math. Phys.* 18, 1235 (1977).
14. F. Guerra and P. Ruggiero, *Lett. Nuovo Cim.* 23, N°15 (1978).
15. J.P. Vigier, *Lett. Nuovo Cim.* 29, 467 (1980).
16. L. Mackinnon, *University of Essex Preprint*, submitted to *Found. Phys.* (1983).
17. L. Mackinnon, *Found. Phys.* 11, 907 (1981).

THE ROLE OF THE QUANTUM POTENTIAL IN DETERMINING PARTICLE TRAJECTORIES AND THE RESOLUTION OF THE MEASUREMENT PROBLEM

B.J. Hiley

Physics Department
Birkbeck College, University of London
London WC1E 7HX, England

ABSTRACT

The method of calculating particle trajectories, using the quantum-potential model, is reviewed and some pertinent details for the two-slit interference experiment, the Aharonov-Bohm effect, and barrier penetration are presented. The implications of this model are discussed and contrasted with the orthodox interpretation of quantum mechanics and with the classical paradigm. The model is then applied to the act of measurement and it is shown how, if irreversibility is included, no collapse problem arises.

1. INTRODUCTION

In this paper I want to review a simple model of quantum phenomena first introduced by de Broglie (1923, 1956) for the single particle and subsequently rediscovered and generalized, to include many-body systems, by Bohm (1952). In this later version, it was assumed that the particles had a stable, autonomous, local existence with a well-defined, though perhaps unknown, position *and* momentum. The particles interacted with each other and the apparatus in the usual way but, in addition, they were subjected to a new field $\psi(\underline{r}_1,\underline{r}_2,\ldots,t)$. This field was to be regarded not as a probability field, although it could be used to obtain information about probabilities, but as the source of a real force which acts on the particles through a new kind of potential which we call the quantum potential. When this potential is negligible, the behaviour of the particles becomes purely classical, so that it is the appearance of the quantum

G. Tarozzi and A. van der Merwe (eds.), Open Questions in Quantum Physics, 237–256.

potential that determines the quantum behaviour.

In his two earlier papers, Bohm had already indicated that this model could reproduce all the known non-relativistic quantum effects, including the stationary states, scattering, etc. The model contained features which, at the time, were considered to weigh against it, but the recent series of optical cascade correlation experiments that culminated in the results of Aspect, Dalibard, and Roger (1982), together with the quark confinement problem, tend to indicate that these "undesirable" features of the model may be very necessary and indeed are already implicit to some degree in orthodox quantum mechanics. For these and other reasons, our group at Birkbeck College has re-examined and explored further the consequences of the model. For example, in the one-body system, the actual particle trajectories have been calculated for the two-slit interference experiment (Philippidis, Dewdney, and Hiley (1979)), and the results are presented in Sec. 2. In Sec. 3 and 4 the Aharonov-Bohm effect (Philippidis, Bohm and Kaye (1982)) and barrier penetration (Dewdney and Hiley (1982)) are reviewed.[1]

Careful examination of each of the above examples shows that the quantum potential produces radically new features that are unlike those produced by a classical potential, a fact that we will bring out in detail later in this paper. Perhaps it is worth summarizing some of the main features here. Firstly, the interaction of particles with each other is *non-local*, and, secondly, it is not fixed once and for all, but in general is determined by the "quantum state" of the *whole system*. Thus, the classical mode of analysis of a system into separate parts whose relationships are independent of the state of the whole is no longer generally valid. Rather, separable and independent parts are now seen to be a special and contigent feature, which is dependent on the whole in a way that cannot even be described in terms of the parts alone. In this respect, the theory was in agreement with the approach of Bohr (1961), but the key difference was that the quantum potential interpretation provided a physical notion of how this wholeness may be brought about in terms of the actual movement of particles under the action of the quantum potential, whereas in Bohr's formulation, such questions are expressly ruled out as having no meaning.

The treatment of the measurement problem is a particularly significant application of the model. The uncertainty principle arises as a result of the observed particle suffering a disturbance that is unpredictable and uncontrollable (in essence, because the particles constituting the apparatus are moving with random thermal motion). But it is not only the particle properties that change under the interaction. The ψ field itself changes since it must satisfy the Schrödinger equation, which now contains the interaction between particle and apparatus, and it is this change that makes it impossible to measure position and momentum together. (See Bohm (1952).)

The third significant feature is that a consistent treatment can be given to what is called, in the usual interpretation, the "collapse of the wavefunction." The orthodox theory either remains silent on this problem or, following von Neumann, suggests that there are two distinct processes occurring, one using the unitary Schrödinger equation, while the other occurs during a measurement process, when one wave function is replaced by another with no physical explanation of how this comes about. Wigner (1965) has never been happy with this situation and proposed that it is ultimately the consciousness of an observer that is involved. Everett (1957), on the other hand, has given an interpretation in which both the "collapse" and the dependence of the result on the consciousness of an observer are avoided. In this interpretation, there is a multiplicity of universes, each corresponding to one of the possible outcomes of the measurement. However, as Bell (1971) points out, this is an extreme case of a lack of economy of concepts involving the assumption of a non-denumerable infinity of universes, with a corresponding non-denumerable infinity of observers within them.

In the quantum potential model, we show that all the packets of the multi-dimensional wave function that do not correspond to the actual result of measurement have no effect on the particle, not only at the moment immediately after its interaction with the measuring apparatus is over, but also for all times from then on. (This is seen to follow from the irreversibility of the random thermal motions of the particles constituting the apparatus.) Consequently, such packets can be dropped from further discussion. Some of the implications of this will be discussed in Sec. 5.

2. THE QUANTUM POTENTIAL MODEL AND THE TWO-SLIT INTERFERENCE PATTERN

We start by assuming that a particle has a well-defined but unkown position and momentum, and that the successful use of the wave function has its origins in some additional potential that acts on each particle. To find the form of this potential we simply write the wavefunction $\psi = Re^{iS/\hbar}$ (with R and S real) in Schrödinger's equation for a single particle and separate the resulting expression into its real and imaginary parts. From the real part we find:

$$\frac{\partial S}{\partial t} + \frac{(\nabla S)^2}{2m} + V + Q = 0, \tag{1}$$

where V is the classical potential and Q the quantum potential

$$Q = \frac{-\hbar^2}{2m} \frac{\nabla^2 R}{R}. \tag{2}$$

When Q can be neglected, Eq. (1) reduces to the classical Hamilton-Jacobi equation for a single particle. If we use the fact that in the Hamilton-Jacobi theory $m\underset{\sim}{v} = \underset{\sim}{\nabla}S$, we can immediately obtain trajectories of individual particles by integrating the phase of the wave function. Philippidis, Dewdney, and Hiley (1979) have illustrated the use of this method by calculating the trajectories of particles passing through a pair of Gaussian slits. The initial condition for each trajectory is determined by the position of the particle as it passes through one of the slits in a direction perpendicular to the plane containing the slits. The ensemble of trajectories, which are shown in fig. 1, exactly reproduces the standard interference pattern.

To see how the probabilities arise, we must now examine the imaginary part of the Schrödinger equation that results from the substitution $\psi = Re^{iS/\hbar}$. This can be written as

$$\frac{\partial P}{\partial t} + \underset{\sim}{\nabla}\left(P\frac{\underset{\sim}{\nabla}S}{m}\right) = 0, \tag{3}$$

with $P = R^2 = \psi^*\psi$. In this form the equation is seen to be an equation for the conservation of probability. If we then assume that the probability of finding a particle on a given trajectory at the slits is $P_i = R_i^2 = \psi_i^*\psi_i$, where ψ_i is the initial Gaussian distribution at one of the slits, then the conservation equation (3) will ensure that the "interference" pattern at the screen must give the correct intensity distribution.

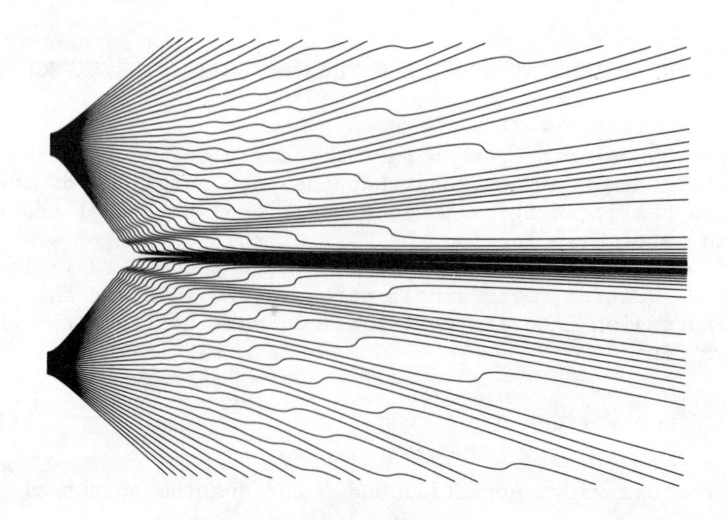

Fig. 1. Particle trajectories for two-slit interference. [From Philippidis, Dewdney & Hiley (1979).]

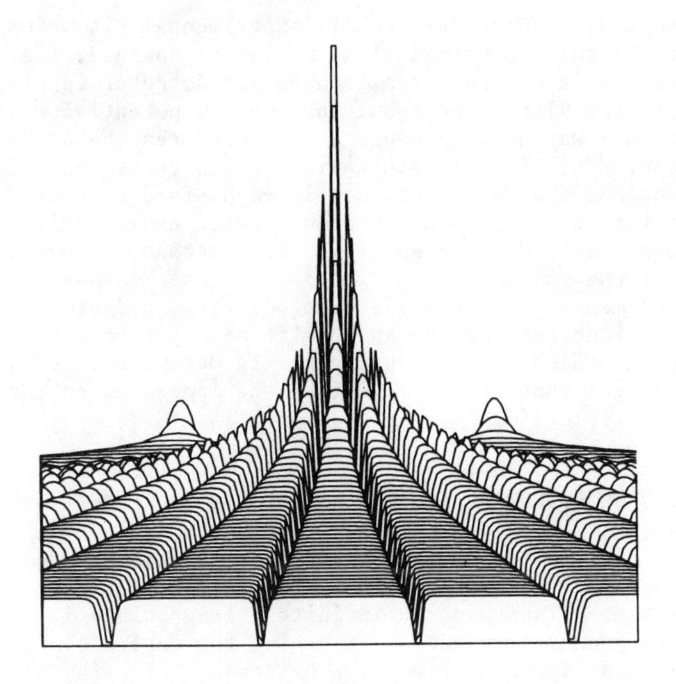

Fig. 2. The quantum potential for two Gaussian slits.
[From Philippidis, Dewdney & Hiley (1979).]

In Fig. 2, the time averaged quantum potential is shown. Here the position at the slits coincides with the two parabolic peaks in the background, the foreground being the position where the "fringes" are formed. If we examine the trajectories in Fig. 1, we see that initially they fan out from each slit in a manner consistent with diffraction from a single slit, and "kinks" develop. Comparing the position of the "kinks" with the quantum potential shown in Fig. 2, we see that they coincide with the troughs in the quantum potential. The kinks arise because when a particle enters the region of a trough it experiences a strong force which accelerates the particle rapidly through the trough onto a plateau region where the forces due to the quantum potential are weak. In consequence, most of the trajectories run along the plateau regions giving rise to the bright fringes, while the troughs coincide with the dark fringes.

An immediate glance at Fig. 2 shows that the quantum potential is unlike any known classical potential. It does not appear to have a localised source in the sense that a Coulomb force has its source in a charged particle. Nevertheless, it carries information about the slit-widths and their distance apart and depends on the momentum of the particle. That is, it carries information about the overall experimental situation. Furthermore, if this experimental situation is changed, i.e., one of the slits is closed, or a transparent detector is placed behind one of the slits, the resulting quantum potential will change in such a way as to produce a very different behaviour in each particle. This is consistent with an important feature of quantum phenomena that was continuously emphasised by Bohr (1961), namely that the word "phenomenon" should refer exclusively to "observations obtained under specified circumstances, including an account of the *whole experimental arrangement*." Thus, if the overall experimental situation is changed, i.e., a device is introduced to find out through which slit each particle passes, one no longer obtains a fringe pattern. In our model, it is the quantum potential that explicitly links the apparatus to the individual particle.

3. THE QUANTUM POTENTIAL AND THE AHARONOV-BOHM EFFECT

Philippidis, Bohm, and Kaye (1982) have applied the quantum potential model to the Aharonov-Bohm effect. It will be recalled that this effect arises when an infinitely long solenoid is placed in the geometric shadow just behind the screen containing the two slits, as shown in Fig. 3. The reason for using an infinite solenoid placed in this position is to ensure that the electron trajectories do not pass through the region containing the magnetic field. Even though one's physical intuition suggests that this field will therefore not produce any effect, Aharonov and Bohm (1959), and indeed the earlier work of Ehrenberg and Siday (1949), showed that standard quantum mechanics predicts that the fringe system will be shifted by an amount that depends on the magnetic field trapped in the solenoid, even though the electrons do not pass through this field. Since the original proposal, a number of beautiful experiments have confirmed the effect (see Tonomura *et al.* (1982) and references therein).

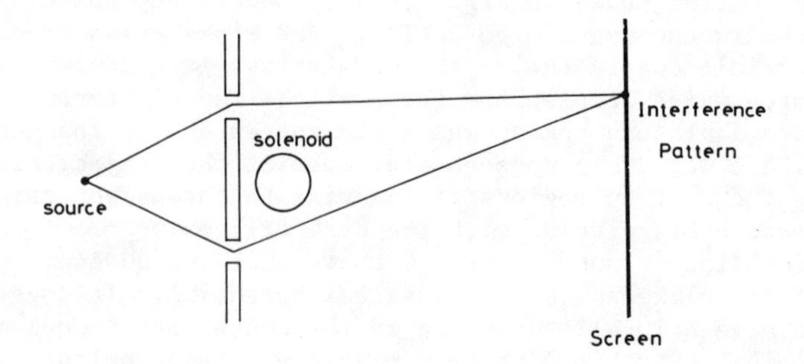

Fig. 3. Experimental arrangement for the Aharonov-Bohm effect.

Although it is clear that the effect exists, there remains a controversy as to how to interpret the result. Aharonov and Bohm (1959) argued that, since the vector potential was non-zero in the region of the trajectories, it is the existence of this vector potential that is directly responsible in producing the change. Thus, they argued that the vector potential possessed a physical significance in quantum theory in contrast to the role it plays in classical physics, where it is regarded as a mere mathematical construct used only to simplify calculations, the real physical effects being produced by the fields themselves.

Belinfante (1962) and De Witt (1962) disagreed with the idea of giving physical significance to gauge dependent quantities like the vector potential and reformulated the problem in a gauge invariant manner. But this had the effect of introducing a non-local element into the orthodox theory. However, this non-locality goes against the spirit of quantum theory which assumes all *field* quantities will change in a local way.

The quantum potential model throws a new light on the problem. The modified Hamilton-Jacobi equation now becomes

$$\frac{\partial S}{\partial t} + \frac{(\nabla S - e/c\mathbf{A})^2}{2m} + \phi + Q = 0, \tag{4}$$

while the corresponding equation of motion reads

$$m\frac{d\mathbf{v}}{dt} = e\left[\mathbf{E} + \frac{1}{c}(\mathbf{v} \times \mathbf{B})\right] - \nabla Q. \tag{5}$$

The quantum potential is again given by

$$Q = \frac{-\hbar^2}{2m}\,\frac{\nabla^2 R}{R}, \tag{6}$$

where R is obtained from the wavefunction $\psi = Re^{iS/\hbar}$ which is, in turn, obtained from the solution of the appropriate Schrödinger equation.

Once again, the quantum potential will contain information concerning slit widths, their distance apart, and the momentum of the electrons, as in the case of the two-slit experiment, but in this case it also carries additional information concerning the magnetic flux in the solenoid. In Fig. 4 the quantum potential is displayed for a magnetic flux of $\Phi = (ch/e)\pi/2$. One can immediately see that the presence of the magnetic flux shifts the quantum potential to the right. The corresponding trajectories are shown in Fig. 5.

Thus, we see it is the quantum potential acting locally on the particles that produces the shift in the pattern. Of course, this potential is derived locally from the vector potential, using the wave-function solutions of the appropriate Schrödinger equation. In classical physics the quantum potential is

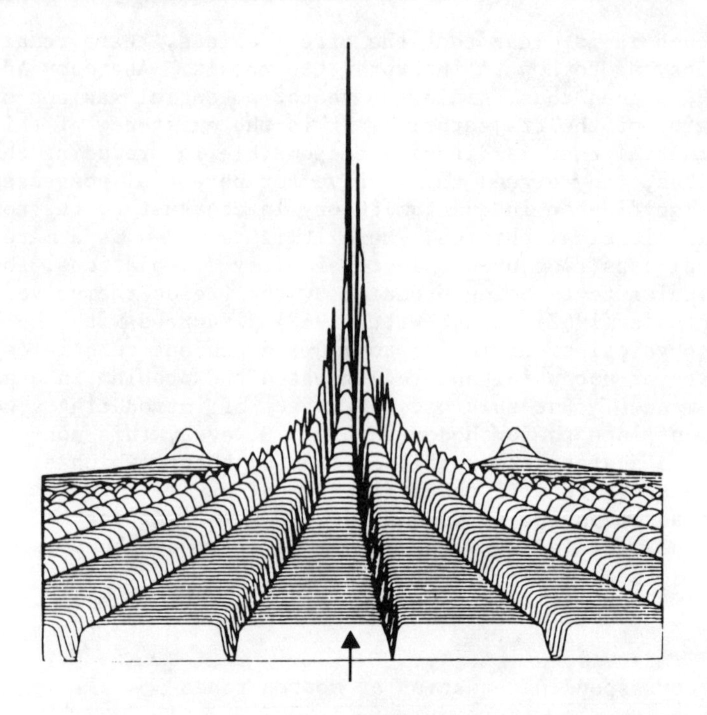

Fig. 4. Quantum potential for Aharonov-Bohm effect. [From Philippidis, Bohm and Kaye (1982).]

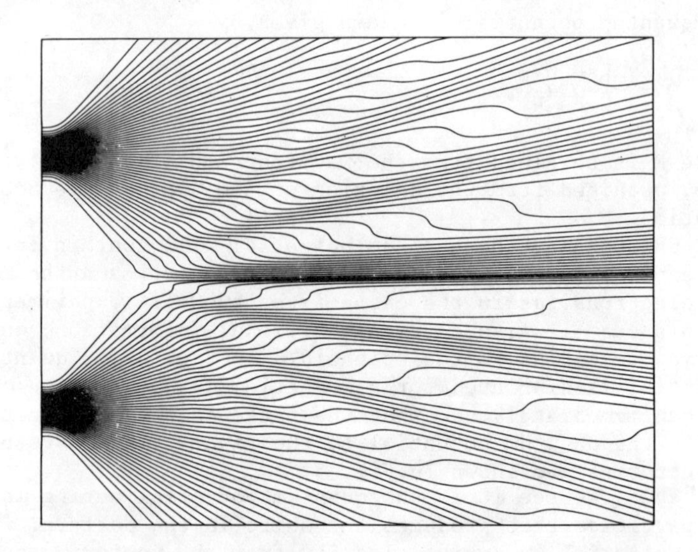

Fig. 5 Trajectories for electrons in Aharonov-Bohm effect. [From Philippidis, Bohm and Kaye (1982).]

negligible, so the vector potential can play no active role, which is exactly what we would expect. Clearly, the quantum potential is gauge-invariant so there is no need to postulate (along the lines of De Witt) that the flux has an immediate nonlocal action on electrons which never enter the region where a magnetic field is nonzero. Thus, since the quantum potential contains information about the global environment of the electron, we see once again how the quantum potential brings out the physical meaning of Bohr's notion of the wholeness of the form of the experimental conditions and the content of the experimental results.

4. THE QUANTUM POTENTIAL AND TIME-DEPENDENT SCATTERING FROM BARRIERS

In both of the above examples, the quantum potential has been time-averaged in the sense that it has been calculated for a steady stream of particles incident on the two slits. Dewdney and Hiley (1982) have examined the time dependence of the quantum potential by considering the case where a single particle approaches a barrier. At this stage an important new feature must be emphasised, namely, the production of a single particle involves a macroscopic piece of apparatus such as an electron gun, and the process needed to produce the electron will involve the quantum potential produced by the interaction of the apparatus with the electron. The electrons emerge with a definite momentum and a definite position, both of which are unknown, together with a wave function which will be in the form of a packet. It is the wave function, when analysed in terms of $Re^{iS/\hbar}$, that determines the subsequent trajectory of the particle, each particle being acted upon locally by the quantum potential as defined in Eq. 2.

At this point, a brief discussion as to the origins and functioning of the quantum potential is needed. In a very early similar approach by Madelung (1926), it was assumed that the wave function had some hydrostatic origin and the particle was "pushed around" by a mechanical force, i.e., the quantum potential was similar to a classical potential. However, our work clearly indicates that there are very basic differences between the two potentials. Apart from the difference we have already indicated in Secs. 2 and 3, a further difference emerges by noticing that multiplication of the wave function by a constant does not change the quantum potential. The quantum potential can therefore still be large even when the wave function is small. This means that its effects do not necessarily fall off with the distance. For example, the complex pattern of plateaus and valleys in the quantum potential is still appreciable at a considerable distance from the slits. This is something that is totally unexpected from the classical point of view.

The essential new feature of the quantum potential is therefore that only the *form* of the Schrödinger ψ fields counts, and not the intensity. So the force arising from this potential is not like a mechanical force of wave "pushing on" a particle with a pressure proportional to the wave intensity. Rather, it acts more like an information content (recall that "to inform" means literally "to form within"). We may make an analogy here to radar waves that guide a ship. These do not push the ship mechanically. If the ship is on automatic pilot it may then be regarded as a self-active system, with its activity directed by radar waves containing information about the whole surroundings. Similarly, we can think of each electron as being surrounded by a field of information concerning the whole environment and the electron can be regarded as having a complex inner structure which responds to this information via the quantum potential.

In some ways this is reminiscent of Feynman's idea when he suggested that we can regard a point in space-time as being an enormously complicated thing. Each point has to remember with precision the values of indefinitely many fields describing indefinitely many elementary particles and has to have data inputs and outputs connected to neighbouring points. It also has to have a little arithmetic element to satisfy the field equations so that it might just as well be an analog computer. (See Finkelstein (1969).)

But all of this may yet be too close to a mechanical analogy when what is needed is something much more subtle. Indeed, de Broglie (1964) has suggested an idea which is very close to Thom's morphogenetic field (1974). However, at this stage we prefer to leave the precise nature of this process open. (For further discussion see Bohm and Hiley (1975), Hiley (1980), and Vigier (1982).)

With these possibilities in mind, let us return to consider a particle incident on a barrier. We will assume the information field takes the form of a Gaussian packet. In Fig. 6 we illustrate the time-dependent behaviour of the quantum potential for such a particle incident on a one-dimensional square barrier when the incident energy of the particles is half the barrier height. The main effect of the quantum potential is to modify the shape of the barrier. In the initial stage of development, the quantum potential is the usual inverse parabola associated with a free Gaussian packet. As time progresses, its shape becomes more complex, developing a series of ridges and valleys. Immediately in front of the barrier the quantum potential is always negative, thus reducing the effective barrier height in this region. Inside the barrier several "pockets" appear as a result of reflections off the rear of the barrier. These pockets disappear as the energy of the incident particles increases. (See for example Fig. 7.)

The trajectories corresponding to the quantum potential illustrated in Fig. 6 are shown in Fig. 8. Each trajectory corresponds to a different initial position in the original Gaussian wavepacket. The trajectory corresponding to the mid-point of the packet originates at 0.5 in Fig. 8. Thus we see that it is the particles at the extreme front of the packet that actually penetrate the barrier, whereas those towards the rear are reflected at, or even before the barrier. As expected, there are particles trapped inside the barrier, some being reflected out eventually, others being transmitted. (Further details for barrier penetration and scattering from wells can be found in Dewdney and Hiley (1982).)

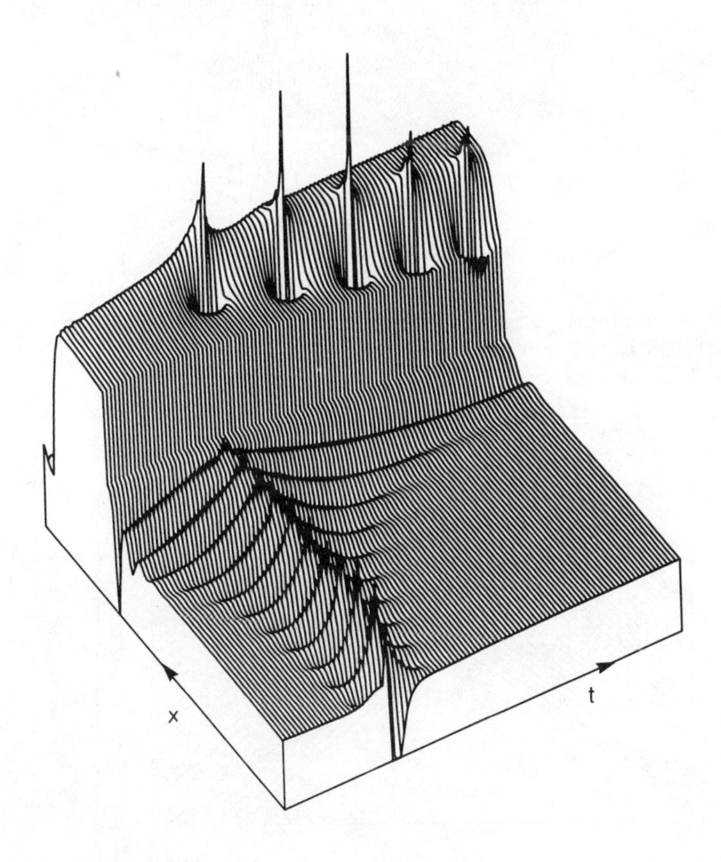

Fig. 6. The quantum potential for a barrier height equal to half the energy of the incident particle. [From Dewdney and Hiley (1982).]

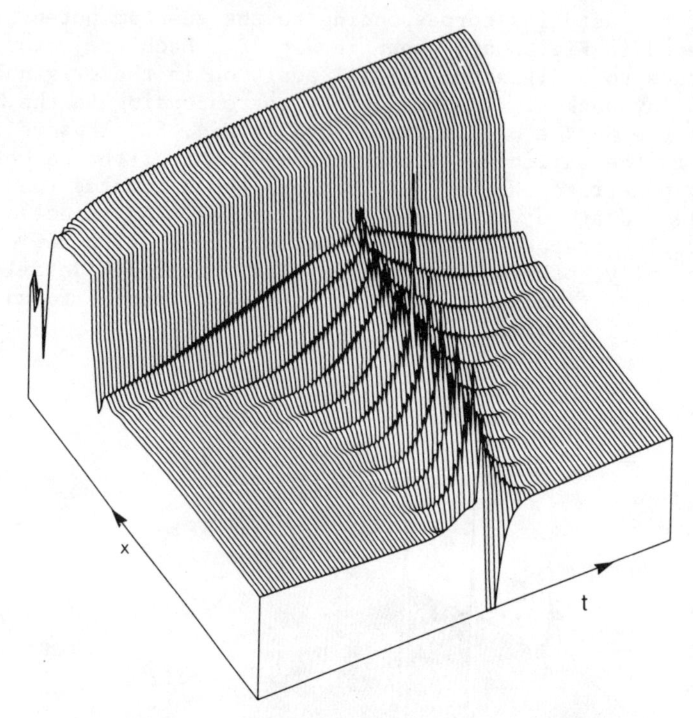

Fig. 7. The quantum potential for barrier height equal to energy of incident particle. [From Dewdney and Hiley (1982).]

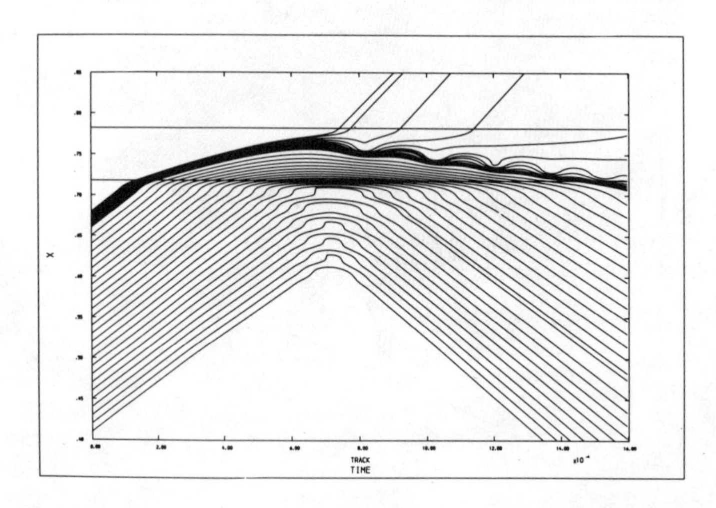

Fig. 8. Particle trajectories for penetration of barrier height equal to half the energy of the incident particles. [From Dewdney and Hiley (1982).]

5. THE QUANTUM MECHANICAL PROCESS OF MEASUREMENT

We shall now consider the process of measurement in terms of the quantum potential interpretation (more details appear in Bohm and Hiley (1983)). In doing this, we find it convenient to divide the process into two stages:

(i) Separation of possible states of the quantum system, through an in principle reversible interaction with an apparatus.

(ii) Registration of an actual result of the measurement in a further *irreversible* interaction.

Let us now consider a wave packet $\psi(\underline{r})$ with a finite width incident on a single crystal as shown in Fig. 9.

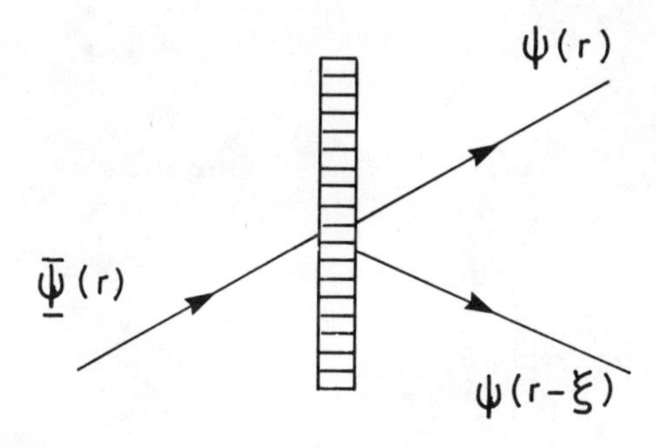

Fig. 9. Beam splitter.

As a result of the interaction with the crystal, two wave packets emerge and separate so that the final wave function is

$$\Psi(\underset{\sim}{r}) = c_1\psi(\underset{\sim}{r}) + c_2\psi(\underset{\sim}{r} - \underset{\sim}{\xi}), \tag{7}$$

where $\underset{\sim}{\xi}$ is the deviation of one of the packets. Once $\underset{\sim}{\xi}$ is much greater than the width of the packet, the two wave packets no longer interfere. The probability of finding the particle in each of the two packets is $|c_1|^2|\psi(\underset{\sim}{r})|^2$ and $|c_2|^2|\psi(\underset{\sim}{r} - \underset{\sim}{\xi})|^2$, respectively. The actual particle must be in one of the packets, and once the packets have separated, the particle remains in that packet since there is negligible probability of crossing the space between the packets (where the probability is essentially zero). It should be noted that in the one-body problem the quantum potential acts locally on the particle, and therefore there is no contribution to the quantum potential from the packet not containing the particle. Thus, as long as the packets do not overlap we can say that the particle behaves as if it were solely in *one* of the two possible final states.

We cannot yet completely ignore the empty packet before a registration takes place, because it is possible (by using appropriately placed singly crystals) to deflect the two packets and cause them to overlap again. This overlap will then allow us to form a state in which the identity of the previous two states is lost. Thus we have not yet explained the irrevocability of an experiment in determining which of the packets will contain the particle. What we need to explain is that after a registration, the packet not containing the particle will never have any effect again.

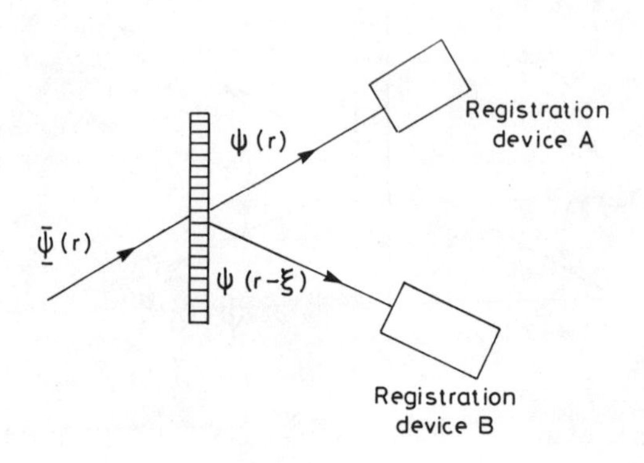

Fig. 10.

The initial total wave function of the whole system will then by given by

$$\psi_T^0(\underset{\sim}{r}, \underset{\sim}{y}) = [c_1\psi(\underset{\sim}{r}) + c_2\psi(\underset{\sim}{r} - \underset{\sim}{\xi})]\phi_A^0(\underset{\sim}{Y}_a)\phi_B^0(\underset{\sim}{Y}_b).$$

The particle entering the single crystal from the left will only trigger one registration device, leaving the other unaffected (Fig. 10). Suppose the particle interacts with device A so that the wave function of A becomes $\phi_A^f(\underset{\sim}{Y}_a)$. In general the wave function of the particle will also change, but we can simplify the discussion without losing the essential features by supposing that its final wave function $\psi^f(\underset{\sim}{r})$ is not correlated to the complex movements of the internal coordinates of the apparatus. A similar result is assured if the particle interacts with B. Consequently, we can write the final total wave function as

$$\psi_T^f = d_1\psi^f(\underset{\sim}{r})\phi_A^f(\underset{\sim}{Y}_a)\phi_B^0(\underset{\sim}{Y}_b) + d_2\psi^f(\underset{\sim}{r} - \underset{\sim}{\xi})\phi_A^0(\underset{\sim}{Y}_a)\phi_B^f(\underset{\sim}{Y}_b). \quad (9)$$

Once again the two wave packets do not overlap. Consequently, the quantum potential will only depend on the wave packet that contains the particle. It remains to be shown that the combined wave packets $\psi^f(r)\phi_A^f(Y_a)\phi_B^0(Y_b)$ and $\psi^f(r - \xi)\phi_A^0(Y_a)\phi_B^f(Y_b)$ can *never* be made to overlap in the future. The key question is then "Under what conditions can we ensure that this overlap is never possible?" Once we have established this result, we are then justified in dropping the packet not containing the particle and leaving it out of the discussion from this point on.

In order to answer this question we note that a registration device is a macroscopic thermodynamic system which is capable of undergoing an irreversible change initiated by its interaction with the particle (i.e., a Geiger counter or scintillation counter for a charged particle or a He_3 counter for neutrons, etc.). This process ultimately produces a distinct macroscopically observable result which indicates where the particle is. A typical registration device will involve storing energy in the device and the passage of the particle will trigger a discharge of some form until no more energy is left stored in the system. We can, for simplicity, think of this as represented by the discharge of a capacitor somewhere in the device.

The original "quantum state" of the device will be represented by the wave function $\phi_A^0(\underset{\sim}{Y}_a)$, which corresponds to a surplus of electrons on one of the plates of the capacitor. The final "quantum state", represented by $\phi_A^f(\underset{\sim}{Y}_a)$, will correspond to the reduction of the surplus electrons. $\phi_A^0(\underset{\sim}{Y}_a)$ will correspond to these electrons in one region of configuration space, while in the discharged state $\phi_A^f(\underset{\sim}{Y}_a)$ they will appear in a very different region of configuration space, since the actual physical movement of electrons has taken place over a macroscopic distance (in this case the distance between the capacitor plates), so that the two registration device wave functions will be well separated in configuration space. Thus when the incident particle triggers device A, the incident particle and the apparatus particles are acted upon by the quantum potential derived from $\psi^f(\underset{\sim}{r})\phi_A^f(\underset{\sim}{Y}_a)\phi_B^0(\underset{\sim}{Y}_b)$. While there is no contribution from the other packet $\psi^f(\underset{\sim}{r} - \underset{\sim}{\xi})\phi_A^0(\underset{\sim}{Y}_a)\phi_B^f(\underset{\sim}{Y}_b)$. (The opposite result holds if B is triggered).

If we are to obtain an overlap of both these packets, then not only must we overlap $\psi^f(\underset{\sim}{r})$ and $\psi^f(\underset{\sim}{r} - \underset{\sim}{\xi})$, but we must overlap $\phi_A^f(\underset{\sim}{Y}_a)$ and $\phi_A^0(\underset{\sim}{Y}_a)$ and also $\phi_B^0(\underset{\sim}{Y}_b)$ and $\phi_B^f(\underset{\sim}{Y}_b)$. This means that we must return the triggered devices to their original state. We could do this by recharging the discharged capacitor using an external source of energy.[2] But then we would have to include in our discussion the wave function of the source. And, of course, here a similar irreversible process would take place, guaranteeing for the extended total system that there could be no return to an overlap of the two extended wave packets, so that the wave packet corresponding to the situation that had *not occurred* could never again contribute to the quantum potential acting on the particle. Thus, we may drop this wave packet from any further discussion.

All of this is reminiscent of what is called the "reduction of the wave packet," so one could ask what has been gained by this analysis. In the orthodox view the wave function is claimed to give the "most complete description of the state of the system," and if the Schrödinger equation of particle plus registration device is solved, one obtains a solution of the form of Eq. (9). But this equation does not correspond to a state with the registration devices A and B themselves in a well-defined macroscopic state (i.e., either fired or not fired and *not* a linear combination as found in Eq. (9)). Therefore the linear laws of quantum mechanics cannot account for the actual fact. (See Wigner (1963).)

The practical use of quantum mechanics is not inhibited by this fact, since physicists simply strike out the contribution of the eigenstates which do not correspond to the observed eigenvalue. Bell and Nauenberg (1965) have discussed this process in detail, and they refer to the procedure as being "very ethical," but they conclude that if the possible interference in Eq. (9) is *not* destroyed, then, while the quantum mechanical description as used in practice is not wrong, it is certainly incomplete. The quantum-potential model amplifies this view by showing that by introducing particles with definite positions and momenta and treating the wave function in a different way (i.e., *not* as the most complete description of state of the system), we are able to avoid this conclusion. In doing this, we have introduced a new role for the wave function, namely, it is a field of information which acts on the particle through the quantum potential. Indeed, the analysis of the measurement problem indicates that we have two types of information: *active* and *inactive*.[3] The packet containing the particle carries active information, while the empty packet in Eq. (9) carries inactive information.

This is essentially what happens in classical situations involving probability distributions. Here, after a new observation is made, we simply discard the previous probability function. We can always do this classically, because in this domain information is not active. In quantum theory, a further problem has arisen because of the general activity of information in this context. We have solved this problem by showing that, through irreversibility in the process of registration, the information in packets not corresponding to the actual result of the measurement ceases to be active, and so can be treated from there on as if it were the kind of information that is in classical probability theory. It is this change of the character of the information that is implied (though in a way that is not very clear) by the "collapse" of the wave function.

We have used here an irreversible process based on a discharge of energy stored in condenser plates, which led to a permanent and clear separation of the position of the electrons before and after the process has taken place. As we have already indicated, this was only a convenient illustration. We could have used other processes, such as those involving random diffusion, or radioactivity, to accomplish a similar aim, but this would have been mathematically more complex. However, the principle of irreversibility would work in a similar way.

Let us apply these ideas for the Schrödinger cat paradox. Here, the wave function will eventually contain a liner superposition of wave functions corresponding to a live cat and to a dead cat. But the cat is (at least initially) a complex living organism, maintained by irreversible processes, which would be very different according to whether the bullet entered it or not. So, as with the condenser plates, there will be no interference between these two wave functions, and, therefore, the information not corresponding to the actual state of the cat will become inactive, so that it can be dropped from all subsequent discussions. (Clearly, it would be a miracle if all the parts of the wave function corresponding to a dead cat were to re-arrange themselves to produce a wave function that would significantly overlap that of a wave function corresponding to a live cat.)

It is clear from the above that the quantum information potential enables us to discuss actuality as independent of the observer (e.g., as in the case of the cat). It can do this because actuality is determined by the movements of the particles constituting a system (which cannot be discussed at all in the usual interpretation). So, it can distinguish between the packets that correspond to actuality and those that do not. Indeed, as we have seen, all packets represent information only, and those that can never affect the particle represent inactive information. On the other hand, if, as in the usual interpretation, the wave function represents *all* that can be said about the system, there is no feature in a wave function covering a number of possible results of a measurement that can represent which of these is the one that is actually present, independently of the activity of an observer. (This is indeed the basis of the Everett interpretation, in which *all* these results are taken as actual, but each present in its own special universe to its own special observer.) By showing that the wave function contains active information, parts of which may become inactive through the very laws of development of this information, we are able thus to obtain an in principle unique description of independent actuality.

Though the wave function in principle combines active information relevant to the whole universe, in typical relations it factorizes, as discussed in Bohm and Hiley (1975). So, beyond certain bounds, the rest of the universe (including the conscious observer) does not matter significantly in discussing specific systems. Of course, if we were to study consciousness itself, as based on the material structure of the brain and nervous system, the question would have to be reopened. Indeed, there are good reasons for expecting that quantum theory, and therefore the notion of a quantum information potential, would be relevant here (Bohm, 1960). Thus, it may well be that in our mental processes, the quantum information potential is significant (as is, for example, suggested by the fact that information regarded as correct is active in determining our behaviour, while, as soon as it is regarded as incorrect, it ceases to be active). The quantum theory may then play a key part in understanding this domain. But (as has been pointed out earlier) there does not seem to be any strong reason to suppose that the human mind is playing a significant part in the kind of experiments ordinarily done in physics.

Finally, one can see a certain possibility of generalization of these notions. For, evidently the definiteness of actual events does not depend necessarily on experiments carried out by physicists in their laboratories. One may reasonably propose that, because the entire universe is almost everywhere engaged in irreversible processes, there is a constant tendency for the parts of its wave function not corresponding to actuality to constitute inactive information, which is no longer relevant. So, through irreversibility of development, the universe is able to determine itself spontaneously as being (at least on the larger scale) in definite and unique evolution (in which it also spontaneously divides into well-defined sub-wholes and sub-sub-wholes, etc., along lines that we have discussed earlier in Bohm and Hiley (1982). This notion has a certain relationship to the work of Prigogine (1980), who, in another context, shows that irreversible processes can give rise to a more or less determinate organisation of events and structures in the microscopic level.

6. SUMMARY AND CONCLUSIONS

The quantum information-potential approach is able to deal with all the problems of principle that may be raised in the quantum context. It provides an intuitive and imaginative understanding, which gives insight into the new properties of matter implied by quantum potential. Such insight may help the search for new physical and mathematical ideas, which could take us beyond the limits both of relativity and of quantum theory as these are now known.

ACKNOWLEDGEMENTS

I should like to thank David Bohm, Chris Dewdney, and Chris Philippidis for their many helpful discussions. Without their considerable efforts much of the work contained in this paper would not have been possible.

NOTES

1. Trajectories have also been obtained by Galehouse (1981), Hirschfelder and his colleagues (1974), and Prosser (1976), but they did not use the quantum potential approach.
2. In principle, it is of course possible for such a reversal to take place spontaneously, but the probability that this would happen is so small that it is unlikely to happen, even over the age of the universe (e.g., as unlikely as it is for water to boil on ice). If we accept such an explanation in thermodynamics, it seems that we should also do so here.
3. This information has nothing to do with the observer, no matter who or however well or badly informed the observer may be.

REFERENCES

Y. Aharonov and D. Bohm, *Phys. Rev. 115*, 485 (1959).
Y. Aharonov and D. Bohm, *Phys. Rev. 130*, 1625 (1963).
A. Aspect, J. Dalibard, and C. Roger, *Phys. Rev. Lett. 49*, 1804 (1982).
A. Baracca, D. Bohm, B.J. Hiley, and A.E.G. Stuart, *Nuovo Cimento 28B*, 453 (1975).
F.J. Belinfante, *Phys. Rev. 128*, 2382 (1962).
J.S. Bell and M. Nauenberg, in *Preludes in Theoretical Physics*, A. de Shalit *et al.*, eds.(North Holland Amsterdam, 1965), p. 279.
J.S. Bell, CERN Preprint TH 1424 (1971).
D. Bohm, *Phys. Rev. 85*, 166, 180 (1952).
D. Bohm, *Quantum Theory* (Prentice-Hall, New Jersey, 1960), Chap. 8.
D. Bohm and B.J. Hiley, *Found. Phys. 5*, 93 (1975).
D. Bohm and B.J. Hiley, *Found. Phys. 12*, 1001 (1982).
D. Bohm and B.J. Hiley, *Found. Phys. 14*, 255 (1984).
N. Bohr, *Atomic Physics and Human Knowledge* (Science Editions, New York, 1961).
L. de Broglie, *C. R. Acad. Sci. 177*, 506, 548, 630 (1923).
L. de Broglie, *Une tentative d'intérpretation causale et non linéaire de la mécanique ondulatoire: la théorie de la double solution* (Gauthier-Villars, Paris, 1956).
L. de Broglie, *La thermodynamique de la particule isolée* (Gauthier-Villars, Paris, 1964).
B. De Witt *Phys. Rev. 125*, 2189 (1962).

C. Dewdney and B.J. Hiley, *Found. Phys. 12*, 27 (1982).
W. Ehrenberg and R.E. Siday, *Proc. Phys. Soc. (London) B62*, 8 (1949).
A. Einstein, B. Podolsky, and N. Rosen, *Phys. Rev. 47*, 777 (1935).
H. Everett, *Rev. Mod. Phys. 29*, 454 (1957).
D. Finkelstein, *Phys. Rev. 184*, 1261 (1969).
D.C. Galehouse, *Int. J. Theor. Phys. 20* 787 (1981).
B.J. Hiley, *Ann. Fond. de Broglie 5*, 75 (1890).
J.O. Hirschfelder, A.C. Christoph, and W.E. Palke, *J. Chem. Phys. 61*, 5435 (1974).
J.O. Hirschfelder and K.T. Tang, *J. Chem. Phys. 64* 760 (1976); *65*, 470 (1976).
E. Madelung, *Z. Phys. 40*, 332 (1926).
C. Philippidis, C. Dewdney, and B.J. Hiley, *Nuovo Cimento 52B*, 15 (1979).
C. Philippidis, D. Bohm, and R.D. Kaye, *Nuovo Cimento 71B*, 75 (1982).
I. Prigogine, *From Being to Becoming* (Freeman, San Fransisco, 1980).
R.D. Prosser, *Int. J. Theor. Phys. 15*, 169, 181 (1976).
R. Thom, *Modèles mathématiques de la morphogenèse* (Union Générale d'editions, Paris, 1974).
A. Tonomura, T. Matsuda, R. Suzuki, A. Fukuhara, N. Osakabe, H. Umezaki, J. Endo, K. Shinagawa, Y. Sugita, and H. Fujiwara, *Phys. Rev. Lett. 48*, 1443 (1932).
E.P. Wigner, *Am. J. Phys. 31*, 6 (1963).
E.P. Wigner, *The Scientist Speculates*, I.J. Good ed. (Putnam, New York, 1965).
J.-P. Vigier, *Astron. Nachr. 303* 55 (1982).

WHEN IS A STATISTICAL THEORY CAUSAL?

Trevor W. Marshall

36 Victoria Avenue
Didsbury
Manchester M20 8RA

ABSTRACT

It is shown that Laplace defined the concept of causality rather than determinism, and that the former concept is compatible with, and indeed demands, a stochastic description of the intraction of a system with its environment. We examine a number of statistical physical theories, both classical and quantum, to see whether they conform with the Laplacian notion of causality. We conclude that all formulations of quantum theory, including those of Bohm and of Nelson, are acausal.

1. INTRODUCTION

In one sense the whole of science is a search for causal explanations. That is not meant as a grand philosophical statement, but rather as a humble attempt to sum up the psychological motivation of the individual scientist. Occasionally a bold idealist, like Heisenberg (1) or Pauli (2), renounces this search for causality as sterile and outdated. Yet, even among strong supporters of the Copenhagen school, it is generally recognized that some form of causality is still necessary. Read Born's letters to Einstein (3), or Rosenfeld's (4) rather dogmatic attempts to convince us that the 'statistical causality' of quantum mechanics is the same as that of thermodynamics.

I am going to offer you a definition of causality -- not a precise one, but possibly a more precise one than you have been offered previously. Among twentieth-century physicists, my definition comes, I would say, from Max Planck (5), and is called

G. Tarozzi and A. van der Merwe (eds.), Open Questions in Quantum Physics, 257–270.

by him "strict causality". But it can be traced back fairly directly to Laplace's statement of 1814, and I will therefore call it 'neo-Laplacian causality'. At the same time I will try to show you that causality *must* be statistical. Indeed it must be more: it *must* be stochastic. (For a discussion of the distinction between 'statistical' and 'stochastic' see Brush (6).) Naturally I do not expect everyone to accept my definition, but I invite anyone finding himself in that position to give us an alternative.

So let us look at what Laplace said in 1814. First, note that he did *not* say "Give me the positions and velocities of all the molecules and I will tell you the future of the world". That is a cruel and libellous caricature which has come to be called "Laplacian determinism". It bears little relation to Laplacian causality, and even less to modern forms of it such as I am offering you. What he did say was:

> Nous devons donc envisager l'état présent de l'univers comme l'effet de son etat antérieur, et comme la cause de celui qui va suivre. *Une intelligence qui pour un instant donné connaîtrait toutes les forces dont la nature est animée et la situation respective des êtres qui la composent*, si d'ailleurs elle était assez vaste pour soumettre ces données a l'analyse, embrasserait dana la même formule les mouvements des plus grands corps de l'univers et ceux du plus léger atome: rien ne serait incertain pour elle, et l'avenir comme le passé serait présent à ses yeux. L'esprit humain offre, dans la perfection qu'il a su donner à l'Astronomie, une faible esquisse de cette intelligence. Ses découvertes en Mécanique et en Géométrie, jointes à celle de las pesanteur universelle, l'ont mis à portée de comprendre dans les mêmes expressions analytiques les états passés et futurs du système du monde. En appliquant la même méthode à quelques autres objets de ses connaissances, il est parvenue à ramener à des lois générales les phénomènes observés, et à prévoir ceux que des circonstances données doivent fair éclore. Tout ces efforts dans la recherche de la vérité tendent à le rapprocher sans cesse de l'intelligence que nous venons de concevoir, mais dont il restera toujours infiniment éloigne. Cette tendance propre à l'espèce humaine est ce qui la rend supérieure aux animaux, et ses progres en ce genre distinguent les nations et les siècles et font leur véritable gloire. (7)

The crucial points at which this statement differs from its caricature cited above lie in (i) the title of the essay, (ii) the use of the conditional, and (iii) the recognition, in the sentence I have italicized that, to predict the future, you need not only "la situation des êtres" but also "toutes les forces dont la nature est animée". I submit that there is no evidence that Laplace thought he knew "toutes les forces", or that they were all of the instaneous action-at-a-distance type which were used in Newtonian mechanics. Such stupidities are to be found rather in the present day search for a Grand Universal Theory based on instantaneously collapsing wave packets.

The one feature needed, in my view, to make Laplace's statement of causality applicable to the physics of our time, is the concept of the *field*, as it developed through the nineteenth century, culminating in Einstein's famous paper of 1905. The recognition that "les forces dont la nature est animée" are mediated by fields propagating through space with finite velocities requires us to make two adjustments to Laplace's statement: (a) Any system, no matter how large and how complicated, interacts, through fields propagating across the boundaries, with its environment. No system may any longer be considered as isolated, and in particular the concept of a "universe" has to be abandoned. For a determinist this is catastrophic, but for Laplace there is no real problem, because he has already recognized that you need both forces and particle coordinates. (b) Because in modern terms the "future" means the set of events contained within the future light cone, certain other adjustments have to be made in describing the way in which the future of a given system is caused by its past.

2. THE PRINCIPLE OF CAUSALITY

I now try to make more precise this neo-Laplacian notion of causality, and to show that the requirement that the time evolution of a physical system be causal implies that it must be modelled by a certain subclass of stochastic processes. In other words, not all statistical theories may be considered causal.

We consider a system which is described by both particle variables $\underset{\sim}{x}_1, \underset{\sim}{x}_2, \ldots$ and field variables $\underset{\sim}{f}_1(x), \underset{\sim}{f}_2(x), \ldots$. We suppose that these variables may be divided into two sets called o-variables ("o" is for "observable" or "ordinary") and e-variables ("e" is for "environmental", "external", or "extra"). The field variables are "les forces dont la nature est animée", while the particle variables are "les êtres qui la composent". The "o" variables are those which present-day technology can measure more or less accurately, while the "e" variables, those for which "l'esprit humain" can give only "une faible esquisse", include vacuum and subquantum fields coming either from the spatial boundaries of the system or from a finer structure of the system than is covered by the "o" variables.

The principle of causality states *in part* that

$$\text{past}^{(o)} + \text{past}^{(e)} \rightarrow \text{future}\ (\text{past}^{(e)}),$$

where $\rightarrow$ means "determines" or "inevitably and uniquely gives rise to". We call this part of the principle *weak locality*. It means that a given "past" (strictly the o-components of the past variables) determines a whole family of possible "futures", according to which particular set of values the e-variables took in the past. For the description to qualify as causal, there must be a *non-negative probability density function* $\rho(\text{past}^{(e)})$, such that the mean value of the future variables is determined by a relation of the type

$$\langle\text{future}^{(o)}\rangle = \int \text{future}^{(o)}(\text{past}^{(e)})\rho(\text{past}^{(e)})d(\text{past}^{(e)}).$$

We will then write

$$\text{past}^{(o)} \rightsquigarrow \langle\text{future}^{(o)}\rangle,$$

where $\rightsquigarrow$ means "causes". This concept of causation implies a direction of time for all processes except those in which all the variables are o-variables. Such processes are called *deterministic*, and for them only the concept of weak locality is applicable.

Note that, because we are analysing, as did Laplace also in practice, a finite system rather than "l'univers", this means that it is no longer sufficient to know "les forces dont la nature est animée" simply "pour un instant donné". For such a system, even Laplace's "intelligence" would need *all* the past field values in order to determine the future values.

3. DIFFUSION THEORIES

We now look at our first family of statistical theories in order to see whether they can be said to provide a causal description of the phenomena.

(a) Macroscopic Diffusion Theory

Consider a cloud of fine raindrops settling in a still atmosphere in a force field $K(x)$. If the density of drops is $f(x)$, then

$$\frac{\partial f}{\partial t} = D\frac{\partial^2 f}{\partial x^2} + \beta\frac{\partial}{\partial x}[K(x)f],$$

where D and β are constants related to the size of the drops and the temperature of the atmosphere. For very small drops $\beta \rightarrow 0$, and this reduces to the equation of heat conduction.

This theory is deterministic, because the past values of f determine the future values uniquely. It is therefore a statistical theory which is not stochastic. Such theories are (or were in "classical" times) manifestly inadequate and temporary. They cry out for some kind of e-variables to be introduced.

(b) Random Walk Theory

Einstein (8) and Smoluchowski (9) independently proposed such a modification. They proposed that the position x of an individual drop satisfy the stochastic differential equation

$$\dot{x} = \beta K(x) + g(t).$$

where g is a white-noise gaussian random function with $\langle g(t_1)\rangle = 0$ and

$$\langle g(t_1)g(t_2)\rangle = 2D\delta(t_1-t_2).$$

This is a causal theory with the o-variable x and the e-variable g whose probability is specified by the above autocorrelation. (Remember that the process is gaussian, so the whole distribution is specified by its first two moments.)

But this classification is valid only with a nonrelativistic notion of locality. From a relativistic point of view, this theory is not causal, because it is not weakly local. For example, the Green function of the differential equation for the case $K = 0$ is

$$f(x,t|0,0) = \frac{1}{2(\pi Dt)^{\frac{1}{2}}} \exp\left(-\frac{x^2}{4Dt}\right),$$

which gives a non-zero probability for the particle to reach points with $|x| > ct$. This is because the theory is not dynamical. The stochastic differential equation is first order and contains no reference to the mass of the particle. (For the distinction between dynamical and kinematical theories of Brownian motion, see E. Nelson (10).) Hence no relativistic modification is possible. There are standard theorems (11) which show that relativistic Markov processes in configuration space cannot exist. Though it is the δ-function autocorrelation which makes the process Markov, the non-local behaviour we have noted is a characteristic feature of any first-order stochastic differential equation. For any autocorrelation of g the Green function will be non-vanishing for some $|x| > ct$.

(c) Uhlenbeck-Ornstein theory (12)

In this case (with units such that $m = 1$), x satisfies a second-order stochastic differential equation

$$x = \dot{p}, \quad \dot{p} = -p + K(x) + g(t).$$

With a white-noise gaussian g, an averaging procedure gives, for the density function $\Phi(x,p,t)$,

$$\frac{\partial\Phi}{\partial t} = \frac{\partial^2\Phi}{\partial p^2} + \frac{\partial}{\partial p}(p\Phi) - K(x)\frac{\partial\Phi}{\partial p} - p\frac{\partial\Phi}{\partial x}.$$

We again have a stochastic process which is norelativistically causal but relativistically acausal. However, if we put in the relativistic mass dependence, $m = (1 + p^2)^{\frac{1}{2}}$, we obtain

$$(1 + p^2)^{\frac{1}{2}}\dot{x} = p, \quad \dot{p} = -f(p) + K(x) + g(t) .$$

where $f(p) \sim p$ as $p \to 0$ is some suitable generalization of Stokes damping. Then $\Phi(x,p,t)$ satisfies

$$\frac{\partial\Phi}{\partial t} = \frac{\partial^2\Phi}{\partial p^2} + \frac{\partial}{\partial p}[f(p)\Phi] - K(x)\frac{\partial\Phi}{\partial p} - \frac{p}{(1 + p\ )^{\frac{1}{2}}}\frac{\partial\Phi}{\partial x},$$

and now the Green function $\Phi(x,p,t|0,p_0,0)$ is zero for $|x| > t$. Our conclusion is that any relativistic version of Brownian motion must be in phase space rather than configuration space.

(d) Stochastic Electrodynamics (13)

This is rather similar to the last example, but is based on the idea that, for electrons in atoms, Lorentz damping is more appropriate than Stokes damping:

$$\dot{\underset{\sim}{x}} = \underset{\sim}{p}, \quad \dot{\underset{\sim}{p}} = -\frac{2e^2}{3c^3}\ddot{\underset{\sim}{p}} + \underset{\sim}{K}(\underset{\sim}{x}) + e\underset{\sim}{g}(t).$$

To get the correct atomic dimensions we need a highly non-white noise

$$\int \langle g_i(t_1)g_j(t_2)\rangle\, e^{i\omega(t_1-t_2)}dt_1 = C\omega^3\delta_{ij},$$

with C a constant of order h. The density of points in phase space is best expressed as a density of orbits $\phi(\xi_1\ \xi_2\ \xi_3)$ where (ξ_1,ξ_2,ξ_3) are the classical action variables. We find that ϕ satisfies a Fokker-Planck equation (14a) of the form

$$\frac{\partial\phi}{\partial t} = \sum_{i,j}\frac{\partial}{\partial\xi_i}[A_{ij}(\xi)\frac{\partial\phi}{\partial\xi_j} + B_i(\xi)\phi],$$

where the 'diffusion coefficients' A_{ij} and the 'drift coefficients' B_i are obtained by integrating certain functions along the orbit.

In this case a relativistic generalization, displaying relativistic causality, follows the same lines as in the previous example, with the improved feature that there is an obvious and well-known relativistic form for the Lorentz damping, and that the above noise spectrum is already relativistically invariant.

(Indeed - see Marshall (14b) - it is the only electromagnetic spectral density having this property).

A theory based on such a real zero-point electromagnetic field has proved capable of explaining a number of results which were previously thought to require a quantum explanation (13). Nevertheless it conspicuously fails in some areas. For example, though it predicts a stable hydrogen atom of approximately the same size as the Bohr atom, it does not give us anything like the correct line spectrum. Hence, though it is a causal theory, it is not at present a serious competitor with quantum mechanics.

4. THEORIES OF ENTROPY INCREASE

(a) The Boltzmann H-Theorem (1872 version)

By assuming that atoms interact by "hard-sphere" collisions, Boltzmann, making the famous *Stosszahlansatz* assumption, derived the following transport equation for the one-particle phase-space distribution $\phi(\underset{\sim}{x},\underset{\sim}{v},t)$:

$$\frac{\partial\phi}{\partial t} = -\underset{\sim}{v}.\mathrm{grad}\;\phi \;+$$

$$+\int[\phi(\underset{\sim}{x},\underset{\sim}{v}'_1)\phi(\underset{\sim}{x},\underset{\sim}{v}'_2) - \phi(\underset{\sim}{x},\underset{\sim}{v})\phi(\underset{\sim}{x},\underset{\sim}{v}_2)]\sigma(\underset{\sim}{v},\underset{\sim}{v}_2,\underset{\sim}{v}'_1,\underset{\sim}{v}'_2)d\underset{\sim}{v}_2 d\underset{\sim}{v}'_1 d\underset{\sim}{v}'_2\;,$$

where $\sigma(\underset{\sim}{v},\underset{\sim}{v}_2,\underset{\sim}{v}'_1,\underset{\sim}{v}'_2)$ is the collision cross-section for collisions between atoms with velocities $\underset{\sim}{v}$ and $\underset{\sim}{v}_2$ giving outgoing atoms with velocities $\underset{\sim}{v}'_1$ and $\underset{\sim}{v}'_2$. From this transport equation he deduced that the quantity

$$H = \int\phi\log\phi\;d\underline{x}d\underline{v},$$

which he identified as $-k$ times the entropy, is a monotonically decreasing quantity. The only time-varying quantity is ϕ, so the theory is deterministic. Furthermore, in the limit of point atoms, the theory is local in both the non-relativistic and the relativistic sense. However, like the first of our examples, it is statistical but not stochastic. Criticisms of the theory, by Loschmidt, and later by the English school of kinetic theorists, forced Boltzmann to view ϕ as a probability density rather than a frequency. The version of the theory that emerged in the *Gastheorie* (1896) is stochastic and laid the foundation for the modern theory which is

(b) The Boltzmann-Gibbs-Ehrenfest H-Theorem (15)

This is based on the phase space density $\Phi(\underset{\sim}{x}_1,\underset{\sim}{p}_1\ldots\underset{\sim}{x}_N,\underset{\sim}{p}_N)$ of all the N atoms of the system. The deterministic equations of motion are

$$\dot{x}_i = \{H, x_i\}, \quad \dot{p}_i = \{H, p_i\},$$

where { } denotes a Poisson bracket, and the Hamiltonian H now includes the interatomic potential involved in all the collisions. Then Φ satisfies the Liouville equation

$$\frac{\partial \Phi}{\partial t} = \{H, \Phi\}$$

and one can then show that the H-function

$$\mathcal{H} = \int \Phi \log \Phi \, dx_1 dp_1 \ldots dx_N dp_N$$

is a constant. However, the distribution Φ may be "coarse-grained" by dividing the phase space into cells of finite volume δ. We denote these cells $\delta_1 \, \delta_2 \ldots$ and denote by Φ_i the integral of Φ over δ_i. Then the coarse-grained modification of $\mathcal{H}$, namely,

$$\mathcal{H}^c = \sum_i \Phi_i \log \Phi_i$$

decreases monotonically with t. In his discussions with the English school in the early 1890s, Boltzmann indicated a mechanism of "molecular disordering" of the N-particle system through interaction with its environment. Such an interaction makes the formal coarse-graining into an objective physical process. The deterministic equations of motion have to be modified to

$$\dot{x}_i = \{H, x_i\}, \quad \dot{p}_i = \{H, p_i\} + \mu(x_i, t),$$

where μ are e-variables.

In this formulation the atomic collisions are not directly responsible for entropy increase. Rather, they provide what modern ergodic theory calls a 'mixing' mechanism which spreads any initial Φ, with increasing uniformity as t increases, over the whole available phase space. It is the interaction with the environment, through the molecular disordering force, which produces the entropy increase. Thus we again see the crucial role played by "les forces dont la nature est animée" in establishing the causal direction of time (see Marshall (16) for further discussion of this point).

5. QUANTUM THEORIES (ONE-PARTICLE VERSIONS)

(a) Copenhagen

According to von Neumann, any operator of the set

$$f_{mn} = S(x^m p^n), \quad (m,n = 0,1,2\ldots),$$

is an observable. Here S denotes the symmetrized product, for example

$$S(x^2p) = \frac{1}{3}(x^2p + xpx + px^2).$$

The o-variables of the theory are the quantities

$$\langle r|f_{mn}|r\rangle,$$

where $\{|r\rangle;\ r = 0,1,2\ldots$ is a complete set of energy eigenstates, while the e-variables are

$$\langle r|f_{mn}|s\rangle \qquad (r \neq s).$$

The time evolution of the system satisfies weak locality:

$$i\hbar\dot{f}_{mn} = Hf_{mn} - f_{mn}H,$$

but, as in all our previous stochastic examples, there is interaction between the o- and e-variables. The theory is not causal, as is illustrated by the non-existence of a positive definite phase space density giving the "correct" marginal distributions in x and p. (For example, the Wigner distribution gives negative probability for certain regions of phase space.)

(b) Bohm's Hidden-Variable Theory (Instantaneous Version)

Here the particle position $\underset{\sim}{x}_1$ and the wave function $\psi(\underset{\sim}{x})$ are both o-variables, and there are no e-variables, so the theory is deterministic. It is also weakly local and the time evolution is given by

$$i\hbar\dot{\psi} = H\psi, \quad \dot{\underset{\sim}{x}}_1 = \frac{h}{m}\,\mathrm{grad}\,S|_{\underset{\sim}{x}=\underset{\sim}{x}_1}$$

$$m\ddot{\underset{\sim}{x}}_1 = -\mathrm{grad}[V - \frac{h}{2m}\{\nabla^2R + (\nabla R)^2\}]_{\underset{\sim}{x}=\underset{\sim}{x}_1},$$

where $\psi = e^{R+iS}$. In my view this (later) version of Bohm's theory, which is the version advanced by Bohn and Hiley (17) and by Bohm in his recent book (18) suffers from the same incompleteness as all deterministic statistical theories, such as macroscopic diffusion theory and Boltzmann's 1872 H-theorem. Indeed, when Bohm advanced the original version of his theory in 1952, he fully recognized that a deterministic theory cannot be causal (19) and that this rather naive theory, which is essentially that originated by de Broglie in 1927 has certain very unsatisfactory "magical" connections between $\underset{\sim}{x}_1$ and $\psi(\underset{\sim}{x})$, namely

$$\underset{\sim}{x}_1 = \frac{h}{m} \text{grad } S \Big|_{\underset{\sim}{x}=\underset{\sim}{x}_1} \quad \text{and} \quad P(\underset{\sim}{x}_1) = e^{2R(\underset{\sim}{x}_1)}.$$

These connections mean that the trajectory of a given particle depends, in a non-local manner, on all the other possible trajectories which it may have taken. As Feynman (20) points out, this is the core of quantum non-locality. The only claim Bohm's theory (instantaneous version) can make to greater realism than the Copenhagen theory is that it does allow us to draw such trajectories. For me, however, the magical connections required mean that its "trajectories" are no more real than Feynman's. But Feynman's analysis at least has the virtue of simplicity. Compare Feynman and Hibbs' discussion (21) of the two slit experiment with that of Phillipidis, Dewdney and Hiley (22)!

(c) Bohm's Hidden-Variable Theory (1952 version) (19)

When he first advanced his theory in 1952, David Bohm was fully aware of these imperfections in the instantaneous form of the theory. He saw that, if $\underset{\sim}{x}_1$ and $\psi(x)$ are to be considered independent entities, so that, for example, as Vigier (23) suggests, a particle may be detached from part of its carrier wave, then the theory must contain a mechanism which explains the magical connections between them. He proposed that the relation $\underset{\sim}{x}_1 = \frac{h}{m}\text{grad } S$ was established by a subquantum fluctuating force, and the relation $P(x) = e^{2R(\underset{\sim}{x}_1)}$ through collisions with macroscopic objects. For present purposes both mechanisms may be classified as o-variables. They play the same role in this theory as do the molecular disordering forces in the modern version of Boltzmann's H-theorem. The "magical" connections are now causally established, but only after a finite time interval. Bohm suggests that this interval is of the order of 10^{-13} seconds, but, clearly, in any relativistic version of the theory, the time taken to establish the relation $P(\underset{\sim}{x}_1) = e^{2R(\underset{\sim}{x}_1)}$ must be proportional to the spatial extent of the system. This means that, in order to qualify as "causal", the equilibrating mechanism has to be a local one.

It seems to me that Bohm's theory between 1952 and 1980 has evolved in the same way as Boltzmann's between 1896 and 1872, that is towards acausality but with the wrong time-direction!

(d) Nelson's Stochastic Theory (10)

This has a great deal of formal resemblance to Bohm's theory. Writing, as above, $\psi(\underset{\sim}{x}) = \exp[R(\underset{\sim}{x}) + iS(\underset{\sim}{x})]$, the particle position $\underset{\sim}{x}$ satisfies the stochastic differential equation

$$\dot{\underset{\sim}{x}}_1 = \underset{\sim}{b}(\underset{\sim}{x}_1) + \underset{\sim}{g}(t),$$

where $\underset{\sim}{g}(t)$ is a white noise with

$$\langle g_i(t_1) g_j(t_2) \rangle = \frac{h}{m} \delta_{ij} \delta(t_1 - t_2),$$

and the drift vector $\underset{\sim}{b}$ is given by

$$\underset{\sim}{b}(\underset{\sim}{x}) = \frac{h}{m} \text{grad}[R(\underset{\sim}{x}) + S(\underset{\sim}{x})].$$

As in Bohm's theory, $\underset{\sim}{x}_1$ and $\psi(\underset{\sim}{x})$ are o-variables, and there is an e-variable $\underset{\sim}{g}(t)$. So we are considering a weakly local stochastic theory.

But also, as in Bohm's (instantaneous) theory, the magical connection $P(\underset{\sim}{x}_1) = e^{2R(\underset{\sim}{x}_1)}$ has to be imposed. No attempt is made to explain how this connection between two otherwise independent physical entities is established. As pointed out by Ghirardi *et al* (24), and also by Grabert *et al* (25) this means that, unlike the diffusion processes with which Nelson tries to compare it, this process cannot be split into subprocesses satisfying different sets of initial conditions. In other words the particle "trajectories" display the same nonlocal dependence on each other as in Bohm's instantaneous theory. It should, however, be noted that, by introducing one e-variable, namely g(t), one of the "magical" connections, namely $\dot{\underset{\sim}{x}}_1 = \frac{h}{m} \text{grad}\, S$, has been removed. This means the Nelson stochastic theory occupies an intermediate position between Bohm's 1952 theory and Bohm's instantaneous theory and gives us a clue as to what modifications are needed to make it causal.

It should be noted, however, that this process, like Einstein's diffusion process, cannot have a proper relativistic version as long as it remains Markov in configuration space. This only becomes possible after introducing a term in $\ddot{\underset{\sim}{x}}_1$ into the stochastic differential equation, so that the process becomes one in phase space.

6. SUMMARY

The definition of causality I have offered in sections 1 and 2 is based on "holding fast absolutely", in the manner advocated by Einstein (26), to local interactions. This means insisting on the normal, macroscopic notions of "past" and "future". I do not believe any change in this definition is possible without making a radical change, of 1905 proportion, in our model of space and time.

With such a definition, all forms of quantum theory, including the so-called "statistical interpretations", are equally acausal. The only microphysical theories which are causal are those which incorporate the ideas of stochastic fields and particles executing Brownian-type motions in phase space. Stochastic electrodynamics provides one example of such a theory, but its inadequacies and contradictions with experiment are now quite manifest.

On the other hand it seems to me that the central challenge to nonlocal theories, such as quantum theory, presented by (27) Einstein, Podolski, and Rosen, remains unanswered. Experiments performed up to this time do no more than "verify" certain models derived from quantum theory. They fall short, by a very long way, from refuting, as they claim to do, all local realistic theories, as is shown in the contributions by Emilio Santos and Franco Selleri to this volume.

There is, of course, a vast area yet to be explored. In particular, our experience with quantum field theories shows us that, especially in the relativistic domain, one-particle theories are totally inadequate. But our progress will be limited to a very inadequate formal type, without any genuine understanding, unless we maintain contact with some form of causality. As I have argued, this means insisting on statistical theories that incorporate genuine (that is non-negative) probability measures and on some definite notion of what is meant by "future" and "past". For the moment there is no serious competitor to Einstein in this latter area, and I, for one, do not expect to see such a competitor emerge.

ACKNOWLEDGEMENT

My attendance at the Bari workshop was made possible by a leave of absence from my post as a Lecturer in Mathematics at the University of Manchester. I also gratefully acknowledge a travel grant from the University of Manchester.

REFERENCES

1. W. Heisenberg, *Physics and Philosophy* (Harper and Row, New York, 1958).

2. C.G. Jung and W. Pauli *The Interpretation of Nature and the Psyche* (Pantheon, New York, 1955).

3. M. Born and A. Einstein, *Letters 1916-1955* (Macmillan, London, 1971).

4. L. Rosenfeld, *Science Progress 163,* 393 (1953), Section 2.

5. M. Planck, *Wege zur physikalischen Erkenntnis* p.223ff (Hirzel, Leipzig, 1944).

6. S.G. Brush, *The kind of motion we call heat* (North Holland, Amsterdam, 1976), Vols. 1 and 2.

7. P.S. Laplace, *Essai philosophique sur les probabilités* 1814, edition des "Maîtres de la pensée scientifique" (Gauthiers-Villars, Paris, 1921), pp.3-4.

8. A. Einstein, *Ann. Physik 17*, 549 (1905).

9. M. Smoluchowski, *Ann Physik 21*, 756 (1906).

10. E. Nelson, *Dynamical theories of Brownian motion* (Princeton Univ. Press, 1967).

11. R. Hakim, *J. Math. Phys. 9*, 1805 (1968).

12. G.E. Uhlenbeck and L.S. Ornstein, *Phys. Rev. 36*, 823 (1930).

13. E. Santos, article in this volume.

14a. T.W. Marshall, *Physica A 103*, 172 (1980).

14b. T.W. Marshall, *Proc. Camb. Phil. Soc. 61*, 537 (1965).

15. D ter Haar, *Elements of statistical mechanics* (Constable, London, 1955), Appendix I.

16. T.W. Marshall, *Eur. J. Phys. 3*, 215 (1982).

17. D. Bohm and B.J. Hiley, *Found. Phys. 5*, 93 (1975).

18. D. Bohm, *Wholeness and the implicate order* (Routledge and Kegan Paul, London 1980).

19. D. Bohm, *Prog. Theor. Phys. 9*, 273 (1953).

20. R.P. Feynman, *Int. J. Theor. Phys. 21*, 467 (1982).

21. R.P. Feynman and A.R. Hibbs, *Quantum mechanics and path integrals* (McGraw Hill, New York, 1965).

22. C. Philippidis, C. Dewdney and B.J. Hiley, *Nuovo Cimento 52B*, 15 (1979).

23. P. Gueret and J.P. Vigier, *Found. Phys. 12*, 1057 (1982).

24. G.C. Ghirardi, C. Omero, A. Rimini and T. Weber, *Riv. Nuovo Cimento vol. 1*, N.3 p.1 (1978).

25. H. Grabert, P. Hänggi and P. Talkner, *Phys. Rev. A 19*, 2440 (1979).

26. A. Einstein, "Autobiographical Notes" in *Albert Einstein:*

philosopher scientist ed. P.A. Schilpp (Tudor, New York, 1951).

27. A. Einstein, B. Podolsky and N. Rosen, *Phys. Rev.* *47*, 777 (1935).

RADIATION DAMPING AND NONLINEARITY IN THE PILOT WAVE INTERPRETATION OF QUANTUM MECHANICS

M.C.Robinson

Departamento de Física, Universidad de Oriente,
Cumaná 6101, VENEZUELA

ABSTRACT:

It is claimed that orthodox quantum mechanics (OQM) is not a scientific theory, but a set of computational recipes. Fundamental differences between the pilot wave interpretation (PWI) and OQM, and their experimental consequences are summarized. PWI is modified for a bound particle by accepting only real eigenfunctions of Schrödinger's time independent equation, in which case the particle is stationary with the electrostatic and quantum force canceling each other. The effect of radiation damping leads to a nonlinear term proportional to $\partial^2 \ln(\psi/\psi^*)/\partial t^2$ in the first approximation. It is shown that the stationary states are stable with the resonant frequencies given by Bohr's relation, $\omega_{nm} = (\mathcal{E}_n - \mathcal{E}_m)/h$.

In order to place in better perspective the possible progress in the pilot wave interpretation (PWI) that we (C.E.Aveledo, D.Bonyuet, L.A.Lameda, and I) have made in Cumaná, I should like to make a few general remarks concerning orthodox quantum mechanics (OQM).

To begin with, there can be no doubt that, over the last 80 years or so, OQM has been an extremely useful tool which has permitted the calculation of certain quantities to a part per million or even better, and the development of devices such as transistors and lasers. However, I claim, in essential agreement with Einstein, Schrödinger, and de Broglie, that OQM is only a tool; that is, a series of recipes for the manipulation, interpretation, and reinterpretation of mathematical formulae--a

G. Tarozzi and A. van der Merwe (eds.), Open Questions in Quantum Physics, 271–282.

sort of cook book tecnology.

The discussion (or, more truthfully, the bitter debate) begins when it is claimed that OQM is a scientific theory; i.e., that OQM can be formulated in accordance with the precepts accepted in practice over the last century by any competent scientist, excluding certain historical aberrations such as the rejection of the atomic theory by Mach, Ostwald, and their followers. If we overlook, in the first approximation, certain differences in detail, language and style, we find surprizingly similar descriptions of the implicit or working philosophy of modern science in the writings of Lenin (1908), Planck (1932), Einstein (1918, 1921, 1927, 1933, 1936, 1949), Popper (1935), Bohm (1957), and Bunge (1967, 1972, 1973, 1981). For our purposes, the main features of the working philosophy of modern science are the following:

(1) A scientific theory is materialistic, or for those who shy away from this term, realistic; that is, a scientific theory concerns itself only with matter in motion and does not accept explanations based on spirits, divine intervention, or supernatural mental powers which can, for example, reduce wave packets (von Neumann, 1932).

(2) A scientific theory is reductionist; that is, it reduces qualitative differences to quantitative differnces; it seeks to explain laws at a "higher" level as emerging from laws at a more basic level. Examples of reductionism are thermodynamics in terms of statistical mechanics; chemistry in terms of molecular physics; optics in terms of electrodynamics; biology in terms of biochemistry; etc.

(3) A scientific theory is logical; that is, its theorems (or laws) are deducible in accordance with the rules of formal logic from a set of consistent fundamental laws, called postulates or axioms, formulated in terms of primitive (irreducible or nondefinable) concepts. Examples of primitive concepts are mass, charge, position, time, electric field, and magnetic field. Examples of physical postulates are Maxwell's equations in a vacuum. A postulate in a scientific theory does not contain terms such as "measurement" or "observation" which are highly complex concepts belonging to the pinnacle of science, not to its foundations.

(4) A scientific theory must be testable; that is, it must be possible to perform experiments or make observations which may agree or disagree with the logical cosequences of the theory. But here we must be careful, remembering that no experiment or observation is theory free, and no theory is philosophy free. Our experimental observations acquire meaning only after being processed theoretically; strictly speaking, we have at our disposal only data, never "facts".

(5) Finally, despite the occasinal and not very convincing objections of Bunge and Popper, a scientific theory must be based on strict determinism; that is if, under supposedly identical

conditions, different effects are observed, then in accordance with the maxim laid down by Popper some 50 years ago, it is necessary to look for the unobserved differences in the conditions-- in other words, we must look for the hidden variables.

I claim that since the mid-ninteenth century, science, including the quantum mechanics that Planck, Einstein, de Broglie and Schrödinger attempted to build, has progressed in accordance with these doctrines, which, following Bunge, we may call scientific materialism . In fact, no matter what faith or ideology he may publically profess, any moderately educated and mentally stable individual today accepts, in practice, the rules and regulations of scientific materialism.(1)

If we now analyse OQM as a scientific theory we find it to be an amalgam of the incoherent, the incredible, and the incrompehensible. It is evident to anyone prepared to admit the nudity of the king that the postulates of OQM are mutually contradictory; with internal agreement limited to special cases (at the very most). In particular, the postulate of the reduction of the wave packet violates Schrödinger's equation, Dirac's equation, the Klein-Gordon equation, Maxwell's equations, and perhaps every other equation of physics-- a little detail which Heisenberg (1930) and von Neumann (1932) reluctantly, and very briefly, half admitted, and which all the standard text books conveniently forget. Moreover, the reduction of the wave packet contradicts laboratory practice. For example, as de Broglie (1964) has pointed out, every measurement of momentum or velocity at the atomic level terminates with a measurement of position, so that from the view point of OQM the momentum of an electron has never been measured and probaby never will be. It follows, of course, that according to OQM there is absolutely no experimental evidence for the validity of Heisenberg's uncertainty principle.

As for the entire quantum theory of measurement it has nothing to do with real measurements performed in the laboratory; no sane experimentalist ever uses it.

The fact is that the postulates of OQM, as it is used in real applications, are never truly formulated. In reality, all of us, orthodox and heretic alike, are forced in practice to switch back and forth between one interpretation of the wave function, $\psi(\vec{r},t)$, and another. When it suits us, a particle is a particle is a particle with a definite position, momentum, mass, charge, size, shape, and internal structure in which case $\psi(\vec{r},t)$ and its Fourier transform, $\phi(\vec{p},t)$ determine the statistics of an ensemble of particles. At other times it suits us to regard $\psi(\vec{r},t)$ and $\phi(\vec{p},t)$ as representing charge clouds in real and momentum space. In other words, a quantum particle can supposedly be everywhere and move with all possible velocities, all at the same time; but despite this awe inspiring turbulence, the "cloud" remains absolutely motionless when the "particle" is in a stationary state. To camouflage this incredible model, the orthodox physicist invokes the "second commandment," forbidding us any intuitive picture of the mathematical formalism. At the same time he proclaims that the mathematical formalism consists of experimentally verified "relations only between observable quantities." (Heisenberg, 1925). In brief, OQM can be formulated as the art of observing the unimaginable.

At this point I claim, repeat and emphasize that we are still completely without any scientific explanation of the spectrum of black body radiatiom, of the photoelectric effect, of the specific heat of solids, of the spectrum of hydrogen, or of any other of the phenomena which we can treat quantitatively with the aid of OQM. For example, not only can we not explain the spectrum of a hydrogen atom with the aid of Scrödinger's (or Dirac's) equation; but, from a scientific viewpoint, we are obliged to consider the spectrum of a hydrogen atom as an experimental violation of these equations, as Scrödinger himself realized immediately after formulating wave mechanics.

However, as Einstein (1936) suggested, if it is possible to make accurate calculations with the aid of QM, it is highly plausible that QM has captured something of the truth. So, let us begin at the beginning, writing down Scrödinger's equation for a single particle in the manner of de Broglie (1927) and Bohm (1952).

$$-\frac{\partial S}{\partial t} = \frac{(\nabla S)^2}{2M} + V(\vec{r}) - \frac{h^2}{2M}\frac{\nabla^2 R}{R}, \qquad \text{(1a)}$$

$$\frac{\partial}{\partial t}(R^2) + \nabla\cdot[\frac{\nabla S}{M} R^2] = 0, \qquad \text{(1b)}$$

where $R(\vec{r},t)$ and $S(\vec{r},t)$ are real functions and

$$\psi(\vec{r},t) = R(\vec{r},t)\exp\{iS(\vec{r},t)/h\}. \qquad \text{(1c)}$$

Eqs.(1a) and (1b) are as true or as false as Schrödinger's equation; and I fail to see how it is possible to accept the validity of Scrodinger's equation (at least in particular cases) and that of the Hamilton-Jacoby equation in classical mechanics without interpreting

$$Q = -\frac{h^2}{2M}\frac{\nabla^2 R}{R} \qquad \text{(1d)}$$

as the quantum potential energy.

If Eq.(1a) is regarded as the quantum mechanical generalization of the Hamilton-Jacoby equation (i.e., the energy

$\mathcal{E} = -\partial S/\partial t$ and the momentum $\vec{p} = \nabla S$) then Eq.(1b) automatically becomes an equation of continuity for R^2. Therefore, if R^2 initially equals the probability density of position of a "normal" ensemble, then R^2 will continue to equal the probability density at all future times. Now, Bohm (1953) and Vigier (1953) have made the plausible conjecture that, even if initially the probability density of an "abnormal" ensemble is not given by R^2, the ensemble will tend to "normality" with time because of stochastic processes. Nevertheless,I claim that,in at least two cases, it is possible to construct or find "abnormal" ensembles, vis., when R^2 ($= |\Psi|^2$) does not equal the probability density of position (Robinson 1969a, 1969b, 1978, 1980). Furthermore, the abnormality of these ensembles appears to be experimentally detectible.

Next we note that in the PWI certain dynamic variables, namely, linear and angular momentum, total energy, position, and functions of position such as potential energy, are given by the relation

$$F = \mathrm{Re}\,\frac{\Psi^* F_{op}\,\Psi}{|\Psi|^2}, \tag{2a}$$

where F is the variable and F_{op}, the corresponding mathematical operator of OQM (Robinson, 1982). Therefore, in "normal" ensembles, the expectation value,

$$\begin{aligned} \langle F\rangle &= \int F\,|\Psi|^2\,d^3 r \\ &= \mathrm{Re}\int \Psi^* F_{op}\Psi\, d^3 r \\ &= \int \Psi^* F_{op}\,\Psi d^3 r. \end{aligned} \tag{2b}$$

Thus the PWI offers a clear and logical explanation of the almost mystical statement that in quantum mechanics dynamical variables are replaced by mathematical operators.

It is important to remember that , within the PWI, eqs.(2a) and (2b) do not hold for the square of the linear and angular momentum, nor for the kinetic energy. For a particle in a stationary state, the square of the orbital angular momentum is not equal to $l(l+1)h^2$, the magnitude of the linear momentum may be exactly zero, and

$$-\frac{h^2}{2M}\int \Psi^* \nabla^2 \Psi\, d^3 r = \langle K\rangle + \langle Q\rangle,$$

which generally is different from $\langle K\rangle$, as in OQM.

It follows that, in general, Heisenberg's principle is not

valid in PWI. However, as Andrade e Silva (1967) has pointed out, eqs.(2a) and (2b) are valid for the square of the linear momentum in the special case of a free particle, initially localized, after the dispersion of the wave packet. Thus Heisenberg's uncertainty principle should be replaced by a relation proposed by de Broglie:

$$(\Delta x)_i (\Delta p)_f \geq h/2,$$

valid only for free particles, which we may call a scatter relation, to use Popper's expression.

According to the PWI, it should therefore be possible in theory and in practice to measure the simultaneous values of the position and momentum with a precision surpassing the limits set by Heisenberg's principle (Robinson, 1969a, 1980). The experiment I have proposed is, in fact, a combination of two experiments suggested by Heisenberg (1930) that had previosly been discussed by Schrödinger (1955) and Popper (1967) who both arrived at similar conclusions concerning the restricted validity of the uncertainty principle (See also Popper, 1935).

Now let us consider the stationary state of a charged particle in a central force field. The wave function is usually written in the form,

$$\psi_{nlm}(\vec{r}) = N_{nl} R_{nl}(r) P_l^m(\theta) \exp(im\phi), \tag{3a}$$

so that

$$S = mh\phi - \mathcal{E}_{nl} t \tag{3b}$$

and

$$L_z = (\vec{r} \times \nabla S)_z = mh \tag{3c}$$

(Bohm, 1952). If m = 0, the particle moves in a circular orbit with an angular frequency,

$$\omega = mh/(Mr^2 \sin^2\theta). \tag{3d}$$

We are thus still confronted with the problem that has haunted physics since the days of Rutherford, of a charged particle supposedly moving in a closed orbit without radiating. One way to avoid this problem is to impose the constraint that the spatial part of the wave function be real. (For an alternative point of view see Gutkowsky, Mole, and Vigier, 1977).

For example, instead of eq.(3a) we could write

$$\psi_{nlm}(\vec{r}) = N_{nl} R_{nl}(r) P_l^m(\theta) \begin{cases} \cos m\phi & (m < 0) \\ \sin m\phi & (m > 0) \end{cases} \tag{3e}$$

If the spatial part, $\psi_n(\vec{r})$, of the wave function is real, then it can be readily shown that the particle is stationary, with the quantum and electrostatic forces exactly canceling each other. This is the tentative model that we are trying to analyse in Cumaná (Robinson et al, 1983).

To test the stability of the stationary particle, we consider the effect of a small electric field

$$\vec{E} = \hat{k}E_0 \sin \omega t ,$$

applied at t = 0, on the charged particle whose wave function is $\Psi^0_n(\vec{r},t)$, where n refers to more than one quantum number. In order to simplify the notation, we suppose

$$\omega \simeq \omega(n',n) = (\mathcal{E}^0_{n'} - \mathcal{E}^0_n)/h,$$

so that, to first order in E ,

$$\Psi(\vec{r},t) \simeq \psi^0_n(\vec{r}) \exp(-i\mathcal{E}^0_n t/h) + E_0 C(t)\psi^0_{n'}(\vec{r}) \exp(-i\mathcal{E}^0_{n'}t/h), \quad (4a)$$

where, according to the usual theory of pertubations,

$$ih\dot{C} = \mathcal{H}_{nn'} \sin \omega t \exp i\omega(n',n)t \quad (4b)$$

and

$$\mathcal{H}_{n'n} = -\int \psi^0_{n'}(\vec{r})(er \cos \theta)\psi^0_n(\vec{r})d r. \quad (4c)$$

In the PWI the particle should now oscillate under the influence of the quantum field and therefore radiate in accordance with Maxwell's equations. For velocities, $|\dot{\vec{r}}| \ll c$, the radiation reaction is given by

$$\vec{F}_r = \frac{2e^2}{3Mc^3}\frac{d^2\vec{p}}{dt^2} = \tau \frac{d^2}{dt^2}(\nabla S), \quad (5a)$$

where the total derivative,

$$\frac{d}{dt} = \frac{\partial}{\partial t} + \vec{v}\cdot\nabla = \frac{\partial}{\partial t} + \frac{1}{M}\nabla S\cdot\nabla . \quad (5b)$$

It can be shown that, to first order in E_0,

$$\frac{d^2}{dt^2}(\nabla S) \simeq \frac{\partial^2}{\partial t^2}(\nabla S) = \nabla[\frac{\partial^2 S}{\partial t^2}], \quad (5c)$$

so that we can write

$$\vec{F}_r \simeq -\nabla\mathcal{R}, \qquad (6a)$$

where the "radiation reaction potential,"

$$\mathcal{R} = -\tau \frac{\partial^2 S}{\partial t^2} = \frac{ih\tau}{2} \frac{\partial^2}{\partial t^2} \ln[\frac{\Psi}{\Psi^*}]. \qquad (6b)$$

To take into account the radiation reaction, we can now modify (1a) to give, for the perturbed system,

$$-\frac{\partial S}{\partial t} = \frac{(\nabla S)^2}{2M} + V(\vec{r}) - \frac{h^2}{2M}\frac{\nabla^2 R}{R} - \tau\frac{\partial^2 S}{\partial t^2} + E_0 H \sin \omega t \qquad (7)$$

which, together with (1b), is equivalent to the nonlinear Schrödinger equation,

$$ih \frac{\partial \Psi}{\partial t} = [-\frac{h^2}{2M}\nabla^2 + V(\vec{r}) + E_0 H \sin \omega t + \frac{ih\tau}{2}\frac{\partial^2}{\partial t^2} \ln \frac{\Psi}{\Psi^*}]\Psi . \qquad (8)$$

We see that, when the particle is in a stationary state described by a real function, $\psi_n(\vec{r})$, Eq.(8) reduces to the usual linear equation. It is hardly necessary to recall at this point that Einstein, de Broglie, and Schrödinger suggested that quantum processes be described by nonlinear equations, although their proposals went far beyond our equation (8). We should however, mention that Andrade e Silva et al. (1960) argued that only nonlinear equations could explain correctly the observed atomic and molecular spectra.

We shall now sketch an approximate solution of Eq.(8) to first order in E_0. From (4a) we have, for sufficiently small E_0,

$$\frac{\Psi}{\Psi^*} = \exp(-2i\mathcal{E}_n^0 t/h)\, \frac{1 + E_0 C(\psi_{n'}^0/\psi_n^0)\exp\{-i\omega(n',n)t\}}{1 + E_0 C^*(\psi_{n'}^0/\psi_n^0)\exp\{i\omega(n',n)t\}} \qquad (9a)$$

$$\simeq \exp(-2i\mathcal{E}_n^0 t/h)(1 + 2iE_0\beta\psi_{n'}^0/\psi_n^0), \qquad (9b)$$

where

$$\beta = \mathrm{Im}[C\exp\{-i\omega(n',n)t\}]. \qquad (9c)$$

After expanding $\ln(\Psi/\Psi^*)$ to first order in E_0, we find

$$S = -i(h/2)\ln(\Psi/\Psi^*) \simeq -\mathcal{E}_n^0 t + \beta h E_0(\psi_{n'}^0/\psi\ ), \qquad (10a)$$

$$\vec{p} = \nabla S \simeq \beta h E_0 \nabla(\psi^0_{n'}/\psi^0_n), \tag{10b}$$

$$\mathcal{R} \simeq \tau \frac{\partial^2 S}{\partial t^2} \simeq -\tau\beta h E_0(\psi^0_{n'}/\psi^0_n). \tag{10c}$$

After substituting Eq.(10c) in (8), time-dependent perturbation theory gives:

$$ih\dot{C} = (\mathcal{H}_{n'n} \sin \omega t - \tau\ddot{\beta} h) \exp\{i\omega(n',n)t\} \tag{11}$$

instead of the usual expression (4b). To solve (11), we introduce the function,

$$\alpha = \mathrm{Re}\,[C \exp\{-i\omega(n',n)t\}], \tag{12a}$$

and we can show that

$$\dot{\alpha} = \omega(n',n)\beta. \tag{12b}$$

After a little algebraic manipulation, we find that (11) reduces to

$$-\tau\dddot{\alpha} + \ddot{\alpha} + \omega^2(n',n)\alpha = -\{\mathcal{H}_{n'n}\,\omega(n',n)/h\} \sin \omega t. \tag{13}$$

After finding the convergent solution of (13) which satifies the initial conditions, $C = 0$; $\vec{p} = \nabla S = 0$, at $t = 0$, we use (12b) to obtain:

$$\beta = \exp[\frac{-\tau\omega_a^2 t}{2}]\,\frac{\mathcal{H}_{n'n}}{h[\{\omega^2(n',n) - \omega^2\}^2 + \tau^2\omega^6]}\,(A \sin \omega_b t + B \cos \omega_b t)$$

$$+ \frac{\omega\,\mathcal{H}_{n'n}}{h[\{\omega^2(n',n) - \omega^2\}^2 + \tau^2\omega^6]}\,(F \sin \omega t + G \cos \omega t) \tag{14a}$$

where A, B, F, and G are functions of $\omega(n',n)$, ω, and τ, and where

$$\omega_a = \omega(n',n)\{1 - 2\tau^2\omega^2(n',n)\}, \tag{14b}$$

$$\omega_b = \omega(n',n)\{1 - (5/8)\tau^2\omega^2(n',n)\}. \tag{14c}$$

If the particle is an electron, $\tau = 6.3 \times 10^{-24}$ s, so that even for $\omega(n',n) \simeq 10^{16}\,s^{-1}$ (ultraviolet),

$$\omega_a = \omega_b = \omega(n',n)$$

with an error of less than 1 part in 10^{14}.

From (14a) and (14b), it is seen that, to first order in E_0, the motion of the particle is that of a forced, underdamped harmonic oscillator. If we wish to use a pictorial analogy, we can think of the particle as being imbedded in a highly elastic "jelly" which resonates at the Bohr frequencies.

The results so far are encouraging. In addition to the above, Aveledo and Lameda have made additional progress during the last few months including the calculation of Einstein's A coeffecient in the special case of a slightly perturbed system. However, we still have to discover whether a more general solution and a more accurate Hamilton will explain phenomena such as "spontaneous" radiation, "transitions," etc. A more difficult problem would seem to be the explanation of magnetic interactions between stationary charged particles. From preliminary research on this problem, we incline to the opinion that under certain conditions the quantum field may be experimentally indistinguishable from the magnetic or electric field.

We wish to emphasize that we do not claim that the model discussed here is correct, even to the a first approximation. However we are convinced that it is a plausible development of the PWI that needs to be explored, without excluding the possibility of some other more radical modification of the PWI including Schrödinger's and/or Maxwell's equations.

At present I am convinced only that, if quantum mechanics does contain elements of truth, then it must be drastically modified and cleansed of orthodox ingredients. However, far more important than the fate of quantum mechanics is the urgent necessity to convert physics into a scientific discipline.

NOTES

1. A more complete summary of scientific materialism would include:

(6) The world and its constituents are undergoing constant change, passing from one state of quasi-stability to another, often in a relatively short time.

(7) The world and its constituents are infinitely complex, or at least effectively so. Thus, our scientific theories should be considered only as approximate and partial descriptions of reality, valid in a limited domain and liable to modification, or even replacement, by a more accurate and general theory.

As far as I am able to understand, scientific and dielectic materialism, are one and the same thing. However, to avoid a pointless discussion, I have used the term "scientific materialism" in this work.

REFERENCES:

Andrade e Silva, J., F. Fer, Ph. Lebuste, and G. Lochak: 1960 Comptes Rendus 25, 2305, 2482, 2662.
Andrade e Silva, J.: 1967, Comptes Rendus 264, 909.
Bohm, D.: 1952, Phys. Rev. 85, 166, 180.
Bohm, D.: 1953, Phys. Rev. 89, 458.
Bohm, D.: 1957, Causality and Chance in Modern Physics, Routledge and Kegan Paul, London.
de Broglie, L.: 1927, J. Phys. Radium 8, 225.
de Broglie, L.: 1964, The Current Interpretation of Wave Mechanics, Elsevier, New York.
de Broglie, L.: 1969, Comptes Rendus 258, 277.
Bunge, M.: 1967, Foundations of Physics, Springer,New York.
Bunge, M.: 1972, Philosophy of Physics, D. Reidel, Dordrecht.
Bunge, M.: 1973, Method, Model and Matter, D. Reidel, Dordrecht.
Bunge, M.: 1981, Scientific Materialism, D. Reidel, Dordrecht.
Einstein, A.: 1918, "Motiv des Forschens," in Zu Max Plancks Geburtstag: Ansprachen in der Deutscher Physikalischer Gesellschaft, Muller, Karlsruhe; English translation: 1954, "Principles of Research," in C. Selig (ed.), Ideas and Opinions, Crown, New York.
Einstein, A.: 1921, "Geometrie und Erfahrung," expanded version of paper, Preuss. Akad. Wiss. Sitzungsber. 123; English translation: 1954, "Geometry and Experience," in C. Selig (ed.), Ideas and Opinions, Crown, New York.
Einstein, A.: 1927, Naturwiss 15, 201; English translation: 1954, "The Mechanics of Newton and Their Influence on the Development of Theoretical Physics," in C. Selig (ed.), Ideas and Opinions, Crown, New York.
Einstein, A.: 1933, On the Method of Theoretic Physics, Clarendon Press, Oxford; Reproduced: 1954, in C. Selig (ed.), Ideas and Opinions, Crown, New York.
Einstein, A.: 1936, J. Franklin Inst. 221, 349.
Einstein, A.: 1949, in P. A. Schilpp (ed.), Albert Einstein: Philosopher-Scientist, Open Court, La Salle, Illinois.
Gutkowsky, D., M. Mole, J.-P. Vigier: 1977, Nuovo Cimento 39, 193.
Heisenberg, W.: 1925, Z. Physik 43, 172.
Heisenberg, W.: 1930, The Physical Principles of the Quantum Theory, University of Chicago Press, Chicago.
Lenin, V. I.: 1908 Materialism and Emperiocriticism, Mir, Moscow.
von Neumann, J:1932, Mathematische Grundlagen der Quantummechanik , Springer, Berlin; English translation: 1955, Mathematical Foundations of Quantum Mechanics, Princeton University Press, Princeton.
Planck, M.: 1932, Where is Science Going? Norton, New York.
Popper, K.: 1934, Logik der Forschung, Springer, Vienna; 2nd English edition: 1968, The Logic of Scientific Discovery, Harper, New York.

Popper, K.: 1967, in M. Bunge (ed.), Quantum Mechanics and Reality, Springer, New York.
Robinson, M. C.: 1969a, Can.J. Phys. 47, 963.
Robinson, M. C.: 1969b, Phys. Lett. 30A, 69.
Robinson, M. C.: 1978, Phys. Lett. 66A, 263.
Robinson, M. C.: 1980, J. Phys. 13A, 877.
Robinson, M. C.: 1982, J. Phys. 15A, 113.
Robinson, M. C., C. E. Aveledo, L. A. Lameda, D. Bonuet: 1983, J. Phys. 16A, 2987.
Schrödinger, E.: 1955, Nuovo Cimento 1, 5.
Vigier, J.-P.: 1953, Comptes Rendus 236, 1003.

STOCHASTIC ELECTRODYNAMICS AND THE BELL INEQUALITIES

Emilio Santos

Departamento de Física Teórica
Universidad de Santander
Spain

ABSTRACT

The basic ideas of stochastic electrodynamics are presented. In the light of these ideas, some general differences between quantum mechanics and local realistic theories are pointed out. The atomic-cascade experimental tests of Bell's inequalities are analyzed and a loophole is reported in the refutation of local realism due to an incorrect substraction of the background which affects mainly the recent experiments by Aspect _et al_. This must be added to the loophole due to the low efficiency of photon detectors. The conclusion is that local realistic theories have not yet been disproved.

1. INTRODUCTION

We are confronting a serious crisis as a consequence of the empirical evidence[1-3] for the violation of Bell's inequalities. The violation is usually interpreted as excluding the possibility of local realistic theories of the physical world. The crisis is more acute after the recent experiment by Aspect and coworkers,[3] which uses time-varying analyzers. In fact, the possibility of a result fulfilling Bell's inequalities in Aspect's experiment was the last hope for some supporters of local realism. In the following I shall present my personal view about the possible solutions to the crisis, but first I shall comment briefly about a number of other solutions proposed.

In the first place, there are people which simply do not see any crisis. Most of these people are supporters of orthodox quantum mechanics. I shall not repeat here the arguments given

G. Tarozzi and A. van der Merwe (eds.), Open Questions in Quantum Physics, 283–296.

elsewhere to show that a real problem exists.[4-6] It is sufficient to mention that quantum mechanics is almost sixty years old and that the controversy about its conceptual foundations not only has not decreased but is at one of its highest peaks at present.

For people who see a real problem in the empirical evidence for the violation of Bell's inequalities there are three possibilities: rejecting realism, rejecting locality, or searching for a loophole in the interpretation of the experiments. Here, realism means the belief that material systems have properties that are independent of whether these are measured or not. In my opinion,[7] this is the last thing one should abandon. It would be preferable even to believe in absolute determinism (nothing can be proved or dispoved empirically if everything is predetermined).

The rejection of locality involves a change in our conceptions of space and time. For instance, signals propagating faster than light or backward in time. That this would violate the spirit of relativity theory is the least that can be said. I shall not comment any further on this possibility, which is actively being investigated by several participants in this workshop.

In my opinion, before rejecting either realism or locality, it is necessary to investigate the existing loopholes in the refutation of local realism via the violation of Bell's inequalities. One such loophole, caused by the low efficiency of photon detectors, has been considered by Clauser and Horne.[8] The existence of that loophole shows that, contrary to a current school of opinion, realistic local theories have not been disproved yet. An experiment has been proposed with the purpose of blocking this loophole,[9] but a new loophole has been pointed out in the proposed experiment.[10] As a consequence, a true test of realistic local theories seems not probable in the near future. The final part of this communication will be devoted to a more detailed analysis of loopholes in atomic cascade experiments, the only ones considered reliable till now as tests for local realistic theories.

In the first part of the communication we present briefly the ideas of stochastic electrodynamics or, more generally, stochastic alternatives to quantum theory. These ideas will be useful as a guide in the search for loopholes in the refutation of local realistic theories.

2. STOCHASTIC ELECTRODYNAMICS

Stochastic electrodynamics (SED) started in the fifties with a series of papers by Braffort and coworkers,[11] although there were several previous expositions of ideas more or less similar to those of SED. SED appeared almost inmediately after the subject of hidden-variable theories, blocked for 20 years by von Neumann's theorem, was revived thanks to the works of de Broglie, Bohm, Vigier and others. In the sixties, SED received a boost from a set of remarkable papers by T.W. Marshall.[12] From that time on an increasing number of authors have been active in the field and important progress has been made. A recent review by L. de la Peña[13] includes a complete bibliography. Other shorter reviews have been written earlier by Boyer[14] and Claverie.[15]

Stochastic electrodynamics is a particular case of a general theory which we may call stochastic theory (ST) of microphysics. ST is just classical physics without the hypothesis that there are isolated systems in the universe. I use the word "without" in order to emphasize that ST does not contain additional, perhaps artificial, hypotheses besides those of classical physics. On the contrary, it eliminates a by no means obvious assumption. The surprising thing is that classical physics, without the hypothesis of isolation (i.e., ST), comes much closer to quantum physics than conventional classical theories, up to the point that it has been considered a potential alternative for quantum theory.

In order to see the changes produced in a classical model when the hypothesis of perfect isolation is removed, we shall analyze the Rutherford planetary model of the atom. According to this model, a hydrogen atom is similar to a solar system with the electron orbiting around the nucleus. However, according to classical electrodynamics the electron should radiate, losing energy, and eventually the atom will collapse. This model is clearly not in agreement with empirical evidence. However, if we do not assume perfect isolation, the Rutherford model much better fits the evidence. In fact, as there are many atoms in the universe, the energy radiated by each atom may eventually arrive at another one. As a consequence, each atom, besides emitting energy, may absorb it, and a dynamical equilibrium should be possible. In this way the stability of atoms could be explained. Besides, we restore time-reversal invariance and homogeneity with time, both properties being absent in the conventional classical model with zero radiation at the initial time.

Without the hypothesis of isolation, we arrive at a model of the hydrogen atom as being formed by a proton, an electron, and a background of electromagnetic radiation filling all of space. The background radiation, coming from a large number of sources, should be viewed as constituting a stochastic field, and the motion of the electron a stochastic process. In this model, although the electron should follow some path with a definite position and velocity at each time, it is impossible to predict these quantities in detail, and one must rely on probability distributions for them. In this way, a stochastic model is obtained of the equilibrium state of the hydrogen atom, which is qualitatively close to the picture given by quantum mechanics.

Stochastic electrodynamics is the theory which attempts to study systems of charged particles immersed in a random electromagnetic radiation field that fills all of space. This theory should be compared with quantum electrodynamics (QED). The inclusion of neutral particles, nuclear forces, and so on would correspond to some generalization of the theory not yet achieved.

In order to allow quantitative calculation, SED needs additional assumptions, in particular to fix the spectrum of the background radiation. The latter is found by introducing the natural hypothesis that the radiation seen by all inertial observers should be equivalent. In this way, we do not introduce priviledged inertial frames, in agreement with relativity theory. It has been shown [13-15] that this hypothesis fixes the spectrum of the radiation, except for a multiplicative constant measuring its intensity. If this constant is identified with Planck constant, SED makes predictions similar to that of QED for several systems. More specifically, the only spectrum which is Lorentz invariant has the form

$$u(\nu) = C \nu^3, \tag{2.1}$$

where $u(\nu)$ is the energy per unit volume and unit frequency interval and C an arbitrary constant. The latter is identified in SED with

$$C \equiv 4\pi h/c^3, \tag{2.2}$$

h being Planck constant and c the speed of light. It can be shown that, after this identification, the background radiation of SED is identical with the zero point radiation of QED, but having a real instead of purely virtual character. We do not comment here on the problems presented by the spectrum (2.1), in particular by the ultraviolet divergence.[13-15]

The background radiation of SED might also be related to the cosmic background radiation of 2.7 K. To put the idea more clearly, SED assumes that everywhere is space there is a random background of electromagnetic radiation whose energy per unit volume and unit frequency interval is given by Planck's (complete) law

$$u(\nu) = (8\pi\nu^2/c^3) \left[\frac{h\nu}{\exp(h\nu/kT)-1} + \tfrac{1}{2} h\nu\right]. \tag{2.3}$$

The first term in brackets arises from the well known microwave cosmic background, the second from the QED zero-point radiation, taken as real. The second part is Lorentz invariant, the first one is not (indeed, it has been used to estimate the velocity of our Galaxy with respect to the radiation). We may say that the cosmic radiation is the thermal part, and the radiation postulated by SED the zero point part, of the total radiation present is space. The former dominates for low frequencies (macroscopic domain), whil the latter is more important for high frequencies (microscopic domain). We assume the thermal component can be observed with macroscopic devices, but that the zero point component, giving rise to what are called quantum phenomena, cannot.

It was believed for some time that SED could be an alternative to QED, but detailed calculations on the hydrogen atom and other systems(13-15) have shown that SED predictions definitely contradict quantum predictions and empirical evidence.

If SED is to be a real candidate for a theory of microphysics, a number of additions and/or modifications are required. The most obvious need is to include the phenomenon of electron-positron pair creation and annihilation. In its present form, SED is at most a theory intermediate between classical and quantum electrodynamics. Indeed, it has been compared to semiclassical radiation theory.(16) Both consider the radiation as classical, but is SED the particles, too, behave classically and one moreover includes the zero point, which is absent in semiclassical radiation theory.

3. THE UNIVERSAL NOISE

Our aim is not to review stochastic electrodynamics, but to see whether the general ideas involved in that theory provide us with a guide for the search of possible loopholes in the empirical refutation of local realistic theories. With this purpose, we shall study more carefully the essential idea of stochastic theory (ST), which is the lack of isolation of material systems.

The concept of an isolated system is one of the cornerstones of physics. Only in statistical physics is it usual to study systems which are not isolated, in order to allow the exchange of heat or particles. However, when the first models of microscopic systems, e.g. atoms, were elaborated, it was assumed that they should be considered as isolated. It is true that classical systems called isolated are considered so as a first approximation only. The lack of absolute isolation is taken into account afterwards as a perturbation. In any case, it is assumed that the perturbations are small if they are due to distant matter. However, there is a general argument which shows that the main perturbation may come from very large distances. This argument is similar to the one used one and a half centuries ago by Olbers in his celebrated paradox. Olbers showed that if stars were homogeneously distributed and space infinite, the night sky should be bright, instead of dark. The argument is as follows. If all systems emit radiation, the observed density of radiation at a point due to a given source will decrease with the square of the distance R from that source to the observer. On the other hand, the number of sources between R and R + dR increases with R^2. Thus the radiation at a point comes mainly from cosmic sources. The expansion of the universe changes this conclusion but the idea of a cosmic noise remains.

The foregoing argument has been given only in order to show that it is consistent with classical physics to assume that every material system is immersed in a sea of noise. This noise should be taken into account even for the socalled isolated systems. The total noise may consist of different types: electromagnetic radiation, gravitational radiation, electron-positron pairs, neutrinos, etc. Each kind of noise should be represented by an adequate stochastic field filling all of space. In this way we replace the vacuum of classical physics by a universal background of noise, which gives a picture not too different from that of modern quantum field theory.

The details of stochastic theory are not yet developed. Only the part corresponding to electromagnetic radiation has been studied under the name of SED. However, from the qualitative picture given, a number of consequences can be derived. Our experience with SED suggests that the assumption of relativistic invariance might fix the stochastic properties of each type of noise except for some multiplicative constant measuring its intensity. Assuming that a kind of equilibrium exists between the different types of noise, it would not be surprising if this constant were universal (the same for all types of noise) and identifiable with Planck's constant.

The spectrum of the radiation of SED is such that the effect of the radiation is negligible at low frequencies (macroscopic domain), but very important at high frequencies (microscopic domain). We may assume this to be true for every type of noise, and we arrive at the conclusion that the universal noise is important only in the microscopic domain, while in the macroscopic domain conventional classical theories are valid.

A number of quantitative predictions can be made. The universal noise corresponds to the zero point energy of quantum field theory, but considered real instead of virtual. Consequently, we associate an energy $\frac{1}{2}h\nu$ with each degree of freedom of the corresponding field. A material system immersed in this noise may absorb and emit radiation, perhaps of several types. It will therefore appear as if the energy of microsystems is not conserved. A fluctuation in energy associated with a motion of frequency ν may be of the order

$$\Delta E \simeq \frac{1}{2} h\nu , \qquad (3\text{-}1)$$

and it may last for a time interval of the order

$$\Delta t \simeq \nu^{-1} . \qquad (3\text{-}2)$$

The violation of energy conservation therefore fulfils

$$\Delta E \, \Delta t \simeq h/2 \,, \tag{3-3}$$

which agrees with the well known result of quantum theory.

Due to the universal noise, a particle cannot be at rest in an external potential well. It will move in a random way. The particle motion might be coupled with many different frequencies of the noise. Assume that ν is the frequency of the motion for which the coupling is largest. The energy of the particle will then be of the order of the energy associated with a normal mode of the noise having the same frequency , i.e., $\frac{1}{2}h\nu$. We may thus write

$$\tfrac{1}{2}\, m \,\langle \dot{x}^2 \rangle \simeq \tfrac{1}{2}\, h\nu \,, \tag{3-4}$$

where one-dimensional motion is assumed for simplicity. On the other hand, the amplitude of the motion will be of the order of the velocity times half the period, i.e.,

$$\langle x^2 \rangle \simeq \langle \dot{x}^2 \rangle \; (2\nu)^{-2} \tag{3-5}$$

Elimination of ν between these two equations gives

$$\langle x^2 \rangle \, \langle m^2 \dot{x}^2 \rangle \simeq h^2/4 \,. \tag{3-6}$$

This means that if we attempt to confine a particle in a region of size Δx the noise produces a motion with typical momentum $h/\Delta x$. Again, the result is in qualitative concordance with the quantum uncertainty relations.

This is enough to show that the assumption of a universal noise whose properties are relativistically invariant, gives rise to a theory, ST, which has some similarity with quantum theory. When more detailed calculations are made, some predictions are found to be in harmony with quantum mechanics and therefore with experiment, while a disagreement with both is obtained in other cases. It remains to be seen whether the general idea of ST can be mantained and a modification of current models used could save the theory. It should be mentioned that ST has some relation to the ideas of L. de Broglie and others on the existence of waves and particles with a mutual interaction.[17] In ST the waves are represented by the universal noise. The idea of a subquantum thermostat, represented in ST by the noise, has also been advocated by de Broglie.[18]

4. STOCHASTIC THEORY VERSUS QUANTUM THEORY

At present, the agreement of quantum theory with empirical evidence seems perfect. No contradiction has been found between the predictions of QM and experiments. If a competing theory, like ST, is to achieve a comparable status, it should agree with quantum theory rather closely. On the other hand, if we wish to maintain local realism we need a definite departure from quantum mechanics, as shown by Bell's theorem. Worse than that, experiments seem to have refuted all local realistic theories, including ST. Is there any escape from this situation?

In order to use the ideas of ST in searching for a loophole in the refutation of local realistic theories we must analyze what general differences might be expected between ST and quantum theory. It is easy to see that the differences will be of a qualitative rather than a quantitative nature. For instance, if we should develop a model which predicts for hydrogen a line spectrum differing from that of quantum mechanics (as given by the formulas of Balmer, etc.), then the model is disproved by empirical evidence. However, ST might predict the same formula but with a qualitatively different interpretation of what a spectrum is.

A specific example will be enlightening. We consider a molecule not acted on by external forces and neglect its intrinsic angular momentum. According to QM the ground state of the molecule will have spherical symmetry and zero angular momentum. According to ST, the molecule will be acted on by the universal noise, which will produce a random motion. The molecule will be changing its orientation in space, all directions being equally probable due to the isotropy (on the average) of the noise. This could be viewed as a state of spherical symmetry. At any time, the molecule will have a given angular momentum but, as the axis of rotation changes at random, the average angular momentum will be zero, in agreement with the predictions of QM. However, QM predicts an angular momentum that is strictly zero, i.e., dispersionless, whilst the dispersion is positive in ST.

We shall not analyze at this moment whether experiments can distinguish between the predictions of QM and ST in this example. Indeed, a more detailed study of the predictions of ST would be needed. We shall focus on theoretical aspects and start by observing that the main difference between the predictions of QM and ST for a molecule is that the former does not include any noise. This gives rise to the paradoxes typical of QM.

For instance, a naive interpretation of QM suggests that a molecule in the ground state does not rotate, the angular momentum being zero with probability one. However, this seems difficult to reconcile with the prediction of QM that all orientations of the molecule in space have the same probability. The solution of orthodox QM is to say that it is meaningless to speak at the same time about the orientation and the angular momentum of the molecule, these being properties associated with non-commuting observables. This solves the problem from a practical point of view, but it gives rise to formidable epistemological problems, like the difficulty of reconciling QM with realism.

These problems do not appear in ST, where the noise is taken into account. Due to the noise, the molecule is never at rest. Then, there is no contradiction between the fact that the angular momentum is zero on the average and the fact that all orientations in space have the same probability. Indeed, both are quite natural consequences of the isotropy of the noise.

If we adhere to the QM picture of a molecule in the ground state as having angular momentum zero with no dispersion and spherical symmetric, then an intuitive picture in impossible. If, furthermore, the state is regarded as a pure state, in agreement with orthodox QM, then we need a theory of measurement, a reduction of the wave packet, and so on. All difficulties disappear by just adding a noise.

It seems that it is not feasible to develop a stochastic theory which is simple (not contrived) and have predictions agreeing with all those of QM. However, a stochastic theory may be possible whose predictions differ from those of QM in the existence of some amount of noise in the former only. This rather vague statement should be clarified in the future research.

We conjecture that QM may be the remmant of an underlying ST when noise is substracted. Then, how can one explain the truly impressive agreement between QM and experiments?. Should not the noise lacking in QM have been detected in experiments? A partial explanation of why this has not happened may be that the substraction of noise is a standard practice in experimental physics.

5. A LOOPHOLE DUE TO THE SUBSTRACTION OF THE NOISE

We shall now consider atomic cascade experiments, the other empirical tests of Bell inequalities not being conclusive.(1) For the sake of clarity we shall analyze the experiment of Aspect, Grangier, and Roger reported in the first paper of Ref. 2. We choose it because it is the most precise of a series of similar experiments,(19–22) while the other two experiments by Aspect _et al._(2,3) introduce a refinement each.

The experiment deals with an atomic cascade $0 \rightarrow 1 \rightarrow 0$ in the calcium atom, where two photons are emitted. Each photon goes through an analyzer of polarization before reaching a detector. If the angle between the polarizer orientations is ψ, the rate of coincidence counting $R(\psi)$, predicted by quantum mechanics, is

$$R(\psi)/R_0 = 1/4\ (1 + \cos 2\psi)\ , \tag{5-1}$$

where R_0 is the counting rate with the polarizers removed; for simplicity, we have considered ideal polarization efficiencies in Eq.(5-1). Aspect reports a nice agreement of the measured rates with the quantum prediction (5-1) and a violation of Bell's inequality by more than 13 standard deviations. A naïve local hidden-variable (LHV) model is obtained as follows. We assume that the pair of photons emitted by an atom is characterized by the hidden variable λ. Then the probability for coincidence detection should be, according to Bell

$$P(a,b) = \int A(a,\lambda) B(b,\lambda) \rho(\lambda) d\lambda, \quad (5\text{-}2)$$

where $A(a,\lambda)$ is the probability that the first photon of a pair is detected in the first detector when the corresponding polarizer is inserted with an angle a, and similarly for $B(b,\lambda)$ in regard to the second photon.(The first and second photon are distinguished by their wavelengths.) If the polarizers are removed, a relation similar to (5-2) holds, with $A(\lambda)$, $B(\lambda)$ instead of $A(a,\lambda)$, $B(b,\lambda)$. Our model assumes

$$\lambda \in \{0, 2\pi\}, \rho(\lambda) = (2\pi)^{-1}, A(\lambda) = B(\lambda) = \alpha,$$

$$A(a,\lambda) = \beta + \gamma \cos(2a-2\lambda), B(b,\lambda) = \beta + \gamma \cos(2b-2\lambda) \quad (5\text{-}3)$$

The constants α, β, γ fulfil the inequalities

$$\alpha > 0, \beta > \gamma > 0, \quad (5\text{-}4)$$

the third one being a convention, while the first two are necessary for the positivity of the probabilities A and B.

From (5-2) and (5-3) it is straightforward to calculate the ratio between the probability $P(\psi)$ of coincidence detection with the polarizers at an angle $a-b = \psi$ and the corresponding probability P_0 with the polarizers removed; one gets

$$P(\psi)/P_0 = (\beta/\alpha)^2 + \tfrac{1}{2}(\gamma/\alpha)^2 \cos 2\psi. \quad (5\text{-}5)$$

If this probability ratio is identified with the ratio of empirical detection rates, it is easy to see that the LHV prediction given by (5-4) and (5-5) cannot agree with the empirically verified QM prediction (5-1).

A closer examination shows however that it is not so easy to disprove (5-5) with the empirical evidence. Actually, the rates $R(\Psi)$ and R_0 considered by Aspect et al.[2] in (5-1) are not the ones obtained directly, but those which result after a background substraction. If we contemplate a test of QM against LHV models in the spirit of ST, the substraction of the background noise seems impermissible. Note that the main difference between QM and ST is the existence of an universal noise in the latter. If we assume that a part C of the background was improperly substracted in obtaining the rate $R(\psi)$, and similarly for a part D with respect to R_0, we should not identify the ratio (5-5) with (5-1), but ought instead to write

$$\frac{R(\psi) + C}{R_o + D} = (\beta/\alpha)^2 + \tfrac{1}{2}(\gamma/\alpha)^2 \cos 2\psi . \tag{5-6}$$

The empirical result (5-1) and the LHV prediction (5-6) with the constraints (5-4) are compatible for several choices of the constants, for instance

$$C = R_o/4 , \quad D = R_o , \quad \gamma = \beta = \tfrac{1}{2}\alpha . \tag{5-7}$$

Actually, the second equality is not compatible with the data reported by Aspect *et al.*,[2] viz.,

$$R_o = 150 \text{ s}^{-1} , \quad D < 90 \text{ s}^{-1} , \tag{5-8}$$

but other choices are possible. We conclude that there are LHV models which cannot be excluded by the commented experiment.[2]

An alternative form of arriving at the same conclusion is to realize that without background substraction the Bell-type inequalities are not violated. For instance, the first experiment by Aspect *et al.*[2] tests the Freedman inequality,

$$\delta \equiv |R(22.5^o) - R(67.5^o)|/R_o - 1/4 < 0 \tag{5-9}$$

The reported value of δ with background substraction is

$$\delta = (5.72 \pm 0.43) \times 10^{-2} , \tag{5-10}$$

but, without that substraction, the value of δ is

$$\delta \simeq - 0.06 \tag{5-11}$$

which certainly does not violate the inequality (5-9). Similar conclusions can be derived for the other two experiments by Aspect *et al.*[2,3]

It is possible to argue that the substraction of the background is quite a natural procedure and, indeed, if the sustraction is performed, a nice agreement with the quantum mechanical prediction is obtained. The point is that, if we are testing LHV theories, everything in the experiment must be interpreted according to these theories. This is an obvious methodological requirement.

The background is assumed to be due to accidental coincidences, i.e., to joint detection of two photons coming from different atoms. Then, according to LHV theory, the joint detection probability should be given by an expresion similar to (5-2), namely

$$P(a,b) = \int A(a,\lambda)\, B(b,\lambda')\, f(\lambda,\lambda')\, d\lambda\, d\lambda', \tag{5-12}$$

where λ (λ') corresponds to the value of the hidden variable associated with the pair of photons produced in the first (second) atom. The function $f(\lambda, \lambda')$ is a joint probability distribution for two atoms. If we assume that there is no correlation in the excitation and decay of both atoms, then

$$f(\lambda, \lambda') = \rho(\lambda)\ \rho(\lambda') , \tag{5-13}$$

and we obtain that P(a,b) is a constant, independent of a and b, corresponding to a constant background rate of accidental coincidences. It is not difficult to show that this constant rate is the same as that obtained with a delayed coincidence circuit. This should justify the procedure of background substraction used by Aspect <u>et al</u>.[7]

However, it is possible that the excitation and decay of neighbouring atoms is correlated. Then eq. (5-13) does not hold. The maximum correlation is obtained if we assume

$$f(\lambda, \lambda') = \delta(\lambda - \lambda')\ \rho(\lambda) \tag{5-14}$$

If this is put into (5-12), we obtain the same dependence on $a-b = \psi$ for the accidental coincidences in the no-delay channel as for the true ones. Then the background correction should be made by multiplying R_0 and $R(\psi)$ by a fixed factor (smaller than 1) rather than by substraction. The result is that the prediction for Bell-type inequalities such as (5-9) is the same after the correction than before, and no violation of these inequalities is found in Aspect's experiments. It is plaussible to assume that the correct assumption is somewhat intermediate between (5-13) and (5-14), but the essential conclusion does not change.

The usual argument that the rate of accidental coincidences must be the same in a time-delayed channel as in the no-delay channel has no value. In fact, the rate of accidental coincidences is the same for (5-13) and for (5-14) when integrated over ψ. Nevertheless, this does not mean that accidental coincidences must be corrected by substraction. The conclusion is that the violation of Bell's inequalities refutes LHV theories only if the violation takes place without background substraction. As quantum mechanics fits the empirical data only after background substraction, this means that experimental tests of LHV theories are only reliable when the background is very low. In atomic cascade experiments, only the early one by Freedman and Clauser[19] reports a violation of Bell's inequalities, even without background substraction.

In summary, after we have shown that the substraction of the background is an incorrect procedure in the experimental tests of Bell's inequalities, we have the following situation. Amongst the seven atomic cascade experiments[19-22,2,3] performed till now, only one[19] shows a clear violation of the Bell-type inequalities without substraction of the background. On the other hand, also one[20] amongst the seven experiments reports a violation of quantum mechanical predictions. This shows, even if the incorrect substraction of the background would be the only existing loophole,that the refutation of local realistic theories cannot be taken as definite at present.

6. OTHER LOOPHOLES

A well known loophole of all atomic cascade experiments, except the last one by Aspect _et al._,[3] is the possibility of communication between the polarization analyzers in static experiments. Even the last Aspect experiment might be criticized on that basis, due to the fact that the time-varying mechanicsm is not a random one. Thanks to this loophole, the performed experiments allow the possibility of maintaining relativistic locality (or Einstein separability). However we would have to renounce some kind of locality, in the sense that interactions would not decrease with distance, which is not a very pleasant prospect for supporters of local realism.

Much more important is the loophole, reported by Clauser and Horne,[8] due to the low efficiency of photon detectors. This loophole has not been considered as seriously as it should, probably because of the highly artificial nature of the counter example proposed by Clauser and Horne. However, more attractive counter examples could be found. The important point is that all this shows the need for additional assumptions for the derivation of Bell type inequalities in atomic cascade experiments. If an adherent of local realism is confronted with the choice of rejecting either realistic locality or the additional hypotheses, he certainly will prefer the last alternative.

Clauser and Horne[8] derive Bell-type inequalities form local realism plus a "no-enhancement" hypothesis, meaning the assumption that the probability of detecting a photon cannot increase by reason of the photon crossing an analyzer. In the spirit of stochastic electrodynamics (SED), this hypothesis makes no sense, because, according to that theory, photons are only mathematical constructs. The optical signal emitted by an atom is, according to SED, a complex thing, including an electromagnetic wave and a background noise, modified by the emitting atom. The presence of a polarizer should change everything, and it is not easy to know _a priori_ if the probability of a detection event should decrease or increase.

The conclusion of our analysis is: The widespread belief that local realistic theories have been disproved does not fit with the facts. Furthermore, the possibilities of a truly reliable experimental test of these theories in the near future seem remote.[9,10] In consequence, the search for such tests should be a highly recommended activity at this moment.

REFERENCES

1. The situation five years ago was reviewed by J.F. Clauser and A. Shimony, Rep. Progr. Phys. 41, 1881 (1978).
2. A. Aspect, P. Grangier, and G. Roger, Phys. Rev. Lett. 47, 460 (1981); 49 91 (1982).
3. A. Aspect, Phys. Rev. D14, 1944 (1976); A. Aspect, J. Dalibard, and G. Roger, Phys. Rev. Lett. 49, 1804 (1982).
4. F. Selleri and G. Tarozzi, "Quantum Mechanics Reality and Separability," Rivista Nuovo Cimento 4, 1 (1981).
5. B. d'Espagnat, Conceptual Foundations of Quantum Mechanics (Benjamin, Reading, Mass., 1976), 2nd ed.
6. B. d'Espagnat, ed., Foundations of Quantum Mechanics, Proceedings of the International School of Physics Enrico Fermi, Course XLIX (Academic Press, New York, 1971).
7. E. Santos, Anales de Física 78, 1 (1982).
8. F.J. Clauser and M.A. Horne, Phys. Rev. D10, 526 (1974).
9. T.K. Lo and A. Shimony, Phys. Rev. A23, 3003 (1981).
10. E. Santos, submitted to Phys. Rev. A as a comment.
11. P. Braffort, M. Spighel, and C. Tzara, C.R. Acad. Sc. Paris 239, 157 (1954); 239, 925 (1954). See also Ref. 13.
12. T.W. Marshall, Proc. Roy. Soc. (London) 276A, 475 (1963); Proc. Camb. Phil. Soc. 61, 537 (1965); Nuovo Cimento 38, 206 (1965). See also Ref. 13.
13. L. de la Peña, "Stochastic Electrodynamics, its Development, Present Situation and Perspectives. A Tutorial Review" in Stochastic Processes Applied to Physics and other Related Fields, Proceedings of the Latin American School of Physics, Cali, Colombia, 1982 (World Scienctific, Singapore, in press). This review includes a complete bibliography.
14. T.H. Boyer, in Foundations of Radiation Theory and Quantum Electrodynamics, A.O. Barut, ed., (Plenum, New York, 1980).
15. P. Claverie, in Proceedings of the Einstein Centennial Symposium on Fundamental Physics, Bogotá, 1979, S.M. Moore et al., eds. (Universidad de los Andes, Bogotá, 1981).
16. P.W. Milonni, Phys. Rep. (Phys. Lett. C) 25, 1 (1976).
17. L. de Broglie, Une tentative d'interpretation causale et non lineare de la mécanique ondulatoire (Gauthier-Villars, Paris, 1956).
18. L. de Broglie, La thermodynamique de la particule isolée (Gauthier-Villars, Paris, 1964).
19. S.J. Freedman and J.F. Clauser, Phys. Rev. Lett. 28, 938 (1972)
20. R.A. Holt and F.M. Pipkin, Harvard University preprint (1974).
21. J.F. Clauser, Phys. Rev. Lett. 36, 1223 (1976).
22. E.S. Fry and R.C. Thomson, Phys. Rev. Lett. 37, 465 (1976).

NONLOCAL QUANTUM POTENTIAL INTERPRETATION OF RELATIVISTIC ACTIONS AT A DISTANCE IN MANY-BODY PROBLEMS

J.P. VIGIER

Institut Henri Poincaré
Equipe de Recherche Associée au CNRS N°533
11, rue P. et M. Curie, 75231 PARIS CEDEX 05 (France)

ABSTRACT

Relativistic constraints on the behaviour of a system of N scalar particles correlated by causal many-body actions at a distance interactions are analysed in relativistic and quantum mechanics. It is shown 1) that the many-body quantum potential associated to the stochastic interpretation of quantum mechanics yields perfectly causal actions at a distance 2) that the corresponding set of distinguishable particle motions corresponds to Bose-Einstein statistics.

INTRODUCTION

The recent experimental confirmation by Aspect et al. in three successive experiments (1, 2, 3) of the existence of non-local superluminal correlations in Bohm's version (4) of the Einstein-Podolsky-Rosen experiment will evidently deepen the controversy opened by Bell's discovery (5) that quantum mechanics is incompatible with the existence of local hidden variables in the Einstein and de Broglie sense. Physicists and philosophers alike are thus confronted with the need to interpret the new quantum superluminal correlation now established between the results of measurements performed on two singlet-pair photons separated by 12 meters.

As is known, only two relativistic interpretations of this effect have been proposed until now.

The first interpretation, which follows the guidelines of the Copenhagen School, has been developed by Wigner, (6), Costa de Beauregard, (7) et al. (8). It implies retroaction in direct

G. Tarozzi and A. van der Merwe (eds.), Open Questions in Quantum Physics, 297–322.

conflict with one of Einstein's basic assumptions (9,10).

The second interpretation rests on the idea that the observed actions at a distance reflect the action of a modified version of the initial quantum potential model of de Broglie (11) and Bohm (12). Following a proposal of Bohm and Vigier, (13) one can indeed show :

(a) that this quantum potential is a reflection of a stochastic process based on the "zero-point" fluctuations of a subquantal covariant Dirac-type aether (14,15,16) ;

(b) that the new superluminal correlations can be understood as being carried not by individual particles (since no particle can leave the light cone) but by collective (phase-like) motions propagating on the extended "rigid" oscillating elements (17,18,19) that constitute Dirac's aether ;

(c) that for the special case of the two-body problem (both in the spin-0 (20) and spin-1 (21) case) this quantum potential provides an action at a distance which is instantaneous in the center of mass frame of the two bodies, so that one can give a perfectly causal description of the results of the Aspect-Rapisarda experiment (1 , 2 , 3 , 22).

Point (c) is evidently a first step towards a more general problem : Can one provide, within this type of quantum potential interpretation of non-locality, imbedded within the so-called stochastic interpretation of quantum mechanics (SIQM), (23) a consistent causal description of a correlated system of N quantum particles tied together by actions at a distance which appear to act together inseparably in a non-local way ? This question is necessary since, in the words of Bohm and Hiley ("Measurement understood through the quantum potential approach", 1983, unpublished), "in the many-body system, firstly, the interaction of particles with each other is now non-local and, secondly, it is not fixed once and for all, but in general determined by the quantum state of the whole system. Thus the classical model of analysis of a system into separate parts whose relationship are independent of the state as a whole is no longer generally valid."

The aim of the present contribution is to discuss this problem in the simple case of N correlated spin-0 particles; leaving the case of spinning particles for a subsequent publication. We will do this in three steps. In Sections 1 to 3 we discuss the problem of the compatibility of causality with actions at a distance in classical relativistic mechanics. In sections 4 to 7 we then show that the non-local quantum potentials introduced in the N-body case satisfy all the causality requirements imposed by relativity theory. In sections 8 and 9 we integrate this behaviour within the frame of the stochastic interpretation of quantum mechanics and establish that the corresponding set of correlated distinguishable motions yield a concrete realistic justification of Bose-Einstein statistics.

To summarize our work, we shall show how one can generalize this treatment to a system of N spin-0 quantum particles, so that the corresponding many-body quantum potential is the first known real causal non-local physical potential. In the conclusion, we shall briefly discuss the implications of the new result for quantum measurement theory and the nature of quantum reality.

1. CAUSAL RELATIVISTIC N-BODY PROBLEM

According to plan we first summarize the classical relativistic treatment of the N-body problem as presented by Droz-Vincent (24).

As one knows, the essential difficulty tied to this problem is that the N time-like world lines connected by actions at a distance and/or retarded local interactions are not separately described by functions of their own proper times. To explicitly quote Droz-Vincent (25) : "According to the multi-time formalism (25,26), for N point particles have the hamiltonian form

$$\begin{aligned} \partial q_a/\partial\tau_b &= \{q_a, H_b\} , \\ &\vdots \\ \partial p_a/\partial\tau_b &= \{p_a\ H_b\} , \end{aligned} \tag{1.1}$$

where $q_a = q_a(\tau_a \cdots \tau_b)$.

They are generated by the Hamiltonians H_a. The indices $a,b = 1,2\ldots$ *are not* submitted to **Einstein's** summation convention. The unconstrained variables $\ _a,\ _a$ are submitted to the standard Poisson bracket Poisson relations, but q_a do not coincide with the position, in general.

The q_a have the transformation properties of a point in Minkowski space, while p_a have those of a four vector. Whenever possible, the Greek indices are omitted. The signature + − − − is used, $c = \hbar = 1$ and m_a are the masses.

We set

$$P = \sum_a p_a , \qquad z_{ab} = q_a - q_b ,$$

$$P_{ab} = (p_a + p_b)/2 , \qquad y_{ab} = (p_a - p_b)/2 ,$$

We call X_a the vector field generated by H_a in the 8N-dimensional phase space E^{8N}.

The worldlines are defined by a differential system (26)

$$d^2x_a^\alpha/d\tau_a^2 = \xi_a(x_1, \cdots, x_N, v_1, \cdots, v_N) \tag{1.2}$$

where x_a are the positions and

$$v_a = dx_a/d\tau_a .$$

The existence of worldlines requires that the Lie brackets

$[X_a, X_b]$ vanish.

In the case of interactions that are symmetric with respect to particle interchange, this condition is equivalent to

$$\{H_a, H_b\} = 0. \tag{1.3}$$

Equations (1.1) and (1.2) can be considered as equivalent ; however, the correspondence between the variables x, v and the canonical variables q, p rests on the solving of the position equations

$$\{H_a, x_b\} = 0 \quad , \quad a \neq b , \tag{1.4}$$

Equation (1.4) yields x as functions of q,p. Then the generalized accelerations are given by

$$\xi_a = \{H_a, v_a\} , \tag{1.5}$$

where v_a is defined by $v_a = \{H_a, X_a\}$.

Solving (1.4) gives rise to a kind of arbitrariness that does not appear in Gallilean mechanics, where it is possible to take the positions as canonical variables. This is not possible here, (26) but we can reasonably settle the questions by requiring that x_a coincide with q_a on the surface Σ defined by

$$P.z_{ab} = 0 , \tag{1.6}$$

This postulate is mathematically correct, because actually (1.4) is, for each x_a, equivalent to a compatible system of (N - 1) first, order linear partial differential equations, while (1.6), which define Σ, is just equivalent to (N - 1) equations.

Moreover, it is easy to check that Σ is never invariant under the infinitesimal transformations generated from any system of (N - 1) vector fields taken among the X_a.

From a physical point of view, Σ has a simple interpretation : It turns out that, for each point of Σ, all the time coordinates are equal in the center-of-mass frame. In this frame, the Poincaré invariant equation (1.6) reduces to

$$x_a^0 = x_b^0 .$$

Let us call Σ the equal time surface. Note that in general the $\xi_a . v_a$ do not vanish, so the parameters τ_a which permit us, to write the Hamiltonian equations, are generally not the proper times σ_a.

By assuming that $2H_a$ must be identified with the squared mass m_a^2 we fix the relationship :

$$d\tau_a / d\sigma_a = (2H_a / v_a^2)^{1/2} .$$

2. N-BODY POTENTIAL FROM ACTIONS AT A DISTANCE

In Equation (1.2) the generalized accelerations cannot be arbitrarily prescribed. They are submitted to a "predictivity condition" which is stronger (26) than a simple integrability condition *because its role is also to ensure the "individuality" of the particles, namely that each solution* x_a *will be a function of* τ_a *only.* The condition

$$(v_a.\partial_a + \xi_a.\partial/\partial v_a)\xi_b = 0 \quad , \quad a \neq b \tag{2.1}$$

being non-linear, the N-body ξ_a *cannot be written as a simple sum of binary terms*

$$\xi_a = \sum_{b \neq a} \eta_a(x_a, v_a, x_b, v_b) \tag{2.2}$$

where η is the acceleration of a 2-body system.

Expressing the same situation in Hamiltonian language, we see that, if the Hamiltonians have the general form

$$H_a = p_a^2/2 + V_a \tag{2.3}$$

the condition (2.3) cannot be satisfied when the "potentials" V_a are a simple sum of binary potentials, as it is usual in Gallilean mechanics.

Hence, to construct a N-body system from a given binary interaction turns out to be a non-trivial problem in relativistic dynamics.

In this article, we consider only single-potential systems, i.e., all the V_a in (2.3) are taken equal, so that we have

$$H_a = p_a^2/2 + V_a \tag{2.4}$$

The condition (2.3) becomes simply

$$\{V, P_{ab}.y_{ab}\} = 0 \tag{2.5}$$

and the problem is to find V.

This question is completely and explicitly settled for the case, N = 2, because in that case P_{ab} coincides with P, which gives rise to a lot of drastic simplifications. We shall see throughout this paper that all the difficulty (26) for N >2 comes from the fact that $P_{ab} \neq P$.

3. THE EQUAL TIME PRESCRIPTION

Actually (2.5) is a set of (N - 1) linear partial differential equations. We have from (2.4)

$$H_a - H_b = P_{ab}.y_{ab}$$

which shows that among the $P_{ab}\cdot y_{ab}$ we have only (N - 1) independent quantities, with vanishing mutual Poisson brackets.

Thus each solution of (2.5) can be determined by its values on a surface with co-dimension (N - 1) (a submanifold of phase-space determined by (N - 1) equations). The best choice of initial surface seems to be Σ defined by (2.6).

This choice means that, at equal times (with respect to the center-of-mass frame), the potential V is assigned to coincide with some given function. Example : For N = 2, requiring that V coincides with kz^2 at equal times determines that $V = k(z^2-(z.P)^2/P^2)$ on the whole phase space (26).

Naturally, it is essential that Σ *is not* characteristic for the partial differential system (2.5). In other words, Σ is never invariant under the infinitesimal transformations generated by a set of (N - 1) independent linear combinations of the $P_{ab}\cdot y_{ab}$. The proof is presented in Ref. (26).

Finally, the recipe for solving (2.5) is just to prescribe the potential V at equal times.

Of course, Poincaré invariance is ensured by requiring that, on Σ, V coincides with a Poincaré-invariant function. In general this potential has nothing to do with a binary interaction. But it is natural to consider the most simple situation where V is constructed by a sort of composition of binary interactions.

In the general case of a single-potential model, as defined by (2.4) and (2.5), we can exhibit a useful combination of H_a and P^2 which is an interesting constant of the motion.

Let us first notice some identities. On the one hand, we have almost obviously

$$\sum_{a<b} (p_a^2 + p_b^2) = (N - 1) \sum p_a^2 \; . \tag{3.1}$$

On the other hand, as can be easily checked by recurrence, direct computation yields

$$P^2 = \sum p_a + 2 \sum_{a<b} p_a^2 p_b^2 \; , \tag{3.2a}$$

$$4y_{ab}^2 = p_a^2 + p_b^2 - 2p_a p_b \; . \tag{3.2b}$$

Thus

$$P^2 + 4 \sum_{a<b} y_{ab}^2 = \sum p_a^2 + \sum_{a<b} (p_a^2 + p_b^2)$$

and by (3.2) we finally have

$$P^2 + 4 \sum_{a<b} y_{ab}^2 = N \sum p_a^2 \; : \tag{3.3}$$

which holds identically on the whole phase space, whatever the model. Now we can take (3.3) into account. Then, by addition,

$$2\, N\Sigma H_a = N\Sigma p_a^2 + 2\, N^2 V \, , \tag{3.4}$$

Insertion of (3.3) into (3.4) yields

$$2\, N \sum H_a^2 = P^2 + 4 \sum_{a<b} y_{ab}^2 + 2\, N^2 V \, . \tag{3.5}$$

Finally, by substraction of P^2, we obtain the combination

$$2\, N \sum H_a^2 - P^2 = 4 \sum_{a<b} y_{ab}^2 + 2\, N^2 V \, , \tag{3.6}$$

which remains constant in the motion. This is the N-body form of a quantity that we have already met in previous works about two-body dynamics (27,28).

Let us emphasize that *only in the special* $N = 2$ *case*, where $y_{12}.P$ is conserved., it is also possible to substract from (3.2) the contribution of $(y_{12}.P)^2/P^2$ and to obtain a simpler constant of the motion related to the relative energy.

In the case of $N > 2$, it would be useless splitting y_{ab}^2 by substraction of $(y_{ab}.P)^2/P^2$,since $y_{ab}\, .\, P$ is not constant. In other words, the orthogonal decomposition of y_{ab} relative to P provides no simplification.

The decomposition relative to P_{ab} would not be better in general, but, in the *case of equal masses*, all the $y_{ab}.P_{ab}$ vanish which permits the statement that all the y_{ab} are purely spacelike, provided an appropriate assumption is added. This does not allow us to define a conserved relative energy, as we did for N = 2, but permits a discussion involving the sign of V.

4. N-BODY WAVE EQUATIONS

By the most straightforward correspondence principle,

$$p_a^\alpha \to \partial/\partial x_a^\alpha \, , \tag{4.1}$$

$$q_a^\alpha \to \text{multiplication by } x_a^\alpha \, , \tag{4.2}$$

the Hamiltonians become operators acting on a wave function $\Psi(x_1, \ldots, x_n)$.
More precisely, 2Ha becomes a squared-mass operator, and we write the wave equation (27)

$$2\, H_a \Psi = m_a \Psi \, . \tag{4.3}$$

This procedure comes out very naturally since it provides N Klein-Gordon equations when applied to the free particle Hamiltonians $p_a^2/2$.

The Poisson bracket condition (1.3) has now the quantum mechanical counterpart

$$[H_a, H_b] = 0 \quad , \tag{4.4}$$

which insures the compatibility of (4.3). Actually (4.4) is stronger than the compatibility condition, this feature owing to the fact that our picture is supposed to admit a classical analog with worldlines.

5. THE KLEIN-GORDON SYSTEM

We consider a system of N free scalar relativistic particles obeying the laws of Quantum Mechanics. This can be described by a wave function $\Psi(q_1^\mu \cdots q_N^\mu)$ submitted to N Klein-Gordon equations (with $\hbar = c = 1$) i.e.

$$(\Box_i + m_i^2)\Psi(q_1^\mu \cdots q_N^\mu) = 0 \quad , \tag{5.1}$$

with $\Box_i = (\partial/\partial q_i^\mu)(\partial/\partial q_{i\mu})$.

We can now generally write Ψ in the form

$$\Psi = \exp\,(\, Q + iW\,) \ . \tag{5.2}$$

As a next step, in a free N-particles system we can consider the case of a Ψ eigenstate of the operator $P_\mu = i\Sigma\partial_{i\mu}$ and separate thereby the motion of the center of mass, which follows a straight world line in the absence of external interactions. We thus write

$$\Psi = \exp\left(iK^\mu \frac{q_{1\mu}\, m_1 + \cdots + q_{N\mu}\, m_N}{m_1 + \ldots + m_N}\right) \phi\,(z_{ij}^\mu, K^\mu) \tag{5.3}$$

where K is a constant timelike vector, z_{ij}^μ denote the relative coordinates $z_{ij}^\mu = q_i^\mu - q_j^\mu$ and $\Phi = \exp(Q + iW)$ represents the wave function of the relative motion in the center-of-mass frame, so that

$$Q = Q(z_{ij}^\mu,\ K^\mu) \quad , \tag{5.4}$$

$$W = w(z_{ij}^\mu, K^\mu) + K^\mu \ \frac{\sum_i m_i q_{i\mu}}{\sum_i m_i} \ . \tag{5.5}$$

The foregoing magnitudes obviously satisfy the following relations

$$(\partial_1^\mu + \cdots + \partial_N^\mu)Q = 0 \quad , \tag{5.6}$$

$$(\partial_1^\mu + \cdots + \partial_N^\mu)W = K^\mu \ , \tag{5.7}$$

which imply that the relative motions are independent of the position $X = (\sum_i q_i m_i)/(\sum_i m_i)$ of the center of mass in space-time.

If we now introduce eq.(5.2) into eq.(5.1) we obtain two sets of equations for the real and imaginary parts

$$\Box_i Q + \partial_i^\mu Q \partial_{i\mu} Q - \partial_i^\mu W \partial_{i\mu} W + m_i^2 = 0 \quad , \tag{5.8}$$

$$2\, \partial_i^\mu Q \partial_{i\mu} Q + \Box_i W = 0 \quad . \tag{5.9}$$

By introducing in (5.8) the Quantum potential $U_i = -(\Box_i Q + \partial_i^\mu Q \partial_{i\mu} Q)/2$ this equation can be rewritten in the form

$$(\partial_i^\mu W \partial_{i\mu} W)/2 + U_i = m_i^2/2 \quad . \tag{5.10}$$

This equation is formally similar to a set of Hamilton-Jacobi equations describing the motion of each individual particle. This formal resemblance will be examined in the next section.

6. HAMILTON-JACOBI EQUATIONS IN RELATIVISTIC DYNAMICS

The scheme of Predictive Mechanics [25] is essentially a multitime formalism, i.e. the motion equations for N point particles involve N independent parameters τ_i, which are suitable generalizations of the proper times. These equations are generated by N covariant Hamiltonians Hi submitted to the predictivity conditions

$$\{H_i, H_j\} = 0 \quad ; \tag{6.1}$$

which ensures that world lines can be actually defined. The N-body phase space admits canonical coordinates q_i^μ, p_j^ν submitted to the standard Poisson brackets

$$\{q_i^\mu, p_{j\nu}\} = \delta_{ij}\, \delta_\nu^\mu \quad , \tag{6.2}$$

The Hamiltonians are obviously constant in the motion and their numerical values are identified with the masses. Accordingly, the conventional Hamilton-Jacobi theory can be transposed in this covariant scheme, as follows : the Hamilton-Jacobi equation is replaced by a system of N partial differential equations where the (half squared) masses $m_i^2/2$ now replace the energy.
In other words the Hamilton principal function S satisfies [25]

$$H_i(q, \frac{\partial S}{\partial q}) + \frac{\partial S}{\partial \tau_i} = 0 \quad . \tag{6.3}$$

We can also introduce Hamilton's characteristic function W^x, assuming that S is of the form

$$S = W^x - (\sum_i m_i^2 \tau_i)/2 \quad , \tag{6.4}$$

where W does not depend on the variables τ_i. Thus (6.3) becomes

$$H_i(q, \frac{\partial W^*}{\partial q}) = m_i^2/2 \quad . \tag{6.5}$$

Let us now assume that each H_i takes the form of free terms completed with additive interaction terms which are generally non local potentials, i.e.

$$H_i = (p_i^\mu p_{i\mu})/2 + V_i(q_i^\mu, p_i^\mu) = m_i^2/2 \quad . \tag{6.6}$$

In order to solve this system we seek canonical transformations to new variables ($\hat{q},\hat{p}$) which trivialize the motion equation, i.e. such that $\hat{p}$ = const. The generating function that produces the desired transformation is $W^*(q, \hat{p})$. Herewith eq. (6.6) takes the explicit form

$$(\partial_i^\mu W^x \partial_{i\mu} W^x)/2 + V_i(q_i^\mu, \partial_i^\mu W^x) = m_i^2/2 \quad . \tag{6.7}$$

As well known [24] a complete solution of this equation contains 3N non-trivial constants of integration which together with the N m_i form a set of 4N independent constants. The condition $\hat{p}_i^2 = m_i^2$ just reflects the fact that the variables $\hat{q}$, $\hat{p}$ cast the equations of motion in the free-form, so that $\hat{p}_i^2 = m_i^2$ represent first integrals of the motion.
In this space characterized by the independent coordinates q, $\hat{p}$ the generating function W (q, $\hat{p}$) defines a transformation by the formulae

$$p_i^\mu = \frac{\partial W^x}{\partial q_i^\mu}(q_i^\mu, \hat{p}_i^\mu) \quad , \quad \hat{q}_i^\mu = \frac{\partial W^x}{\partial p_i^\mu}(q_i^\mu, \hat{p}_i^\mu) \quad . \tag{6.8}$$

One should stress at this point that equations (6.8) are used to determine $\hat{p}$ and $\hat{q}$ out of <u>unconstrained</u> p and q. Therefore the dependence on $\hat{p}$ in the W^x function should be such that no constraints on p and q should result out of eq.(6.8). One can easily check e.g. that every degeneracy in the $\hat{p}$ dependence inevitably leads to constraints that are undesirable in the frame of Hamilton-Jacobi theory.
Exactly this point turns out to be of great importance in what follows. In fact we intend to show that the originally examined Klein-Gordon system can be considered as a special type of motions imbedded within a general suitable Hamilton-Jacobi system in phase space. But it can be easily seen that such a limitation of canonical phase space motions onto our initial motion cannot be performed in a straightforward way. Indeed, a direct way of mapping could be to impose on W^* a dependence only on the $\Sigma_i \hat{p}_i^\mu$ instead of the general one, and subsequently identify the $\Sigma_i \hat{p}_i^\mu$ with K^μ in W^*. This would guarantee the mathematical analogy between the two systems (KG and HJ) if the transformations of eq.(6.8) could be solved with unconstrained p and q. But it turns out, as can be

easily verified, that the system of eq.(6. 8) is highly singular and admits no solution at all.

A way out of this difficulty goes along the following lines :

(a) One extends the set of variables of the Klein-Gordon system by introducing $k_1^\mu \cdots k_N^\mu$ with the condition $k_1^\nu + \cdots + k_N^\nu = K^\nu$

(b) The phase part of the wave function W depends now on the total set of variables

$$\tilde{W} = K^\mu \frac{q_{1\mu} m_1 + \ldots + q_{N\mu} m_N}{m_1 + \ldots + m_N} + W(q_i^\mu - q_j^\mu, K^\mu) + F(q_i^\mu, k_i^\mu) \quad (6.9)$$

where $F(q_i^\mu, k_i^\mu)$ is completely arbitrary in phase space:with the only boundary condition that it takes the value

$$F(q_i^\mu, k_i^\mu) = 2\pi n \qquad (6.10)$$

on the "Klein-Gordon surface" in phase space.

(c) One can now assume that W has exactly the same dependence as $\tilde{W}$ of eq.(6. 9) provided one sets $\hat{p}_i^\mu = k_i^\mu$. This $W^x(q_i^\mu, k_i^\mu)$ meets the requirement that it yields k_i^μ as functions of the unconstrained p and q canonical variables in eq.(.) and overcomes the singularity of the dependence only on $K^\mu = \sum_i k_i^\mu$. Furthermore

(d) One can easily show that for this specific choice of W the canonical total momentum P of the Hamilton-Jacobi system is equal to the constant timelike vector K on the Klein-Gordon surface. In fact, one can calculate

$$P^\mu = \sum_{i=1}^{n} \hat{p}_i^\mu = \sum_i \partial_i^\mu W^x(q, \hat{p})$$

using the assumed identity $p = k$.

Then one has

$$\partial_i^\mu W^x = \partial_i^\mu \tilde{W} = K^\mu (m_i / \sum_i m_i) + \frac{\partial w}{\partial q_i^\mu}(q_i^\mu - q_j^\mu, K^\mu) + \frac{\partial F}{\partial q_i^\mu}(q_i^\mu, k_i^\mu) \quad (6.11)$$

On the right hand side of eq.(6.11) the last term vanishes identically since F is a constant on the Klein-Gordon surface. Furthermore the $\sum_i \partial_i w$ averages out to zero, since they contain always contributions $\partial_i^\mu w$ and $\partial_j^\mu w$ which are equal and of opposite sign. Thus, we finally get a relation that implies $p^\mu = \hat{p}^\mu$ on the Klein-Gordon surface.

As a consequence, the transformation equation and the mapping of our initial motion by a phase space motion of an N-particles system admits an infinity of solutions, since F(q, p) is arbitrary up to boundary conditions on the Klein-Gordon surface of motion. Furthermore we wish to stress that this arbitrariness does not affect the Klein-Gordon system because the boundary condition ensures the uniqueness of the wave function

$$\Psi = \exp(Q + iW) = \exp(Q + iW + i2\pi n) \, . \qquad (6.12)$$

Thus we are allowed to set

$$W = W^{x} \qquad \text{and} \qquad U = V \tag{6.13}$$

on the Klein-Gordon surface. This also enables us in the following to consider the classical system defined abstractly by the Hamiltonians

$$H_i = (p_i^{\mu} p_{i\mu})/2 + V_i(q_i^{\mu}, p_i^{\mu}) \quad , \tag{6.14}$$

provided one satisfies the predictivity conditions.
It has been shown [25] that this predictivity condition implies that in the Hamiltonian system the potentials cannot be chosen arbitrarily and ensures the existence of world lines. This means that the N particles system

(a) possesses time-like paths,

(b) can be solved in the forward (backward) time direction in the sense of the Cauchy problem.

This condition can be written in the form [25]

$$\{H_i, H_j\} = 0 \quad , \quad \text{for all} \quad i,j : \tag{6.15}$$

where $\{\cdot,\cdot\}$ denote the usual relativistic Poisson brackets.
With $H_i = H_{oi} + V_i$ we have

$$\begin{aligned} \{H_i, H_j\} &= \{H_{oi}, H_{oj}\} + \{V_i, H_{oj}\} + \{H_{oi}, V_j\} + \{V_i, V_j\} \\ &= \{H_{oi}, V_j\} - \{H_{oj}, V_i\} \quad ; \end{aligned} \tag{6.16}$$

since $\{H_{oi}, H_{oj}\}$, $\{V_i, V_j\}$ vanish identically. Writing (6.16) then explicitly yields

$$\{H_i, H_j\} = p_i^{\mu}(\partial V_j/\partial q_i^{\mu}) - p_j(\partial V_i/\partial q_j^{\mu}) \, . \tag{6.17}$$

With eq.(6. 7) we get

$$\partial V_j/\partial q_i^{\mu} = -\tfrac{1}{2}\, \partial_i^{\mu} (\partial_{j\nu} W \partial_j^{\nu} W) = -\partial_{j\nu} W(\partial_i^{\mu} \partial_j^{\nu} W) \quad , \tag{6.18}$$

and respectively

$$\partial V / \partial q^{\mu} = - \partial_{j\nu} W(\partial_j^{\mu} \partial_i^{\nu} W) \quad , \tag{6.19}$$

so that eq.(6. 17) can be rewritten in the form

$$\{H_i, H_j\} = p_i^{\mu} \partial_{j\nu} W(\partial_i^{\mu} \partial_j^{\nu} W) - p_j^{\mu} \partial_{i\nu} W(\partial_j^{\mu} \partial_i^{\nu} W) \, . \tag{6.20}$$

Taking into account that μ and ν are dummy indices, this is immediately transformed into the new form

$$\{H_i, H_j\} = p_i^{\mu} \partial_{j\nu} W(\partial_i^{\mu} \partial_j^{\nu} W) - p_j^{\nu} \partial_{i\mu} W(\partial_j^{\nu} \partial_i^{\mu} W) \, . \tag{6.21}$$

If we now use the general property of equality of the second mixed partial derivatives of many variables functions, i.e.

$$\partial^{\nu}_{j}\partial^{\mu}_{i}W = \partial^{\mu}_{i}\partial^{\nu}_{j}W \quad , \tag{6.22}$$

and the transformation eq.(6.8), i.e. $p^{\mu}_{i} = \partial^{\mu}_{i}W$, we immediately recognize that eq.(6.21) vanishes identically and we finally obtain

$$\{H_i, H_j\} = \partial^{\mu}_{i}W\partial_{j\nu}W(\partial^{\mu}_{i}\partial^{\nu}_{j}W - \partial^{\nu}_{j}\partial^{\mu}_{i}W) = 0 \quad : \tag{6.23}$$

a relation that is satisfied in the whole of phase space.

We have thus succeeded in the search for a covariant and mathematically well-defined mapping which relates the free-particle wave-function with an hypothetical system of classical particles which undergo mutual interactions.

(a) We do not claim that this correspondence is unique. Besides the well-known ambiguities of the covariant hamiltonian formalism (possibilities of an alternative to the equal-time condition) our identification of $\tilde{W}$ in eq.(6.9) with the Hamilton characteristic function present in eq.(6.7) is evidently not mathematically unique. But it is anyway the most natural generalization of the conventional procedure [29] which gives rise to the concept of quantum potential in the non relativistic case.

(b) The <u>global</u> timelikeness of the classical trajectories is not automatically ensured but, in principle, should be checked. As the definition of these world lines rests on the equal-time condition, checking this property would be a good test for the adequacy of the condition.

(c) The interaction present in the classical picture we have constructed is by no means local. Indeed, locality is exceptional in the relativistic theories of action at-a-distance. But causality (in the sense of a Cauchy problem) is guaranteed.

(d) We shall not discuss here the physical meaning of this hypothetical system of classical particles.

The question whether the classical trajectories have a certain meaning out of a suitable averaging procedure on the quantum mechanical system and/or from a stochastic interpretation has an old story in the non relativistic case and is still a subject of controversy, in spite of encouraging results [30]. The present work provides a solid ground for the continuation of this old debate in the domain of special relativity.

Also it might have some connection with recent progress towards a relativistic formulation of the stochastic interpretation [30].

7. DISCUSSION OF PARTICULAR N-BODY PROBLEM SOLUTIONS

At this stage one can only remark that the N-body problem should be solved explicitly in every particular case where one adds to the Quantum Potentials U_l local or non-local interactions Vij.

As shown in particular by Takabayasi [31] the system of N point particles, bound by harmonic non local forces, is important for many reasons, including its relevance to hadronic models. We shall briefly summarize here his reasoning, since it a typical treatment of N-body behavior :
A relativistic motion of the system is represented in classical theory by world lines W^α ($\alpha = 1, 2, \ldots, N$). We consider a one-parameter sequence of spacelike hypersurfaces $\{\Sigma_\tau\}$ and denote the intersections of each Σ_τ with W^α as $x^\alpha_\mu(\tau)$. (The choice of $\{\Sigma_\tau\}$is arbitrary for the present, but will be fixed below). We then define the geometrical center of mass $X_\mu(\tau) = \Sigma x^\alpha_\mu(\tau)/N$ ($\alpha = 1,\cdot\cdot,N-1$) and relative coordinates with $r = 1,\cdot\cdot,{}^\alpha N-1$, $\xi^r_\mu(\tau) = \sqrt{(2/N)}\, \sum_\alpha \cos[\pi r(2\alpha-1)/2N]\,.x^\alpha_\mu(\tau)$, Further we introduce

$$Y^\alpha_\mu = \sqrt{(2/N)}\, \sum_r \sin(\frac{\pi r(2\alpha-1)}{2N}).r\xi^r_\mu \quad ,$$

which satisfy $\Sigma_\alpha(-1)^\alpha Y^\alpha_\mu = 0$ to postulate the action principle $\delta \int_{\tau_2}^{\tau_1} Ld\tau = 0$, with $\dot{x}^\alpha_\mu = dx^\alpha_\mu(\tau)/d\tau$, $(\dot{x}^\alpha)^2 = \Sigma_i(\dot{x}^\alpha_i)^2-(\dot{x}^\alpha_o)^2$ etc... and

$$L = -\,\Sigma_\alpha[-(\dot{x}^\alpha)^2\{\kappa^2(y^\alpha)^2 + m_o^2\} + \kappa^2(\dot{x}^\alpha y^\alpha)^2]^{1/2}. \quad (7.1)$$

From (7.1) we get the definition of the canonical momenta $p^\alpha_\mu(\tau)$ conjugate to $x^\alpha_\mu(\tau)$ and find that they satisfy the following primary constraints :

$$(p^\alpha)^2 + \kappa^2(y^\alpha)^2 + (\kappa^2/m_o^2)(p^\alpha y^\alpha)^2 + m_o^2 = 0 \quad . \quad (7.2)$$

Also (7.1) gives Euler equations, and the 4-momentum of the system is $P_\mu = \Sigma_\alpha p^\alpha_\mu(\tau)$. In the adiabatic force-free limit $\kappa\to 0$, (7.2) tends to $(p^\alpha)^2 + m_o^2 = 0$, meaning that each constituent has the common limiting mass m_o.

We now put the conditions $\dot{x}^\alpha(\tau)\, y^\alpha(\tau) = 0$, that is,

$$p^\alpha y^\alpha = 0 \quad (\alpha = 1,2,\cdots,N) \quad . \quad (7.3)$$

These work as the conditions suppressing the arbitrariness in the identification of the equal τ points on W^α's, besides implying one physical constraint. Then (7.2) simplifies to

$$(p^\alpha)^2 + \kappa^2(y^\alpha)^2 + m_o^2 = 0 \quad . \quad (7.4)$$

In the special case $m_o = 0$, (7.3) and (7.4) directly result from (7. 1) as primary constraints. This means that in this case the theory is invariant under arbitrary transformation of $\{\Sigma_\tau\}$) By (7. 3) and (7. 4) P is ensured to be timelike. From (7.4) we obtain

$$H = K + U = 0. \tag{7.5}$$

where $K = \frac{1}{2}\Sigma_{\alpha=1}^{N}(p^\alpha)^2 = \frac{1}{2}N\,P^2 + \frac{1}{2}\Sigma_{r=1}^{N-1}(\Pi^r)^2$ with $\Pi_\mu^r(\tau) = \sqrt{2/N}\,\Sigma_{\alpha=1}^{N}\cos[\pi r(2\alpha-1)/2N]\,.p_\mu^\alpha(\tau)$ and $U = \frac{1}{2}k^2\Sigma_{\alpha=1}^{N}(y^\alpha)^2 + U_0 = \frac{1}{2}k^2\Sigma_{r=1}^{N-1}r^2(\xi^r)^2 + U_0$ with $U_0 = \frac{1}{2}Nm_0^2$. Essentially H works as a Hamiltonian that determines the τ-evolution of the system. The The potential function U is reexpressed as $U=(1/8)k^2\Sigma_{\alpha,\beta=1}^{N}F_{\alpha\beta}\cdot(x^\alpha(\tau)-x^\beta(\tau))^2 + U_0$ with

$$F_{\alpha\beta} = -(-1)^{\alpha-\beta}\left\{\mathrm{cosec}^2\left(\frac{\pi(\alpha-\beta)}{2N}\right) - \mathrm{cosec}^2\left(\frac{\pi(\alpha+\beta-1)}{2N}\right)\right\}$$

which shows that in our model strong attractive harmonic forces work between 'nearest neighbors' ($|\alpha-\beta| = 1$), weaker repulsive ones between second nearest neighbors ($|\alpha-\beta| = 2$), and so forth [31]. The eqs. (7.3) and (7.4) contain, besides (7.5), (2N-1) constraints, and they further lead to (2N-3) secondary constraints. Thus we have totally (4N-3) constraints, among which H is of first class while other are of second class. It is convenient to introduce $C_\mu^r(\tau) = \sqrt{(N/2)}(\Pi_\mu^r(\tau)-i\kappa r\xi_\mu^r(\tau))$ where $r = 0,\pm 1,\dots,\pm(N-1)$, and $\Pi^{-r}_\mu = \Pi^r_\mu$, $\xi^{-r}_\mu = \xi^r_\mu$. They satisfy the Poisson bracket relations $\{C_\mu^r(\tau), C_\nu^s(\tau)\} = ik'r\delta_{r+s,0}g_{\mu\nu}$ $(k' = Nk)$. Then all the constraints are just represented by

$$\Lambda^r = \frac{1}{2}\sum_{n=r-N+1}^{N-1} C^n(\tau)C^{r-n}(\tau) + \frac{1}{2}N^2m_o^2\delta_{ro} = 0$$

$$(r = 0, 1, 2, \cdots, 2N-2)\,, \quad (\Lambda^0 = NH)\,. \tag{7.6}$$

They satisfy the closed algebra $\{\Lambda^r,\Lambda^s\} = -i\,\kappa'(r-s)\Lambda^{r+s}$, with the definition that $\Lambda^r = 0$ for $r \geqslant 2N-1$.

Classical solutions are given explicitly as

$$C_\mu^r(\tau) = C_\mu^r(0)e^{-ir\kappa'\lambda_o\tau}\,, \quad X_\mu(\tau) = P_\mu\lambda_o\tau + X_\mu(0)\,,$$

where λ_o is a constant, while $C_\mu^r(0)$ are constant complex vectors satisfying the constraints (7.6) : $\Lambda^r(0) = 0$.

Quantization is effected by replac g Poisson brackets by commutators and by putting $\Lambda^r\psi = 0$ $(r = 0, 1, \dots, 2N-2)$, where we assume indefinite metric regarding relative-time degrees.

The N=3 case and the large N case are especially interesting. The former has appropriate properties as a simple trilocal quark model for baryonic states. To consider the latter case we employ the parameter $\sigma = \pi\alpha/N$ and write $x_\mu^\alpha(\tau) = x_\mu(\sigma,\tau)$ Also we define $P_\mu(\sigma,\tau) = (N/\pi)P_\mu^\alpha(\tau)$, so that in the limit $N\to\infty$ we have $\{x_\mu(\sigma,\tau), P_\nu(\sigma',\tau)\} = g_{\mu\nu}\,\delta(\sigma-\sigma')$ and $P_\mu = \int_0^\pi P_\mu(\sigma,\tau)\,d\sigma$ The essential point is that in this limit the multilocal quantity

y^{α}_{μ} just goes over to the local quantity $-\partial x_\mu/\partial\sigma$, and accordingly the alternating harmonic forces to contiguous tension. Since we keep κ' and $\omega = (Nm_o/\kappa')^2$ finite, the Lagrangian (7.1) goes over to

$$L = -\frac{\kappa}{\pi} \int_0^\pi d\sigma[-(x)^2\{(\frac{\partial x}{\partial\sigma})^2 + \omega\} + (x \frac{\partial x}{\partial\sigma})^2]^{1/2}, \quad (7.7)$$

which is exactly the Lagrangian for the 'massive' relativistic string . The constraints go over to $\Lambda^r = (1/2)\Sigma^{\infty}_{n=\infty}C^n C^{r-n} + \frac{1}{2}k'^2\omega\delta_{r0}$ $(r = 0, 1, 2 \ldots \infty)$. The corresponding quantum theory is just the same as the one formerly given by 'detailed wave equation' [42].

All demonstrations given in Secs. 5 and 6 are less satisfactory than the general demonstration of causality given in [20] [21] in the two-particle case. It seems that for $N > 2$ one cannot avoid the introduction of constraints to ensure the causal character of N-body solutions, except in a few particular cases. These constraints however seem quite natural and compatible with all-known physical situations. Indeed, as will be shown in a subsequent publication, the usual wave fields utilized in the many-body problem in configuration space satisfy in general both the unipotential condition and the causality constraints (1.3).

Our causality constraints yield the same basic physical properties as the N=2 case, since one sees
- that the paths are time-like ;
- that since the proper time τ of the center of mass can be written $\tau = (\Sigma^N_{\alpha=1}\tau\alpha)/N$ and if we denote the derivatives $d/d\tau$ we have $\dot{P} = 0$, so that, since $P.z_{ab} = P(x_a - x_b) = 0$, we get $P.(v_a - v_b) = Py_{ab} = 0$.

A relation which shows that no energy can be exchanged between pairs of particles in the rest frame and the actions at a distance binding any pair of particles are instantaneous in the rest frame system ;
- that all the correlated world lines of the system are independent of the initial conditions measured at $t = 0$ for any inertial observer on the corresponding space-like surfaces, so that the collective N-body system of actions at a distance given by the quantum potential yields the same motions for all possible sets of inertial initial conditions. The latter is a crucial property which, as was stressed by J. Bell [33], must be satisfied if one wants to preserve Einstein's definition of causality in relativity theory.

8. STOCHASTIC INTERPRETATION OF THE N-BODY MOTION IN CONFIGURATION SPACE

We conclude this intervention with a brief discussion of the interpretation of the wave equation [29, 30] in the stochastic interpretation of quantum mechanics. Our starting point is just the N-particle generalization in configuration space of the one-particle model discussed by Lehr and Park [34], Guerra-Ruggiero [35], and Vigier [36]. Indeed, let us assume N identical scalar particles imbedded in Dirac's stochastic aether [14, 15]. Any correlated particle motion along their world lines in real space time builds a general world line in a 4N-dimensional configuration space, so that the set of all possible correlated motions can be represented by a fluid in this space-time. These motions are not independent (since the presence of any particle a disturbs the "aether", i.e., the motion of any particle b, and vice versa), and one assumes that we are dealing (as in the one-particle case) with stochastic jumps at the velocity of light (in physical space-time) which transfer any pair a-b from one drift line of flow (in configuration space) to another. Physically this amounts (in the hydrodynamical model of Bohm and Vigier [37]) to the superposition in space-time of N interacting fluids which undergo light-like internal stochastic motions, particle-antiparticle transitions, and possible number-preserving transfers from one fluid to another - so that we have a conserved scalar fluid particle density in configuration space.

Mathematically this model can thus be described (Cufaro-Petroni and Vigier [38]) in a 4N-dimensional configuration space where a pair position is defined by a 4N-component vector X

$$\{X^i\} \equiv \{X_1^\mu, \cdots, X_N^\mu\} \qquad (i = 1, \cdots, 4N) \quad (\mu,\nu=0,1,2,3) \tag{8.1}$$

with $X_1^\mu, \cdots, X_N^\mu$ four-vectors of the position of each body. The metric is defined by

$$g_{ij} = \begin{vmatrix} 1 & 0 & 0 & 0 & & 0 & 0 & 0 & 0 \\ 0 & -1 & 0 & 0 & & 0 & 0 & 0 & 0 \\ 0 & 0 & -1 & 0 & & 0 & 0 & 0 & 0 \\ 0 & 0 & 0 & -1 & & 0 & 0 & 0 & 0 \\ & & & & \ddots & & & & \\ 0 & 0 & 0 & 0 & & 1 & 0 & 0 & 0 \\ 0 & 0 & 0 & 0 & & 0 & -1 & 0 & 0 \\ 0 & 0 & 0 & 0 & & 0 & 0 & -1 & 0 \\ 0 & 0 & 0 & 0 & & 0 & 0 & 0 & -1 \end{vmatrix}$$

so that

$$X = \sum_{i=1}^{N} X_i X^i = g_{ij} X^i X^j = \sum_{i=1}^{N} (X_i^\mu X_{i\mu}) \tag{8.2}$$

If $X_1^\mu(\tau_1), \ldots, X_N^\mu(\tau_N)$ are the trajectories for the N-particles, then the trajectory in configuration space will be an $X^i(\tau_1, \cdots, \tau_N)$. As a consequence of Nelson's equation [39] we can now generalize the differential operators defined by Guerra and Ruggiero [23], for the single-particle case to a system of N identical particles :

$$D = \sum_{i=1}^{N} \partial/\partial\tau_i + b^i\partial_i \,, \quad \delta D = \delta b_i \partial^i - \frac{\hbar}{2m}\Box \,,$$

$$\partial_i = \partial/\partial x^i \,, \quad \Box = \partial_i\partial^i = \sum_{i=1}^{N} \Box_i \quad (i=1,\cdots,4N) \,,$$

$$b_i = DX_i \quad \text{and} \quad \delta b_i = \delta DX_i$$

Now a direct extension of Guerra and Ruggiero's [23] formulas gives the following dependence of δb_i on a density $(X',\tau_1, \cdots ,\tau_N)$:

$$\delta b_i = -\frac{\hbar}{m}\partial_i \log \rho^{1/2} \,, \tag{8.3}$$

where for the density we have as continuity equations

$$\frac{\partial\rho}{\partial\tau_i} = -\partial_{i\mu}(\rho b_i^{\mu}) \,, \tag{8.4}$$

valid for all i with

$$b_i = DX_i \quad \text{and} \quad \partial_i = \frac{\partial}{\partial x_{i\mu}}$$

In our model, as a generalization of the assumption that ρ is independent of the proper time in the one-body case, we advance the physical hypothesis that the total number of particles (i.e., pair in the real space-time) is conserved ; thus we write

$$\sum_{i=1}^{N} \frac{\partial\rho}{\partial\tau_i} = 0 \,, \tag{8.5}$$

so that our continuity equations in configuration space are

$$\partial_i(\rho b^i) = 0 \,. \tag{8.6}$$

We assume as before (Guerra and Ruggiero [35], Vigier [36], Lehr and Park [34]) that our fluid motions are irrotational so that

$$b^i = \frac{1}{m}\partial^i W \,, \tag{8.7}$$

where $\Phi(X^i,\tau_1,\cdots,\tau_N)$ is a phase function and, if we look for a steady state (i.e., proper time-independent) equation,

$$\Phi(X^i,\tau_1,\cdots,\tau_N) = \frac{mc}{2}\sum_{i=1}^{N}\tau_i + W(X^i) \,. \tag{8.8}$$

Now it is clear that (as generally assumed and later demonstrated by Cufaro-Petroni and Vigier [38], Newton's equations for the N-free particles can be written in the compact form

$$(DD - \delta D\,\delta D)X^i = 0 \,. \tag{8.9}$$

Starting from (8.6) and (8.7) and using (8.8), (8.3) and (8.9), we obtain a Hamilton-Jacobi type equation ($R = \rho^{1/2}$) for our N-body system, i.e.

$$(\partial_i \partial^i - \frac{\partial_i W \partial^i W}{\hbar} - N \frac{m^2 c^2}{\hbar}) R = 0 \quad ; \qquad (8.10)$$

which yields for the continuity equation the form

$$2\ \partial_i R\ \partial^i W = R\ \partial_i \partial^i W = 0 \quad . \qquad (8.11)$$

Finally, if we consider (8.10) as the real part and (8.4) as the imaginary part, the total equation for $\Psi = R \exp(iW/\hbar)$ is

$$(\Box - N \frac{m^2 c^2}{\hbar^2}) \Psi = 0 \quad , \qquad (8.12)$$

which is equivalent to relation (8.12). Moreover, the model implies the subsidiary condition (2.4), since in the rest frame of particles a and b we have, by the principle of action and reaction, $P_a = -P_b$, i.e.,

$$P_a P_a - P_b P_b = (P_a + P_b)(P_a - P_b) = 0$$

which are equivalent to (2.4) if R and W are symmetric permutations, so that $U_a = U_b$.

9. THE CAUSAL NON-LOCAL CHARACTER OF QUANTUM STATISTICS

As one knows [40], the usual way of introducing the classical Maxwell-Boltzmann statistics is founded on the assumption of a basic classical property, namely the distinguishability of classical particles. N classical particles are distinguishable, they can be labeled by an index K (K = 1 ... N), and the probability of distributing them in a set of M available states i (i = 1 ... M) (where each state has an equal probability weight 1/M) can be calculated. The probability $P(\{n_i\})$ of having n particles in state i, can then be easily calculated to be

$$P(\{n_i\}) = M^{-n} \frac{N!}{n_1! \cdots n_N!} \quad . \qquad (9.1)$$

On the other hand, one can in principle justify the utilization of quantum Bose-Einstein statistics, by the new property of bosons of being indistinguishable. In that case the probability of distributing N bosons in M states with equal probability weight for each state, can be easily evaluated to be

$$P(\{n_i\}) = \frac{N!(M-1)!}{(N+M-1)!} \quad . \qquad (9.2)$$

Of course the widespread opinion that M.B. and B.E. statistics differ only through the (in)distinguishability of their elements has the well-known merit of presenting this difference in a concise calculable form. It raises however evident epistemological difficulties. Indeed any set of identical localizable particles, which move on different chaotic world lines in the frame of relativistic space-time, are evidently distinguishable in principle ... so that B.E. statistics are at first sight not describable in terms of real subquantal localizable motions i.e. in terms of local hidden parameters. This suggests that bosons are entities which do not move permanently in space-time (in conformity with Bohr's view that micro elements are waves <u>or</u> particles never the two simultaneously) and their stochastic behaviour cannot be understood in terms of subquantal random motions. As we shall now show this judgement is not correct. Locality certainly implies distinguishability, but the reverse is not true ... and one can have new forms of stochastic processes corresponding to distinguishable well localized particles in space-time, if correlated by non-local causal actions at a distance.

To clarify this essential point let us first recall [40] the three postulates which underly classical M.B. statistics i.e.
(1) each statistical element is distinguishable from all others at least by its individual time-like world line ;
(2) all elements are submitted to random <u>local</u> interactions ... so that there is no action at a distance between spatially separated pairs of elements.;
(3) <u>no correlation</u> exists between two colliding molecules, so that the distribution function is independent of the local collision model used.

Now in the author's opinion any attempt to describe quantum statistics as resulting from the existence of real hidden parameters must preserve the principle of distinguishable particles, if it wants to account for quantum phenomena without abandoning the existence of their motion in space-time. Indeed as pointed out by de Muynck [41] each statistical element has internal (e.g. spin, electromagnetic charge etc.) and external (e.g. position, momenta) parameters, so that the notion of identical particles refers to intrinsic properties whereas the notion of distinguishability adheres to external properties, or "dynamically relevant valiables" [40]. In any realistic model, even identical particles become distinguishable in principle by means of their external motion parameters in space-time.

We shall now show that assumptions (2) and (3) of classical statistics, which imply local non-correlated states, are untenable in the frame of non-local hidden variables models. Indeed a set of spinless particles   (K = 1 ,..., N) is described by a single wave-field.

$$\Psi(\cdots,x_\mu^K,\cdots) = R(\cdots,x_\mu^K,\cdots)\ \exp(iS(\cdots,x_\mu^K,\cdots) \qquad (9.3)$$

in configuration space (which represents K real correlated waves moving in real space-time [51]) and their average motions $v_\mu^K = (1/m)\ \partial_\mu^K S$ are no longer submitted to random local correlations but also tied by a permanent non-local action at a distance described by their quantum potential

$$Q = \sum_K \left(\frac{\hbar}{2m}\right) \frac{\Box_K R}{R} = \sum_K Q_K \qquad (9.4)$$

This quantum potential acts instantaneously in their common center of mass' rest frame and implies that this interaction is causal, since their individual Hamiltonians $H_K = (1/m) p^\mu_K p^\mu_K + Q_K \alpha$ satisfy the causality constraints $H_K , H_J = 0$. This fact can be summarized in two new assumptions i.e. :

(2') all statistical elements are submitted to random non-local causal interactions mediated by a causal many-body quantum potential. This implies that our subquantal stochastic model presents striking differences from all those formerly used, since it has been shown, that the quantum potential does not decrease in general with distance. This action at a distance implies a permanent correlation of all colliding molecules with the overall ensemble, so that they cannot be considered as independent (or free) at any time. The definition of their state thus depends on the ensemble as a whole, via the quantum potential Q so that

(3') any pair of "colliding" particles is always non-locally correlated with the ensemble as a whole.

Evidently assumptions (2') and (3') imply that

(a) each individual state is not identical with all others

(b) each state is permanently "disturbed" by its non-local correlation with the ensemble.

If we want to describe in this case the process of probability weighting of such quantum states, we have to assume

(A) that the probability weights ω_i of the states are not in general equal but change at random in the process ;

(B) that they depend on all former possible random motions in phase space which reflect the real stochastic motions associated with a random Dirac-type of subquantal aether [14].

This implies that the probability weights are subject only to the usual probability restrictions

$$0 \leqslant \omega_i \leqslant 1 \qquad \text{with} \qquad \sum_{i=1}^{M} \omega_i = 1 \qquad (9.5)$$

and that, in order to get the total statistics, one has to average over all possible ω_i in all possible configurations.

This clearly justifies the starting point of Tersoff and Bayer [42] which they only attempted to justify with a mathematical economy arguement, i.e. that in the absence of prior knowledge it is more natural (i.e. entails a weaker assumption) to assume all probability weights of the states as arbitrary and random, rather their equal.

The results of the averaging process over a Maxwell-Boltzmann classical statistics with unequal random weights has been performed by Tersoff and Bayer [42]. In order to calculate an expression of the form

$$P(\{n_i\}) = \int_0^1 \dots \int_0^1 d\omega_1 \dots d\omega_M \frac{N!}{n_1! \dots n_M!} \omega_1^{n_1} \cdots \omega_M^{n_M} \delta(1-\sum_{i=1}^{M}\omega_i), \quad (9.6)$$

one has to evaluate this integration in M steps, incorporating the $\Sigma_{i=1}^{M} \omega_1 = 1$ constraint in every step.

The first step is just the δ-function contribution

$$\int_0^1 d\omega_M \; \omega^{n_M} \delta(1-\sum_{i=1}^{M} \omega_i) = (1-\sum_{i=1}^{M-1}\omega_i)^{n_M}. \quad (9.7)$$

The second integration gives

$$\int_0^{1-\Sigma_1^{M-2}\omega_i} \omega_{M-1}^{n_{M-1}} \left[\sum_{L=1}^{M-2} \omega_i) - \omega_{M-1} \right]^{n_M} d\omega_{M-1}$$
$$= (1-\sum_{L=1}^{M-2}\omega_i)^{n_{M-1}+n_M+1} \frac{n_{M-1}!\,n_M!}{(n_{M-1}+n_M+1)!} \quad (9.8)$$

Clearly the first term in this result should be now integrated in the third step, the second being a multiplying constant. Now the third integration can be written as follows :

$$\int_0^{1 \sum^{M-2} \omega_i} \omega_{M-2}^{n_{M-2}} \left[(1-\sum_{L=1}^{M-3} \omega_i)-\omega_{M-2}\right]^{n_{M-1}+n_M+1} d\omega_{M-2} \quad (9.9)$$
$$= (1-\sum_{i=1}^{M-3}\omega_i)^{n_{M-2}+n_{M-1}+n_M+2} \frac{n_{M-2}!\,(n_{M-1}+n_M+1)!}{(n_{M-2}+n_{M-1}+n_M+2)!}$$

which again gives a factor, that produces an integral of the same type in the fourth step, plus a multiplying factor. Of course this procedure can be iterated up to the Mth step which yields the integral :

$$\int_0^1 \omega_1^{n_1}(1-\omega_2)^{n_2+\dots+n_M+M-2} d\omega_1 = \frac{n_1!(n_2+n_3+\dots n_M+M-2)!}{(n_1+\dots n_M+M-1)!} . \quad (9.10)$$

The total result of the integration is now a product of all factors resulting from each one of the M steps, i.e. :

$$\frac{n_{M-1}!\,n_M!}{(n_{M-1}+n_{M+1})!} \cdot \frac{n_{M-2}!\,(n_{M-1}+n_M+1)!}{(n_{M-2})+n_{M-1}+n_{M+2})!} \dots \frac{n_1!\,(n_2+\dots+n_M+M-2)!}{(n_1+n_2+\dots+n_M+M+1)!} \quad (9.11)$$

and one sees that all factors cancel except for the product of $n_1!$ on the numerator and the $(n_1+\dots+n_M+M-1)!$ in the denominator.

We thus obtain, finally :

$$\frac{n_1!n_2!\ldots n_M!}{(N+M-1)!}$$

Of course to write the expression for the probability distribution one has to take into account that the
constraint that was incorporated in the integration limits can be satisfied in (M-1) ! ways, corresponding to the number of permutations of the M-1 weights . The integration result must accordingly be multiplied by (M-1) ! We thus get for P

$$P(\{n_i\}) = \frac{N!}{n_1!..n_M!}\ \frac{n_1!..n_M!}{(N+M-1)!}\ (M-1)! = \frac{N!(M-1)!}{(N+M-1)!}\ , \qquad (9.12)$$

which is in fact equivalent to the Bose-Einstein result. This is independent of the configurations , the latter being the cause for the introduction of the notion of indistinguishability. As Tershoff and Bayer [42] have shown there is thus no absolute necessity to introduce such an intuitively incomprehensible notion, in order to recover Bose-Einstein statistics for a set of particles. However if one indeed justifies the random probability weights by means of non-local causal action at a distance correlations of particles imbedded in Dirac's stochastic covariant aether, one recovers, in addition to the formal mathematical picture, a physical picture that permits also an intuitive understanding of the origins and striking differences of Bose-Einstein quantum statistics with Maxwell-Boltzmann classical statistics. Instead of playing with independent dices (as in M.B. statistics) B.E. statistics results from action at a distance correlated dices. This statistical difference is a manifestation of the fundamental non-local character of the quantum phenomena, a fact established by the non local property of the quantum potential. This physical picture is simple if one accepts the idea that the so-called "vacuum" is really filled with the violent subquantal random motions, as suggested by Dirac [14], Bohm-Vigier [13], and de Broglie [11]. In this model particles disturb their neighborhood and interact non-locally via the stochastic quantum potential.

CONCLUSION

One can conclude this paper with three remarks on the properties and consequences of the action at a distance associated with the quantum potential in the stochastic interpretation of quantum mechanics.

The first is that, if one assumes that these non-local quantum potentials are carried by a real subquantum covariant stochastic medium of the Dirac type, there is no reason to assume that their action extends to infinity (as results from the usual formalism) but, as assumed by Baracca et al. [43], it might be that the "vacuum's chaos" implies that

$$\Psi(\cdots x_a, \cdots x_b, \cdots) \rightarrow \cdots \Psi_a(x_a) \cdots \Psi_b(x_b) \cdots$$

beyond a certain distance, so that one recovers Einstein's locality (and Bell's inequalities) for macroscopic distances.

The second remark is that the process of measurement probably follows the model presented by Prof. Cini in this conference, since the stochastic interpretation implies that one should drop the Bohrian concept of wave packet collapse, so that "empty" waves not containing particles just melt into the "vacuum noise" of Dirac's aether.

The third remark is that the very existence of non-local correlations, varied by a quantum potential which reflects the vacuum distorsion the particle and its wave, implies

- (1) the reality of the de Broglie pilot waves surrounding the particle aspect of matter (microobjects are waves and particles both in Einstein's and de Broglie's views), and
- (2) the possibility to test experimentally the existence of such pilot waves, as discussed actually both in optics [44] and in neutron interferometry [45]. Crucial experiments are evidently forthcoming on that question. If these experiments confirm that de Broglie's waves exist independently of particles, one will not be able to avoid the question of the nature of their subquantal carrier.

In my point of view, after the experimental discovery of action at a distance, the crucial question in physics becomes the problem of the real nature of the chaotic subquantal covariant vacuum reintroduced into physics by Dirac in 1951. I thus cannot resist concluding this report with a quotation of Isaac Newton (founder of modern deterministic-realistic physics) which perfectly describes the problem confronting the tenents of a causal stochastic description of quantum mechanics, which include actions at a distance :

"that gravity should be innate, inherent and essential to matter, so that one body may act upon another at a distance through a vacuum without the mediation of any thing else, by and through which their action and force may be conveyed from one to another, is to me so great an absurdity that I believe no man who has in philosophical matters a competent faculty of thinking, can ever fall into it".

Acknowledgements. The author of this brief and incomplete report on the N-body problem is deeply thankful to Drs Droz-Vincent and Cufaro-Petroni for many discussions, suggestions and material help in its preparation. Common work with Cufaro-Petroni is summarized in Sec. 7 , and essential parts of Droz-Vincent's demonstration are reproduced in Secs. 2 to 6 . Sec. 8 contains a demonstration by Prof. Takabayasi, a well-known pioneer both in this field and in the causal interpretation of quantum mechanics. The author also wants to thank Dr. J.S. Bell for enlightening discussions on the causality problem in N-body configurations and Dr. Kyprianidis for common research on the material in sections 6 and 9.

REFERENCES

1. A. Aspect, P. Grangier and G. Roger, Phys. Rev. Lett., 47, 460, (1981).
2. A. Aspect, P. Grangier and G. Roger, Phys. Rev. Lett., 49, 91, (1982).
3. A. Aspect, J. Dalibard and G. Roger, Phys. Rev. Lett., 49, 1804, (1982).
4. D. Bohm, Quantum Theory (Prentice Hall, Englewood Cliffs, N.J., 1951).
5. J.S. Bell, Physics, 1, 195, (1964).
6. E.P. Wigner, Symmetries and Reflections (M.I.T. Press, Cambridge, Mass. 1971).
7. O. Costa de Beauregard, Phys. Rev. Lett.,
8. H.P. Stapp, Nuovo Cim., B40, 151, (1977).
9. F. Selleri and J.P. Vigier, Nuovo Cim. Lett., 29, 7, (1980) ; A. Garuccio et al., Nuovo Cim. Lett., 27, 60, (1980).
10. J.P. Vigier, Astr. Nachr., 303, 55, (1982).
11. L. de Broglie, Une interprétation causale et non linéaire de la mécanique quantique, Gauthier-Villars, Paris, (1972).
12. D. Bohm, Phys. Rev., 85, 166, 180, (1952) ; Phys. Rev., 89, 458, (1953).
13. D. Bohm and J.P. Vigier, Phys. Rev., 96, 208, (1954) ; Phys. Rev.,109,1882, (1958).
14. P.A.M. Dirac, Nature, 168, 906, (1951).
15. J.P. Vigier, Nuovo Cim. Lett., 29, 467, (1980) ; see also N. Cufaro-Petroni and J.P. Vigier, Found. Phys., 12, 253,(1982).
16. N. Cufaro-Petroni and J.P. Vigier, Found. Phys., 13, 253,(1983).
17. P. Guéret, P. Mérat and J.P. Vigier, Lett. Math. Phys., 3, 47, (1979) ; see also Maric et al., Nuovo Cim. Lett., 29, 65, (1980).
18. J.P. Vigier, Nuovo Cim. Lett., 24, 258, 265, (1979).
19. N. Cufaro-Petroni and J.P. Vigier, in Old and New Questions in Physics, Cosmology, Philosophy, and Theoretical Biology, A. van der Merwe, ed. (Plenum, New-York, 1983).
20. N. Cufaro-Petroni, P. Droz-Vincent and J.P. Vigier, Nuovo Cim. Lett., 31, 415, (1981).
21. N. Cufaro-Petroni and J.P. Vigier, Phys. Lett., 93A, 383,(1983).
22. F. Falciglia, G. Iaci and V.A. Rapisarda, Nuovo Cim. Lett., 26, 327, (1979) ; L. Papalardo and V.A. Rapisarda, Nuovo Cim., A65, 269, (1981) ; A. Garuccio and V.A. Rapisarda, Nuovo Cim., A65, 269, (1981).
23. See references (13) and E. Nelson, Phys. Rev., 150, 1079, (1966) ; W. Lehr and J. Park, J. Math. Phys., 18, 1235,(1977); F. Guerra and P. Ruggiero, Nuovo Cim. Lett., 23, 529, (1978); N. Cufaro-Petroni and J.P. Vigier, Phys. Lett., 73A, 289, (1979), and 81A, 12, (1981) ; Kh. Namsrai, Found. Phys., 10, 353, 731, (1980).

24. P. Droz-Vincent, Ann. Inst. H. Poincaré, 27, 407, (1977) ; Phys. Rev., D19, 702, (1979).
25. P. Droz-Vincent, Ann. Inst. H. Poincaré, 32, 377, (198).
26. P. Droz-Vincent, Nuovo Cim. Lett., 1, 839, (1969) ; Physica Scripta, 2, 129, (1970).
27. P. Droz-Vincent, Rep. Math. Phys., 8, 79, (1975).
28. P. Droz-Vincent, C.R. Acad. Sc. (Paris), A182, (1979) ; For N=2, see, e.g., R.P. Feynman, K. Kislinger and R. Ravnel, Phys. Rev., D3, 2706, (1971) ; Y.S. Kim and M.L. Noz, Phys. Rev., D15, 33 (1977) ; J.F. Gunion and F. Li, Phys. Rev., D12, 3583, (1975) ; H.W. Crater, Phys. Rev., D16, 1580, (1977).
29. L. de Broglie, C.R. Acad. Sc. (Paris), 184, 273, (1927) ; 185, 380, (1927) ; D. Bohm, Phys. Rev., 85, 166, 180, (1952); D. Bohm and J.P. Vigier, Phys. Rev., 96, 208, (1954).
30. W. Lehr and J. Park, J. Math. Phys., 18, 1235, (1977) ; F. Guerra and P. Ruggiero, Nuovo Cim. Lett., 23, 529, (1978); J.P. Vigier, Nuovo Cim. Lett., 24, 265, (1979) ; N. Cufaro-Petroni and J.P. Vigier, Random motions at the velocity of light and relativistic quantum mechanics, Inst. H. Poincaré, preprint (1983), J. Phys. (1983) in press.
31. T. Takabayasi, Prog. Theor. Phys., 43, 1117, (1970) ; T. Takabayasi, Prog. Theor. Phys., 51, 262, 571, (1974).
32. T. Takabayasi, Prog. Theor. Phys., 44, 1429, (1970).
33. J.S. Bell, private communication.
34. W. Lehr and J. Park, J. Math. Phys., 18, 1235, (1977).
35. F. Guerra and P. Ruggiero, Nuovo Cim. Lett., 23, 529, (1978).
36. J.P. Vigier, Nuovo Cim. Lett., 24, 258, 265, (1979).
37. D. Bohm and J.P. Vigier, Phys. Rev., 96, 208, (1954) ; 109, 882, (1958).
38. N. Cufaro-Petroni and J.P. Vigier, Nuovo Cim. Lett., 26, 149, (1979).
39. E. Nelson, Phys. Rev., 150, 1079, (1966).
40. Z. Maric, M. Bozic and D. Davidovic, Randomness and determinism in the kinetic equations of Clausius and Boltzmann. Proceedings of the Boltzmann meeting, Vienna (1981).
41. W.M. de Muynck, Int. J. Th. Phys., 14, 327, (1975).
42. J. Tersoff and D. Bayer, Phys. Rev. Lett., 50, 553, (1983).
43. A. Baracca, D. Bohm, B.J. Hiley and A.E.G. Stuart, Nuovo Cim., 28B, 453, (1975).
44. A. Garuccio, K. Popper and J.P. Vigier, Phys. Lett., 86A, 397, (1981) ; A. Garuccio, V.A. Rapisarda and J.P. Vigier, Phys. Lett., 90A, 17, (1982) ; A. Gozzini, in The wave-particle dualism, S. Diner et al., eds. (Reidel, Dordrecht, 1983).
45. J. Andrade e Sylva, F. Selleri and J.P. Vigier, Nuovo Cim. Lett., 36, 503, (1983) ; F. Selleri, Found. Phys., 12, 1087, (1982) ; H. Rauch, A. Wilfing, W. Bauspiess and U. Bonse, Z. Physik, B29, 281, ((1978) ; see also H. Rauch's report in this book.

Part 3

The Realistic Interpretation of the Wave Function: Experimental Tests on the Wave-Particle Dualism

THIRD KIND MEASUREMENTS AND WAVE-PARTICLE DUALISM

A. Garuccio

Dipartimento di Fisica dell'Universita' di Bari
I.N.F.N. Sezione di Bari, Italy

ABSTRACT
Modern experimental apparatuses allow us to duplicate photons. Some experiments, based on this process, have been proposed for testing the existence of de Broglie's empty wave and the possibility of a description in space and time of microphysics.

1. INTRODUCTION

W.Pauli in an article on measurement theory[1] introduced the following definitions:

M1. A measurement is said to be of the first kind if it leaves the measured state in an eigenstate of the measured eigenvalue.

M2. A measurement is said of the second type if it leaves the measured state in some other eigenstate.

In a recent paper[2], N.Herbert introduced a new kind of measurement defined as follows:

M3. A measurement of third kind is one that duplicates the measured state exactly.

If the measured state is a photon state,it is possible to perform measurements of the third kind using the process of stimulated emission of light in which the emitted photon is

G. Tarozzi and A. van der Merwe (eds.), Open Questions in Quantum Physics, 325–331.

"identical" to the impinging one.

The aim of this paper is to discuss some experiments concerning the foundations of quantum mechanics that are based on this process.

2. WHAT IS A DUPLICATOR?

The first duplicator proposed was a laser gain tube (LGT), which is a laser tube maintained below threshold, i.e., basically a system of atoms in a steady state such that the population of the excited level is greater than that of the lower state. Clearly this system emits light because of the spontaneous emission of excited atoms.

Professor Gozzini[3] proposed the use of a nonlinear process involving three photons, which has been developed in Doppler-free spectroscopy (DFS).

A two-level system g,e, whose energetic separation is $\hbar\omega_o$, is irradiated with two lasers beams having frequencies $\omega_1 = \omega_o + \delta$ and $\omega_2 = \omega_o + 2\delta$, respectively, so as to satisfy the resonance condition $2\omega_1 - \omega_2 = \omega_o$.

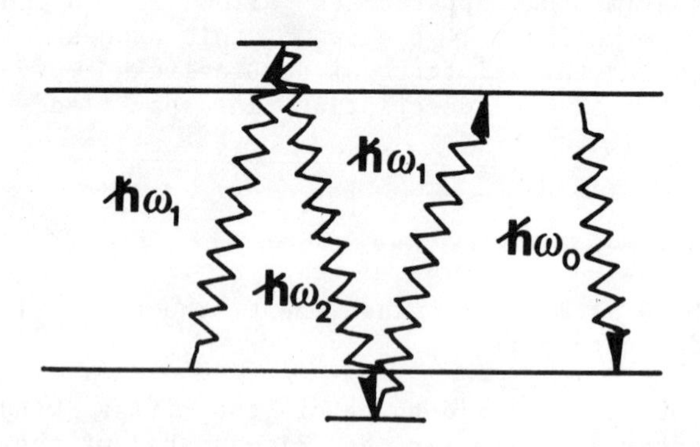

Fig. 1. Doppler-free spectroscopy scheme.

The system can absorb a photon ω_1, emit a stimulated photon ω_2, and absorb a new photon ω_1, passing to the exitated level e. Then it decays from the the level e by a spontaneous emission $\hbar\omega_o$.

A light amplifier utilizing such a process presents with respect to a LGT these advantages:

a) possibility of focussing the beams on a very small volume;

b) low noise level;

c) possibility, by an appropriate choice of δ, of filtering the frequency signal ω_2 from ω_1 and ω_o;

d) possibility to use the emitted $\hbar\omega_o$-energy phothon to discriminate between unamplified photons and amplified photons.

3. DESCRIPTION IN SPACE AND TIME

Let us now consider the application of a third kind measurement to the problem of a the fundamental dualistic nature of elementary entities.

It is possible to imagine that associated with each beam of particles is an "empty wave" that does not carry energy or momentum (Einstein's "Gespensterfelder") or that carries only a very small fraction of the available energy (de Broglie's "pilot wave"). This wave could however produce some physical process, specifically a change in the transition probability or the interference pattern.

If an experiment should prove the existence of a process produced by an "empty wave", then the reality of this wave would be proved.

Let us now discuss some experiments which have been proposed in order to prove the reality of empty waves and thus the possibility of a description in the space and time of microphysics.

4. SELLERI'S EXPERIMENT

The goal of Selleri's experiment[4] is to prove the reality of empty waves via the stimulated emission of light.

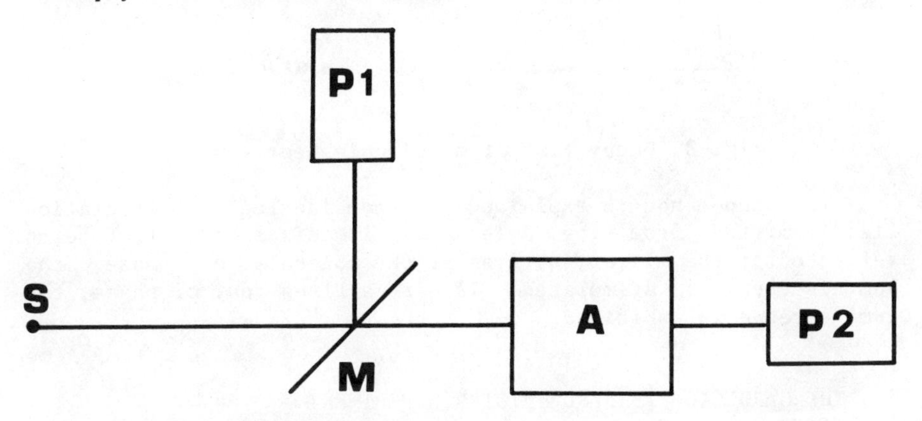

Fig. 2. Set-up of Selleri's experiment.

A very low intensity beam emitted by the source S (see Fig. 2) is split by the beam splitter M into two orthogonal beams. On the reflected beam a photomultiplier P1 is located. The transmitted beam traverses an ~~amplificator~~ amplifier A and impinges on a second photomultiplier P2. The amplifier consists of a number of

atomic systems in an excited level.
It Is known that the photon can be localized(5) in the arm M-P1 or in the arm M-A. Therefore, if P1 reveals the arrival of a photon there are no photons in the transmitted beam, and the empty wave can reveal its existence by generating in A a stimulated emission. In this way, a P1-P2 coincidence rate greater than the accidental background rate would prove the propagation of a zero-energy undulatory phenomenon, transmitted by the beamsplitter M, and the existence of real empty waves.

But can an empty wave influence the stimulated emission of light?

An indirect answer to this question comes from experiments(6) on the decay time of the luminescence of a molecule in front of a metal mirror.
Figure 3 shows the ratio between the decay rate τ_d of molecule at distance d from mirror and the decay rate τ_∞ at infinite distance as a function of the distance; the decay time depends markedly on the distance between the molecule and the mirror.

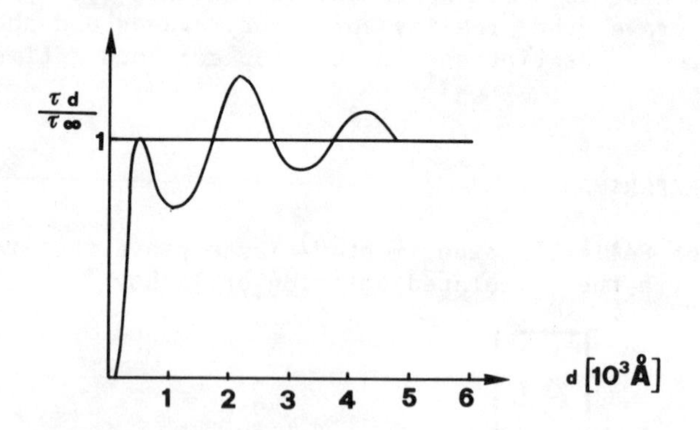

Fig. 3. Decay time of a molecule near a mirror.

The phenomenon is explained by considering the radiation field emitted from the molecule. If this field, after being reflected at the mirror, arrives at the molecule on phase, the luminescence is stimulated, if it arrives out of phase, the luminescence is inhibited.

5. THE GARUCCIO-RAPISARDA-VIGIER EXPERIMENT

The experiment (Fig. 4) proposed by these authors(7) consits of a Mach-Zehnder interferometer in which a mirror is replaced by the beam splitter M2. The photomultipliers PA and PB are placed in the constructive and the destructive interference regions, respectively, and the duplicator A is placed along the arm M1-M2. The source emits a very low-intensity beam and the outputs of photomultipliers PA and PB are analyzed in coincidence with PC.

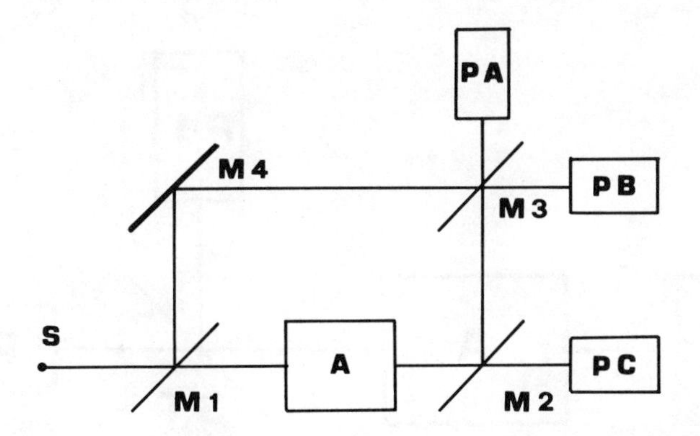

Fig. 4. Set-up of Garuccio-Rapisarda-Vigier experiment.

Let us now suppose that the empty wave does not produce stimulated emission. Under this assumtion, if a photon is reflected from M1, it is detected on PA or PB and we have no coincidence; whereas if it is transmitted , it is duplicated in A and two photons will impinge on M2.

If both photons are either transmitted or reflected from M2, then we have no coincidence; a coincidence occurs only when a photon is transmitted and a photon is reflected.

But if a photon is detected by PC, we can conclude that no photon is traveling along the M1-M4-M3 path. Then, in the Copenhagen interpretation of quantum mechanics, the wave packet has been reduced and no interference should appear in the M3 region (i.e., the number of PC-PA coincidences must be equal to the number of the PC-PA coincidences).

If an empty wave exists, it travels along the M1-M4-M3 path and in M3 overlaps whith the not-empty wave traveling along M1-M2-M3, producing interference (i.e., PA-PC coincidences alone).

6. MARTINOLLI-GOZZINI EXPERIMENT

Martinolli and Gozzini in Pisa performed an experiment with a photon amplifier. The goal of this experiment was to test the behaviour of two photons in the same quantum state when they impinge on a semitransmitting mirror.

The sketch of their apparatus is shown in Fig. 5. The light beam emitted from the source S (a sodium lamp) is attenuated down to one photon at a time in the apparatus (10^6 photons/sec). Then it is amplified by A (a rodhamine cell excited by a laser). The semitransmitting mirror ST splits the amplified beam into two orthogonal beams, which are detected by the two photomultipliers

P1 and P2. The outputs of the photomultipliers are analyzed in coincidence.

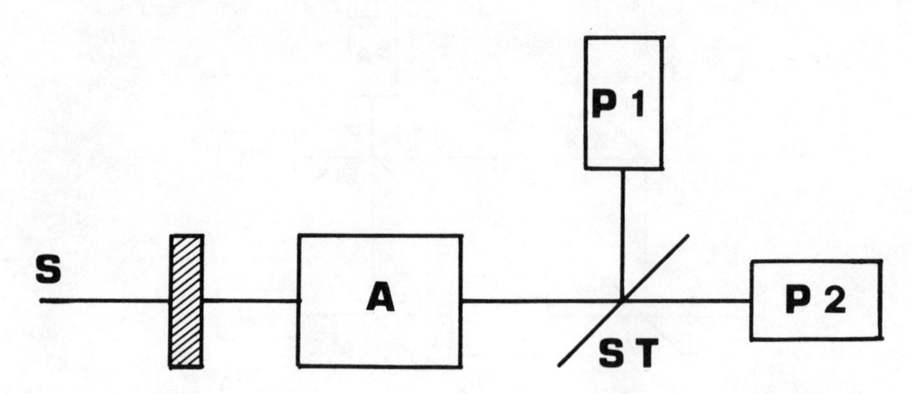

Fig. 5. Set-up of the Martinolli-Gozzini experiment.

Let us now suppose that when a photon impinges on A, it is duplicated and therefore two photons will travel towards the semitransmitting mirror.

Two different results are then possible:

(a) The photons are not only in the same quantum state, but also have the same "hidden variable". Therfore they make the same choice on ST (whether both transmitted, or both reflected) and there are no coincidences between the photomultipliers P1 and P2.

(b) The photons are only in the same quantum state and the P1-P2 coincidence rate is different from zero.

In the actual experiment not all the photons can be duplicated (the amplificacation rate is $\phi = 1.2$) and there is an accidental coincidence background rate. The predicted coincidence number in case (a) therefore is equal to the accidental coincidence number, which in the Martinolli-Gozzini experiment is $N_a = 310 \pm 15$,whereas in case (b) it is $N_b = 1910 \pm 615$. The experiment yields 1071 ± 30 , which is not compatible with either hypoteses (a) or (b).

It is important to note that, in the computation of the coincidence number N_b , Martinolli and Gozzini used the hypothesis that the photons impinging on ST behave like classical particles and that each photon has the independent probability 1/2 of being reflected or transmitted (Fig. 6).
This hypothesis led them to assign the probability 1/4 to the configurations (a) and (b) of Fig. 6 and the probability 1/2 to the configuration (c) and to over-estimate the number of coincidences.

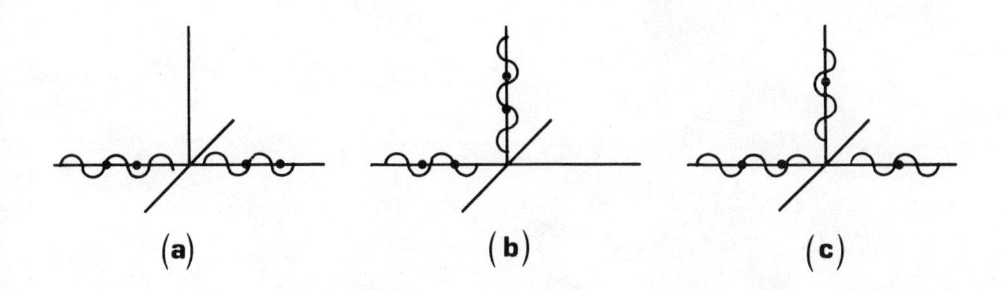

Fig. 6. The three possible distributions of two photons impinging on a semitransmitting mirror.

If one start from the idea that the interaction between the photons and the semitransmitting mirror ST follows Bose-Einstein statistics, then the probability of the configuration (c) is 1/3 and it is possible to prove that the expected value of N_b is equal to the experimental value.

Therefore if further tests should confirm these results, we can conclude that the duplicated photons interact like Bose-Einstein particles.

REFERENCES

1. W. Pauli, in Handbuch der Physik, Vol. V/1, S. Flugge, ed. (Springer, New York, 1957).
2. N. Herbert, Found. Phys. 12, 1171 (1982).
3. A. Gozzini, in The Wave-Particle Dualism, S. Diner, D. Fargue, G. Lochak, and F. Selleri, eds. (Reidel, Dordrecht, 1984), pp. 129-138.
4. F. Selleri, Found. Phys. 12, 1087 (1982).
5. J.F. Clauser, in Coherence and Quantum Optics, L. Mandel and E. Wolf, eds. (Plenum, New York, 1973), pp. 815-819.
6. H. Kuhn, Y. Chem. Phys. 53, 101 (1970).
7. A. Garuccio, V.A. Rapisarda, and J.-P. Vigier, Phis. Lett. 90A, 17 (1982).
8. R. Martinolli (relatore A. Gozzini), "Un esperimento di ottica a intensita′ molto basse", Thesis, Universita′ di Pisa (1980).
9. A. Carbone (relatore F. Selleri), "Studio degli aspetti ondulatori e corpuscolari dei sistemi quantici", Thesis, Universita′ di Bari (1981).

QUANTUM EFFECTS IN THE INTERFERENCE OF LIGHT

L. Mandel

Department of Physics and Astronomy
University of Rochester
Rochester, New York, U.S.A.

ABSTRACT
Interference effects produced by two sources of light are discussed within the framework of the quantum electrodynamic interpretation. Interference should disappear when all the source atoms are in the fully excited state, and this prediction may distinguish between quantum mechanics and various guided wave theories. The introduction of a light amplifier into an interferometer arm cannot unambiguously determine the path of the photon, because of spontaneous emission effects. Quantum mechanics, like classical wave optics, therefore predicts only a reduction of visibility in such interference experiments. In the special case of an interference experiment in which the two light sources consist of two single atoms, two photons cannot be found simultaneously at two positions that are separated by an odd number of half interference fringes. This quantum mechanical prediction amounts to another kind of EPR paradox, and should lend itself to another experimental test of the theory.

1. INTRODUCTION

It has been generally recognized, since the earliest development of quantum mechanics, that the phenomenon of interference presents a certain challenge to the quantum mechanical description of nature. Interference, whether of photons or material particles, tends to violate some of our intuitive notions of physical reality. It also confronts us with two of the fundamental quantum ideas, the probabilistic interpretation and the addition of probability amplitudes for indistinguishable paths. The latter in particular sometimes seems to run counter to intuition and to lead to paradoxes, such as the Einstein-Podolsky-Rosen paradox.

G. Tarozzi and A. van der Merwe (eds.), Open Questions in Quantum Physics, 333–343.

In this paper we shall discuss a number of less well-known features of the optical interference phenomenon, that appear to be explicitly quantum mechanical, in the sense that they have no counterpart expressible in terms of classical light waves. In principle, these features lend themselves to experimental test, although the corresponding experiments are not easy. The experiments may also provide evidence distinguishing between conventional quantum mechanics and some of the other theoretical alternatives that have recently been discussed.(1-5)

2. THE INTERPRETATION OF TWO-BEAM INTERFERENCE

We start by considering a conventional two-beam interference experiment, in which an incident light beam of wavelength λ is split into two by a beam splitter, such as a partly silvered mirror, and the two resulting beams are again superposed at some small angle θ (see Fig. 1). If the path difference introduced in this way is sufficiently small compared with the coherence length of the light, interference fringes of periodicity λ/θ will appear in a plane that is approximately perpendicular to the two beams. These fringes can readily be identified if a photodetector with an entrance pupil much narrower than λ/θ is scanned across the plane.

It is well known that the interference pattern persists at an arbitrarily low level of illumination, when photons enter the interferometer one at a time. The interpretation of this offered by conventional quantum theory is that when a photon encounters a beam splitter it has a non-vanishing probability amplitude of being found in each of the two newly formed light beams, and these two probability amplitudes must be added if we wish to calculate the probability of detecting the photon in the receiving plane. It is these two probability amplitudes, rather than the different photons, that interfere, and this conclusion is summed up in a famous statement by Dirac(6) that "... each photon interferes only with itself ...."

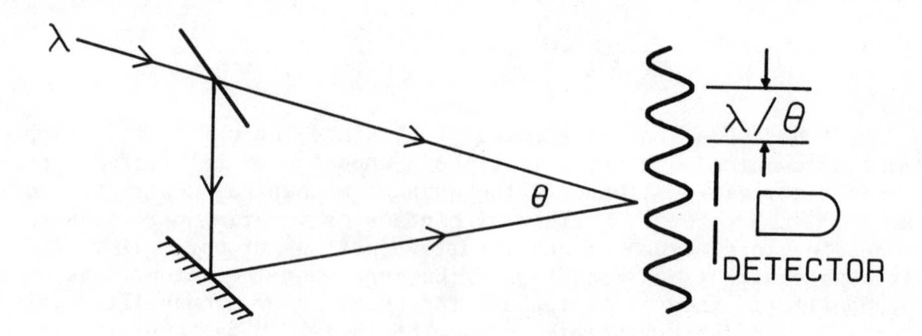

Fig. 1 Outline of a conventional interference experiment.

Although the foregoing experiment and its interpretation pose some intuitive difficulties, these difficulties are compounded in the experimental situation illustrated in Fig. 2, in which the beam splitter has been removed and replaced by a second source. Experiments of this kind have been carried out.(7) We wish to examine the question whether interference fringes will again be observed in the receiving plane if the two sources are completely independent, and if so under what conditions. As the phase of a classical light wave generally performs a random walk in time, it follows that if there is an interference pattern, then its position, which is determined by the phase difference between the two light beams, will drift at random. The result is that any interference effects must be expected to average to zero over a long time interval. However, according to classical wave optics, interference should always be observable in a sufficiently short time, i.e., short compared with the coherence time.

Within the framework of quantum electrodynamics, another condition has to be imposed before the effects become observable: enough photons have to be emitted within a coherence time to make the phase of each beam reasonably well-defined. This also ensures that there are enough photons to define the position of the interference pattern. We may suppose that the coherence time is very large compared with the photon transit time through the interferometer. If, in addition, the photons are so well separated that the average interval between them is very great compared with their transit times, so that with high probability there is only one photon within the apparatus, it may not be quite so obvious whether interference effects are again to be expected. In fact, the existence of interference has been demonstrated experimentally even under these conditions.(7) Let us see how quantum mechanics accounts for it in the absence of a beam splitter.

In order to resolve the interference pattern of spacing λ/θ we require a slit Δx--or a detector with spatial resolution Δx--

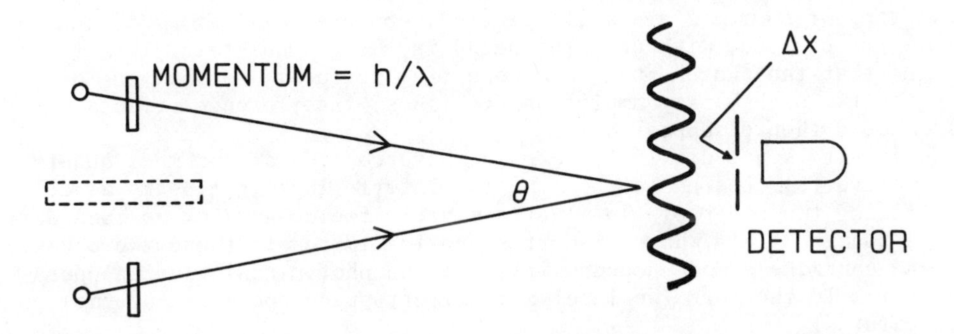

Fig. 2. Outline of an interference experiment with two light sources.

such that

$$\Delta x < \lambda/\theta . \quad (1)$$

This introduces a momentum uncertainty Δp_x for each photon in the same direction, such that $|\Delta x||\Delta p_x| \gtrsim h$, or

$$|\Delta p_x| > (h/\lambda)\theta . \quad (2)$$

But reference to Fig. 2 shows that $(h/\lambda)\theta$ is the transverse component of each incoming photon of wavelength λ. The inequality (2) then makes it impossible to determine from which of the two incident light beams the photon came. In other words, in the act of localizing the position of the photon to an accuracy Δx, we automatically disturb its momentum to the extent that the source becomes indeterminate, because the sources cannot be resolved if we look through the slit Δx. This means that we have to associate two probability amplitudes with each of the two possible, indistinguishable photon paths, and these amplitudes have to be added. The addition of the two probability amplitudes for each photon results in interference, and accounts for the observed effects within the framework of Dirac's statement. But because the superposition state is here created *a posteriori*, the explanation may appear even less attractive intuitively than that relating to Fig. 1.

Nevertheless, because of the existence of two separate sources, it should be possible to subject the quantum mechanical interpretation to experimental test. According to quantum mechanics, interference effects are a consequence of the intrinsic indistinguishability of the two photon paths. They must therefore disappear if it is possible to determine from which source the photon has come. Let us consider a situation in which sources 1 and 2 consist of exactly N_1 and N_2 excited atoms, respectively. After a short time, when a photon has been registered at the detector, we re-examine the two sources (in an energy eigenstate the energy can be measured without much disturbance), and we find that all N_2 atoms of source 2 are still excited, but only N_1-1 atoms of source 1 are excited, with one atom being in the ground state. This tells us that the photon must have come from source 1, so that one of the two probability amplitudes vanishes. Interference effects should then disappear also.

It is not difficult to make these considerations more quantitative.(8) Let us look on the two interfering light beams as defining two modes of the electromagnetic field, and let us consider a quantum field whose Hilbert space is limited to these two modes. We can make a mode decomposition of the photon annihilation operator $\hat{a}$ in the form (we label all Hilbert space operators by the caret ^)

$$\hat{a} = c_1\hat{a}_1 + c_2\hat{a}_2 , \tag{3}$$

in which $\hat{a}_1$ operates only on states belonging to mode 1, and similarly for $\hat{a}_2$. Let the quantum state of the total field be a simple product state $|S\rangle = |\Psi\rangle_1|\Phi\rangle_2$ over the two modes. Then the probability that a photon is absorbed at the detector when the field is in this state is proportional to $\langle S|\hat{a}^\dagger\hat{a}|S\rangle$, and with the help of Eq. (3) we readily obtain

$$\langle S|\hat{a}^\dagger\hat{a}|S\rangle = |c_1|^2\langle\Psi|\hat{a}_1^\dagger\hat{a}_1|\Psi\rangle + |c_2|^2\langle\Phi|\hat{a}_2^\dagger\hat{a}_2|\Phi\rangle + c_1^* c_2\langle\Psi|\hat{a}_1^\dagger|\Psi\rangle\langle\Phi|\hat{a}_2|\Phi\rangle + \text{c.c.} \tag{4}$$

The first two terms are clearly identifiable as the probabilities that a photon is detected from beam 1 and from beam 2, respectively. But the total probability is not just the sum of these two probabilities, and the additional terms show that interference effects are present. However, these terms contribute only if

$$\langle\Psi|\hat{a}_1|\Psi\rangle \neq 0 , \quad \langle\Phi|\hat{a}_2|\Phi\rangle \neq 0 , \tag{5}$$

which rules out Fock states, or states of definite photon excitation. But these are just the states of the electromagnetic field produced in an interaction in which the number of excited and unexcited atoms is well-defined both at the beginning and at the end, so that the source of the photons can be identified. It follows that interference effects should disappear precisely under those conditions in which the source of the photons is identifiable in principle, and *vice versa*. This prediction is explicitly quantum mechanical, has no analog in terms of classical interfering light waves, and should be testable experimentally. However, the experiments are not easy to carry out.

3. USE OF A LIGHT AMPLIFIER

It has recently been suggested that the problem of determining the path of the photon through an interferometer might be solved in principle, without destroying the interference effects, by the use of a light amplifier.(9,10) If such an amplifier had an intensity gain of 2, then an incident photon in an arbitrary quantum state would be expected to give rise to two photons in the same state at the output, one of which could be used to signal the arrival of the incident photon. This mechanism appears to allow one to determine the path of the photon through the interferometer without disturbing the interference pattern, and therefore to violate the uncertainty principle. Actually precise "cloning" of a photon in an arbitrary state is impossible;(11) it is prevented by the phenomenon of spontaneous emission that always accompanies any amplification process,(12,13) and introduces ambiguity in the input to the light amplifier on the basis of the output.

The photon statistics of the light amplifier have recently been treated in some detail.(14) Figure 3 shows plots of the probability $p(n|n_0)$ that n photons emerge from the amplifier when n_0 photons are incident, for an amplifier consisting of a large number of inverted atoms and having an intensity gain of 2. These probabilities exhibit the symmetry property(14)

$$p(n|n_0) = p(n|n-n_0) \quad , \tag{6}$$

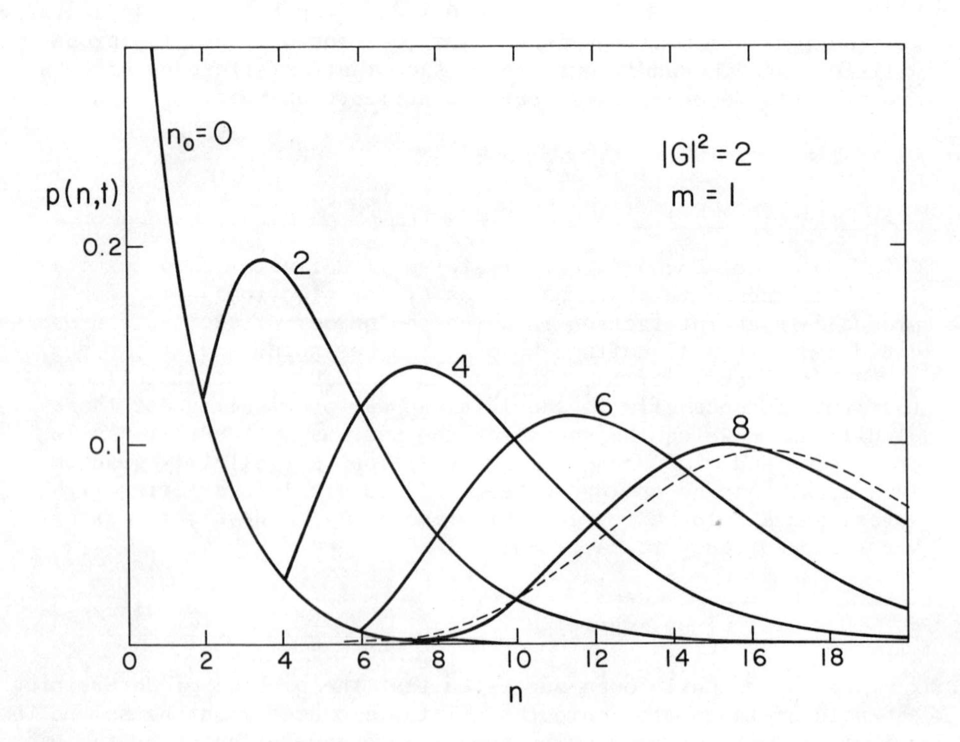

Fig. 3. The probability of the number n of photons at the amplifier output, for various definite photon numbers n_0 at the input. [Reproduced from S. Friberg and L. Mandel, in *Coherence and Quantum Optics V*, L. Mandel and E. Wolf, eds. (Plenum, New York, 1984), p. 470.]

from which it follows that the probabilities that the amplifier produces one photon at the output when one photon is incident at the input, and when no photon is incident at the input, are exactly equal. Because of spontaneous emission, the input to the amplifier cannot be deduced with certainty from the output, and this ambiguity makes it impossible to determine with certainty, by use of a light amplifier, which path a photon followed through the interferometer. According to quantum mechanics,interference effects occur whenever there exists a fundamental uncertainty concerning the path. The light amplifier will therefore not destroy the interference pattern, although it may reduce the visibility of the fringes. In this respect the quantum mechanical prediction is not significantly different from that given by classical optics.

4. INTERFERENCE DETECTION BY PHOTOELECTRIC CORRELATION

In the foregoing we have considered the interference fringes to be detected by a photoelectric detector with a narrow slit that is scanned across the field of view. This method is appropriate for a steady pattern, such as would be produced under the conditions shown in Fig. 1. When the interference fringes move randomly in time, as they would under the conditions in Fig. 2, this is not usually feasible, and one generally has to use two or more detectors simultaneously and to correlate their output. This is the method that was adopted in the experiments in Ref. 7.

Let us then consider the situation illustrated in Fig. 4, in which the light from two independent sources is superposed in some receiving plane, and two detectors located at positions x and x' are used to observe the interference pattern having a periodicity L. The photoelectric pulses from the two detectors, after amplification and pulse shaping, are fed to the two inputs of a coincidence counter that registers a count only when two pulses arrive

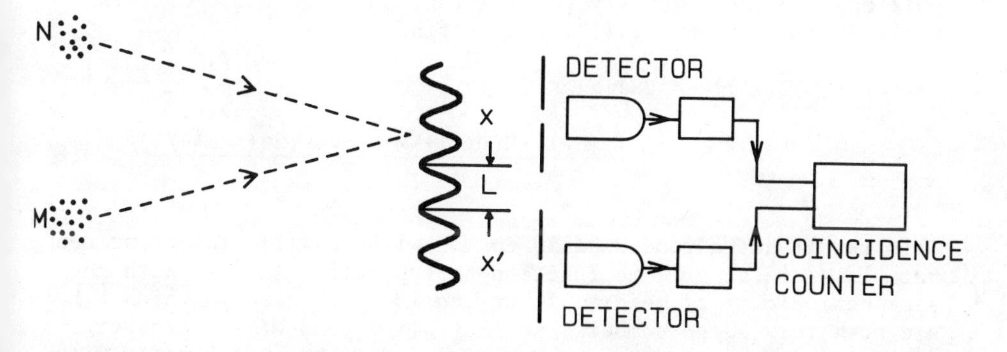

Fig. 4. The technique of photoelectric coincidence counting for detecting interference (fringes).

in coincidence (within some resolving time of the counter). A detailed quantum mechanical treatment of such an experiment has recently been presented,(15) and we shall make use of a few conclusions from it.

Let us first approach the problem from a semiclassical point of view, in which the radiation field is considered to consist of classical light waves with instantaneous light intensity I. The probability that the two photodetectors at positions x and x' will emit two photoelectric pulses in coincidence is then proportional to the average product or correlation(16) $\langle I(x)I x'\rangle$ (the time arguments of I have been suppressed). If the phases of the light waves leaving the two sources are completely random and uncorrelated, and if the sources are of equal strength, one can show that the required correlation is given by(15)

$$\langle I(x)I(x')\rangle = \frac{1}{2\beta}\langle I(x)\rangle\langle I(x')\rangle[1 + \beta\cos(2\pi(x-x')/L)] \ . \tag{7}$$

The coincidence counting rate is therefore a periodic function of the detector separation x-x' with periodicity L, which reveals the existence of the interference fringes. The constant β is the relative modulation amplitude, and it takes the value 1/2 when the light waves leaving the sources are of constant intensity and the value 1/3 when the light waves have fluctuating intensities and obey thermal statistics. The coincidence counting rate is greatest when the detectors are separated by a whole number of interference fringes, and it is least when they are separated by an odd number of half fringes, but it is never zero.

We now contrast these conclusions with those obtained by a full quantum mechanical analysis. Let us suppose that the two sources consist of exactly N and M identical two-level atoms, respectively, all in the same state of excitation except for their phases, which are random. The light intensity $\hat{I}(x)$ is now an operator in a certain Hilbert space, and $\langle\hat{I}(x)\rangle$ denotes the quantum mechanical expectation. The coincidence counting rate is again proportional to the intensity correlation, but with the operators written in normal order,(17) and we find(15)

$$\langle:\hat{I}(x)\hat{I}(x'):\rangle = 2\langle\hat{I}(x)\rangle\langle\hat{I}(x')\rangle \times \frac{[N(N-1)+M(M-1)+NM+NM\cos(2\pi(x-x')/L)]}{(N+M)^2} \ . \tag{8}$$

This result has a good deal in common with Eq. (7). The coincidence rate is again a periodic function of separation (x-x'), with the same periodicity as before. If we consider the case in which N,M are both very large numbers, so that N(N-1) and M(M-1) are very nearly N^2 and M^2, and we let N=M, we recover Eq. (7) exactly with

$\beta = 1/3$. In other words, the quantum field then behaves very much like the classical field when the light obeys thermal statistics. This result may seem difficult to understand at first, because the atoms were not assumed to be in thermal equilibrium. However, because of the random phase assumption, the central limit theorem comes into play, and the field resulting from the superposition of a large number of randomly phased fields obeys Gaussian statistics, just like a thermal field.

5. ANOTHER FORM OF EPR PARADOX

A more interesting consequence of Eq. (8) is obtained if we go to the other extreme, and assume that each source consists of a single atom, or N=1=M. Then Eq. (8) reduces to

$$\langle:\hat{I}(x)\hat{I}(x'):\rangle = \frac{1}{2}\langle\hat{I}(x)\rangle\langle\hat{I}(x')\rangle[1 + \cos(2\pi(x-x')/L)] , \qquad (9)$$

which shows that the modulation of the coincidence counting rate with separation $x-x'$ is now 100%. In particular, the probability of detecting two photons at two points x,x' separated by an odd number of half fringes is zero. This is an interesting and explicitly quantum mechanical prediction, that has no classical counterpart, and can be put to the experimental test. The reason for it can be understood in the following way. When each source consists of one atom, there are only two ways in which a coincidence count can be achieved: Either the photon from source 1 activates the detector at x and the photon from source 2 activates the detector at x', or *vice versa*. These two possibilities have almost equal probability, and they are indistinguishable, so that the corresponding probability amplitudes have to be added. But the two amplitudes are out of phase when $(x-x') = (n+\frac{1}{2})L$, $n=0,\pm1,....$, and therefore sum to zero.

The result embodied in Eq. (9) implies another form of Einstein-Podolsky-Rosen paradox. Let us imagine that the two detectors are well separated in distance by several thousands of interference fringes. Yet the probability that the detector at x' registers a photon at a certain time, namely when the detector at x registers a photon, depends critically on the exact position of the detector at x. At certain positions x the probability is zero, whereas it is non-zero at other positions, despite the fact that the two detectors are far apart. Except for the fact that the paradox relates to positions rather than to polarization orientations, it is reminiscent of the EPR paradox, and its origin again lies in the addition of probability amplitudes and the non-locality of quantum mechanics.

6. LIGHT SOURCES WITH INDEFINITE NUMBERS OF ATOMS

Equation (8) applies to a situation in which each source consists of a definite number of atoms. However, in many experimental situations, such as when the source is in the form of a gas discharge or an atomic beam behind an aperture, the number of atoms is not constant but fluctuates. We can accommodate this situation by averaging the right-hand side of Eq. (8) over the fluctuations of N and M. Let us assume that N and M are independent and obey Poisson statistics. Then

$$\left.\begin{aligned} \langle N(N-1)\rangle &= \langle N\rangle^2 , \\ \langle M(M-1)\rangle &= \langle M\rangle^2 , \\ \langle NM\rangle &= \langle N\rangle\langle M\rangle . \end{aligned}\right\} \quad (10)$$

If we make use of the fact that $\langle\hat{I}(x)\rangle\langle\hat{I}(x')\rangle/(N+M)^2$ is independent of N,M and if we put $\langle N\rangle=\langle M\rangle$, Eq. (8) goes over into the classical Eq. (7) with $\beta = 1/3$. In other words, when the number of atoms fluctuates, the quantum field behaves just like the thermal classical field, no matter how small the average numbers $\langle N\rangle,\langle M\rangle$ may be. At first sight this conclusion may seem surprising, for one might suppose that the case $\langle N\rangle = \langle M\rangle \lesssim 1$ would correspond to a quantum mechanical situation. However, when there is indefiniteness in the number of atoms, the two photons counted by the two detectors in coincidence could have come from the same source, rather than from two sources. Only when each source consists of exactly one atom is that possibility ruled out, and only then does quantum mechanics lead us to Eq. (9).

Although Eq. (9) can, in principle, be tested experimentally, the experiments are not easy to carry out, as the two atoms need to have fairly well-defined positions, and they need to be prepared repeatedly in the same quantum state. Nevertheless, the experiment is attractive, for the outcome appears to distinguish clearly between the quantum mechanical and various other(1-5) predictions.

This work was supported by the National Science Foundation.

REFERENCES

1. L. de Broglie, *Non-Linear Wave Mechanics: A Causal Interpretation* (Elsevier, Amsterdam, 1960); *The Current Interpretation of Wave Mechanics: A Critical Study* (Elsevier, Amsterdam, 1964); *Ann. Fond. L. de Broglie 2,* 1 (1977).
2. L. de Broglie and J. Andrade E. Silva, *Phys. Rev. 172,* 1284 (1968).
3. F. Selleri and G. Tarozzi, *Nuov. Cim. B 43,* 31 (1978); F. Selleri, *Found. Phys. 12,* 1087 (1982).

4. A. Garuccio and J.-P. Vigier, *Found. Phys. 10*, 797 (1980).
5. N. Cufaro-Petroni and J.-P. Vigier, *Phys. Lett. 93A*, 383 (1983).
6. P.A.M. Dirac, *The Principles of Quantum Mechanics* (Oxford University Press, Oxford, 1947), 3rd ed., Chap. 1.
7. R. L. Pfleegor and L. Mandel, *Phys. Rev. 159*, 1084 (1967); *J. Opt. Soc. Am. 58*, 946 (1968); L. Mandel, in *Quantum Optics*, R. J. Glauber, ed. (Academic Press, New York, 1969), p. 176.
8. L. Mandel, *Phys. Lett. 89A*, 325 (1982).
9. A. Garuccio, K. R. Popper, and J.-P. Vigier, *Phys. Lett. 86A*, 397 (1981).
10. A. Garuccio, V. Rapisarda, and J.-P. Vigier, *Phys. Lett. 90A*, 17 (1982).
11. W. K. Wootters and W. H. Zurek, *Nature 299*, 802 (1982).
12. L. Mandel, *Nature 304*, 188 (1983).
13. P. W. Milonni and M. L. Hardies, *Phys. Lett. 92A*, 321 (1982).
14. S. Friberg and L. Mandel, *Opt. Comm. 46*, 141 (1983); and in *Coherence and Quantum Optics V*, L. Mandel and E. Wolf, eds. (Plenum, New York, 1984), p. 465.
15. L. Mandel, *Phys. Rev. A 28*, 929 (1983).
16. L. Mandel, E.C.G. Sudarshan, and E. Wolf, *Proc. Phys. Soc. (London) 84*, 435 (1964); L. Mandel and E. Wolf, *Rev. Mod. Phys. 37*, 231 (1965).
17. R. J. Glauber, *Phys. Rev. 130*, 2529 (1963); *131*, 2766 (1963).

TESTS OF QUANTUM MECHANICS BY NEUTRON INTERFEROMETRY

H. Rauch

Atominstitut der Österreichischen Universitäten
A-1020 Wien

ABSTRACT

Matter wave interferometry got a new impetus due to the invention of perfect crystal neutron interferometry. Well-separated, coherent beams are available which can be influenced by nuclear, magnetic, or gravitational phase shifts. The longitudinal and transverse coherence lengths are determined and related to the properties of the beam. The problem of locality or nonlocality of quantum mechanics is discussed in view of the self-interference property of such a device. Characteristic neutron interferometric measurements are discussed: the 4π-symmetry of a spinor-wave function, the explicit verification of the quantum mechanical spin-superposition law, wide-slit diffraction, the neutron Fizeau effect, and the influence of gravitation. Formulations are chosen to stimulate discussion with theoreticians. The Appendix gives an outline of dynamical diffraction theory, which is the basis of perfect crystal interferometry.

1. INTRODUCTION

The neutron has well-defined particle and wave properties, which have been summarized recently by Ramsey (1982) and Shull (1982) at the 50th anniversary meeting of the discovery of the neutron held at Cambridge in September 1982. Particle and wave properties are coupled together by the well-known de Broglie relation (de Broglie, 1923),

G. Tarozzi and A. van der Merwe (eds.), Open Questions in Quantum Physics, 345–376.

$$mv = \frac{h}{\lambda}, \tag{1}$$

where m is the neutron mass, v the group velocity, h Planck's constant, and λ the wavelength of the associate matter wave ψ. The latter is described by the well-known Schrödinger equation

$$H\psi = i\hbar\frac{\partial\psi}{\partial t} \tag{2}$$

or, for a stationary situation, its time-independent form

$$H\psi = E\psi, \tag{3}$$

where H is the Hamilton operator and E the total energy ($E = \hbar\omega$). The solution for the case of the free motion of the neutron can be written in the form of a wave packet (e.g., Messiah, 1965; Hittmair, 1972):

$$\psi = \frac{1}{(2\pi)^{3/2}} \int A(\vec{k}')\, e^{i(\vec{k}'\vec{r} - \omega t)}\, d\vec{k}', \tag{4}$$

which defines the particle in space $\vec{r}$ and time t. This is a coherent superposition of plane waves having a wave vector $\vec{k}' = (2\pi/\lambda)\,(\vec{v}/v)$ and an amplitude $A(\vec{k}')$.

Neutron interferometry provides a unique tool for the realization of quantum mechanical textbook experiments on a macroscopic scale, which permits the observation of certain properties of the wave function itself. In the past, neutron interferometry based on wave front division (Maier-Leibnitz and Springer, 1962) and on amplitude division (Rauch, Treimer, and Bonse, 1974) have been developed. The former case uses Fraunhofer slit diffraction and has close analogies to classical light interferometers, while the second method is based on the dynamical diffraction phenomenon in perfect crystals, which has been used before in X-ray interferometry (Bonse and Hart, 1965).

The two techniques are to some extent complementary (Fig.1). The first method permits very long (10 m) beam paths, but only a very small beam separation (100 μm), because the slit diffraction at slits providing enough intensity is very small for neutrons having wavelengths in the range of 2 - 20 Å. The second method provides a very good beam separation (5 cm) due to the Bragg-diffraction at the perfect crystal ($0.5 \leq \lambda \geq 3$ Å). But the beam paths are limited to about 10 cm, due to the requirements on the perfectness of the crystal, whose lattice planes have to be parallel throughout the crystal, which favors the use of silicon and the monolithic design of such a device. Most of the experiments have been performed using such a monolithic perfect crystal

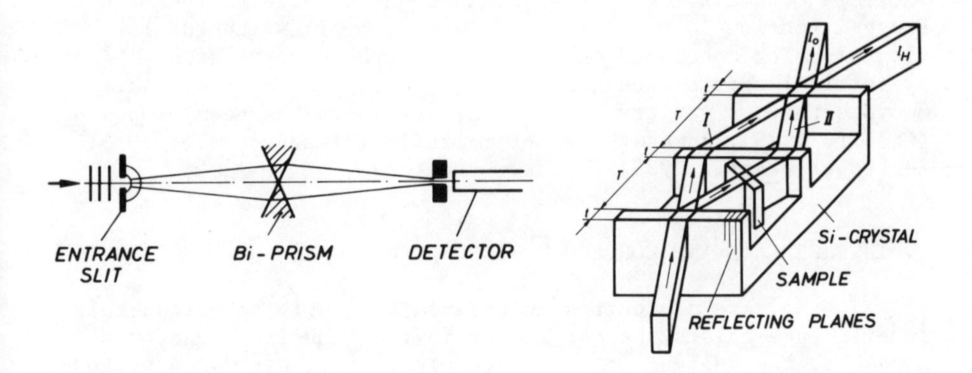

Fig. 1. Principle of wavefront and amplitude division interferometry indicated by a slit Fresnel prism and by a perfect crystal arrangement.

interferometer, and therefore a sketch of the basic theory of its operation will be outlined below.

The diffraction of neutrons at a perfect crystal structure is described by the dynamical diffraction theory, which is treated in the Appendix of this article. Here it is important to notice that one can write down the wave function for the neutron behind a perfect crystal slab as a function of the incoming wave function (Eq.A18). If we proceed to the second and third plate of the interferometer, we can write the wave function behind the interferometer as a sum of wave functions coming from beam paths I and II (Petrascheck, 1976; Bauspiess et al., 1976):

$$\psi_0 = \psi_0^I + \psi_0^{II} = [v_0(y)v_H(y)v_{-H}(-y) + v_H(y)v_{-H}(-y)v_0(y)] \cdot \exp[-2\pi\, i\, y(T+t)/\Delta_0]\psi_e, \qquad (5)$$

where the relevant quantities are defined in the Appendix (Eq. A18 and A19) and in fig.1. Using these definitions, and assuming ideal geometry, an important relation follows:

$$\psi_0^I = \psi_0^{II}, \qquad (6)$$

which means that the wave functions coming via beam paths I and II are equal, and the whole intensity becomes

$$I_0 = |\psi_0^I + \psi_0^{II}|^2 = 4|\psi_0^I|^2. \qquad (7)$$

Particle conservation is guaranteed together with the beam in Bragg-direction $I_H = |\psi_H^I + \psi_H^{II}|^2$. One recognizes the ideal situation for coherence measurements in the case of such a symmetrical triple Laue (LLL)arrangement of the crystal plates. Besides that, a variety of other arrangements has been proposed (Graeff, 1979) and tested experimentally (Kikuta _et al_., 1975; Zeilinger _et al_., 1983).

2. INTERFERENCE AND COHERENCE MEASUREMENTS

In the case of neutron interferometry, only so-called self-interference exists, because, even in the high-flux reactor at Grenoble, the maximum of the available flux within the crystal the interferometer is 1.3×10^3 cm^{-2} s^{-1} (Bauspiess _et al_., 1978), which gives a mean time interval between successive neutrons of about $\overline{\Delta t} = 780$ μs, while the time-of-flight through the interferometer is always much smaller, $t_{TOF} = 35$ μs. Most experiments that need higher beam monochromatism or other types of beam tailoring, or the experiments at our small research reactor in Wien, have a much smaller intensity, and we therefore have the relation

$$\overline{\Delta t} \gg t_{TOF}.$$

For most of the experiments the statement can be made that the following neutron is on the average not yet "born", it is still contained within the uranium nucleus if the observed neutron passes the interferometer. Although the particle properties of the neutron (mass, magnetic moment, etc.) must be transported through the interferometer via beam path I or II, the wave function (Eq. (4)) spreads out across both beams, which are spatially separated up to 5 cm. The wave function is treated as a probability wave, which describes a neutron in a certain ensemble and has the characteristics of the individual neutron (k,ω) and of the ensemble $(A(k))$. The problem of locality and nonlocality (e.g., Wigner, 1970; Pitowsky, 1982; Selleri, 1982) is inherently involved in any interpretation, but here a heuristical approach is given that describes the experiments and their results.

The interference properties of the beam become most easily visible if a phase-shifting material characterized by its thickness D and its index of refraction n (Eq.A8) is inserted into the coherent beams (Fig.1b). This causes an optical path length difference, analogous to the situation in classical optics, and therefore a change of the wave function:

$$\psi_0^{II'} = \psi_0^I e^{-i(1-n)kD} = \psi_0^I e^{-i\lambda b_c ND} = \psi_0^I e^{i\chi}, \qquad (8)$$

where b_c is the coherent scattering length, N is the particle density, and χ denotes the nuclear phase shift defined by this equation. Using this relation one immediately obtains an intensity modulation

$$I_0 = |\psi_0^{I} + \psi_0^{II'}|^2 = 2|\psi_0^{I}|^2 \ (1 + \cos N b_c \lambda D)$$
$$= 2|\psi_0^{I}|^2 \ (1 + \cos\chi). \tag{9}$$

A variation of the phase shift is easily obtained by rotating the samples within the coherent beams. Various imperfections of the crystal, the phase-shifting material, or the neutron beam itself cause a reduction of the complete beam modulation. Variations of the geometry of the crystal or slight rotations of the lattice planes or small vibrations of the whole crystal causes a constant reduction of the contrast. The effects of the phase shifter may be accounted for by a factor describing the variation of the thickness and by a complex index of refraction (Halpern, 1952):

$$n = 1 - \frac{\lambda^2 N}{2\pi} \sqrt{b_c^2 - \left(\frac{\sigma_r}{2\lambda}\right)^2} + i\,\frac{\sigma_r N \lambda}{2\pi}, \tag{10}$$

which accounts for absorption (σ_a) and incoherent scattering (σ_i) processes ($\sigma_r = \sigma_a + \sigma_i$). The neutron beam's imperfections are described by its distribution of k vectors, $W(\vec{k})$; the latter is directly related to the amplitude of the wave packet $A(\vec{k})$ (see Eq. (4)) and measurable from the wavelength distribution function $W(\lambda)$, which depends on the experimental arrangement and often is approximated by a Gaussian function centered around λ_0 and with a full half-width $\Delta\lambda$:

$$W(\lambda) = \frac{2\sqrt{\ln 2}}{\sqrt{\pi}\,\Delta\lambda} \exp\left\{-\left[\frac{(\lambda-\lambda_0)2\sqrt{\ln 2}}{\Delta\lambda}\right]^2\right\}. \tag{11}$$

Inserting Eq. (10) into Eq. (8) and averaging Eq. (9) over the wavelength distribution, one obtains

$$I_0 = \int I(\lambda)W(\lambda)d\lambda$$
$$= 2|\psi_0^{I}|^2 \exp(-\sigma_r ND/2) \exp\left[-\left(\frac{\pi}{2\sqrt{\ln 2}} \frac{D}{D_\lambda} \frac{\Delta\lambda_H}{\lambda_0}\right)^2\right]$$
$$\cdot \left[\cosh(\sigma_r ND/2) + \cos 2\pi \frac{D}{D_\lambda}\right], \tag{12}$$

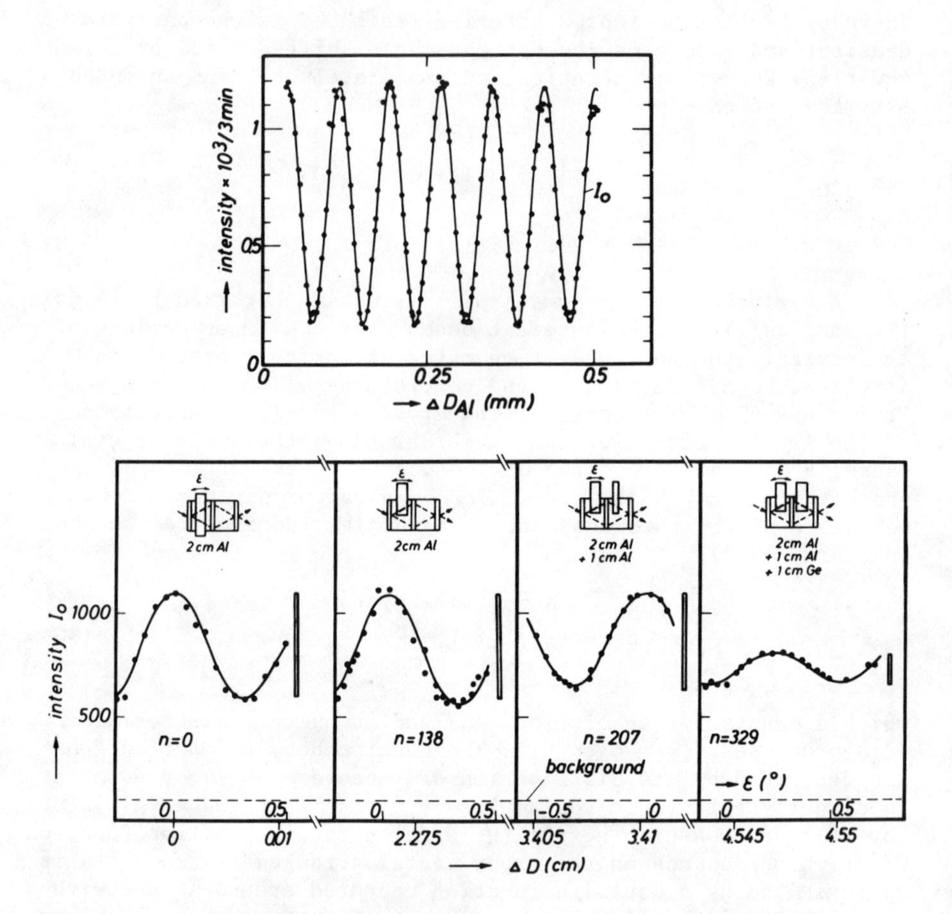

Fig. 2. Characteristic interference pattern for low-order interference, showing the high degree of coherence available (above), and high-order interference observed with a highly monochromatic beam ($\Delta\lambda/\lambda = 10^{-3}$, below) (Rauch, 1979).

where we have introduced the λ-thickness $D_\lambda = 2\pi/(Nb_c\lambda_0)$, which counts the order of the interference $m = D/D_\lambda$. Analogous to classical optics, the influence of the wavelength spread becomes more important at high order.

The experimental results (Fig.2) agree very well with the calculated formulas. For a well-balanced interferometer crystal we obtain for low-order interferences nearly the ideal behavior (Eq. (9), Fig. 2a), and for high-order interference measurements, where thick phase-shifting materials are introduced into the coherent beams, the contrast decreases according to Eq. (12) (Fig. 2b), in agreement with the separately measured wavelength spread $\Delta\lambda/\lambda_0 = 10^{-3}$ (Rauch, 1979). From these results it can be concluded by extrapolation that the longitudinal coherence length is at least $\Delta x_L \sim 500\lambda = 875$ Å, because the wavelength used in the experiment was $\lambda_0 = 1.75$ Å (E = 0.028 eV). It should be noticed again that the coherence pattern depends on the wavelength spread, which is a beam property even though there is only self interference of the neutrons and even though there is a remarkable spreading in space of the wave packet of matter waves (e.g., Messiah, 1965):

$$[\Delta x(t)]^2 = [\Delta x(0)]^2 + [\frac{(\hbar/2m)t}{\Delta x(0)}]^2, \qquad (13)$$

where for slow neutrons the second term easily dominates the first one. The same conclusions concerning the coherence length have recently been drawn by the Columbia-interferometer group after similar experiments, where again the contrast has been observed at high-order of interference (Kaiser _et al._, 1983). The accuracy available for the determination of the nuclear phase shift χ reaches, for an advanced interferometer setup and a high-flux neutron source, about $\Delta\chi/2\pi \simeq 7 \times 10^{-5}$ (Bauspiess _et al._, 1978; Bonse and Kischko, 1982).

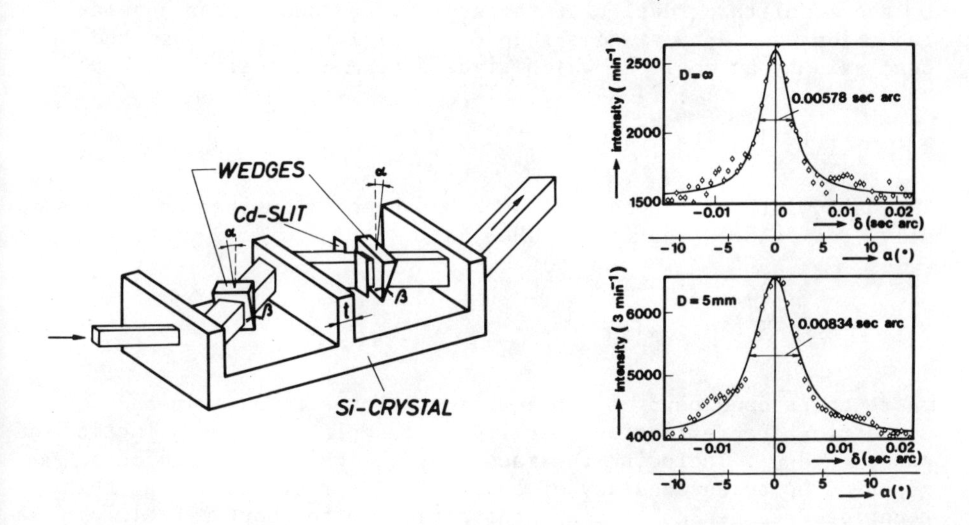

Fig. 3. Sketch of the experimental arrangement and typical results of triple Laue-reflection curve measurements and the results of a wide single-slit diffraction experiment (Rauch _et al._, 1983).

Information about the transverse coherence length can be obtained from the width of the Borrmann fan produced due to mutual interference of the wave fields within the perfect crystal (see Appendix, Eq. A24). The observed interferences, using the two plate interferometers (Kikuta et al., 1975; Zeilinger et al., 1979), show directly that there exists coherence across the whole Borrmann fan. The transversal coherence length can also be deduced from the multiple Laue-rocking curves, where a very narrow central peak exists due to the convolution procedure of the Pendellösung diffraction curves (Eq. A20 and A2), Bonse et al., 1977). The width of the central peak is in the order of 0.001 sec of an arc, which defines a correspondingly small Δk_T value, which gives, together with the uncertainty relation $\Delta k_T \Delta x_T \gtrsim 1/2$, the transverse coherence length. Figure 3 depicts the experimental arrangement and typical results showing the central peak broadening due to Frauenhofer diffraction at a macroscopic slit having a width of D = 2.5 mm (Rauch et al., 1983). The broadening of the zero-order slit diffraction peak is given according to the Frauenhofer formula as (e.g., Shull, 1969)

$$\Delta\beta = 0.888\ \lambda/D. \tag{14}$$

A transverse-coherence length of $\Delta x_T \simeq 6.5$ mm can be deduced from these data. These experiments could be performed in this way due to the monolithic design of the crystal and due to the angular reduction caused by a wedge-shaped material rotated around the beam axis by an angle ξ, which gives a beam deflection of

$$\delta = [2(1-n)\mathrm{tg}\,\frac{\beta}{2}]\,\sin\alpha. \tag{15}$$

The half-width of the central peak has been calculated analytically (Petrascheck and Rauch, 1983) as

$$\Delta y_H = \frac{2.1}{A} = \frac{2.1\,k\cos\Theta_B}{2\pi b_c N t}, \tag{16}$$

where the second part of the equation belongs to an even and symmetric Laue reflection of silicon at a plane having a lattice constant d_{hkl}. Including the fact that the half-width can be measured up to an accuracy of about 1%, one recognizes that this technique reaches an angular sensitivity up to about 2.5×10^{-5} sec of an arc, which may have advances for fundamental physics application.

The diffraction at narrower slits ($\sim$ 10 μm) has been measured before (Shull, 1969), using a double-crystal arrangement, and by Zeilinger et al. (1982), using long-wave length neutrons and a long optical bench, which has also been used by Gähler et al. (1981) to observe the diffraction at a thin (100 μm) absorbing

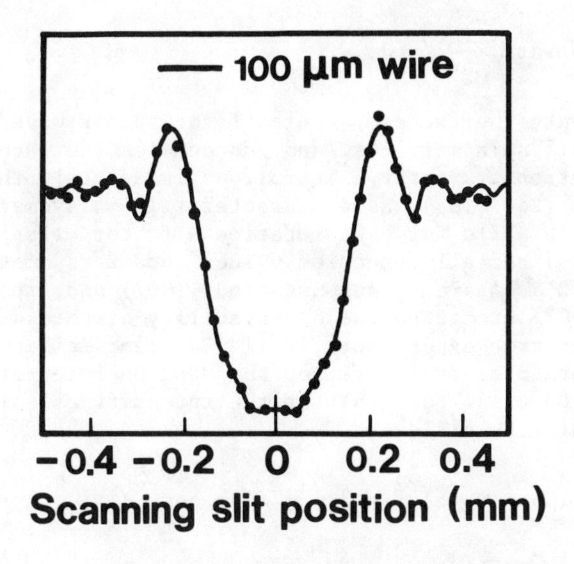

Fig. 4. Diffraction pattern of a thin absorbing wire, observed with a slit-diffraction interferometer (Gähler *et al.*, 1981)

wire whose diffraction pattern is shown in Fig. 4 compared to the quantum mechanical prediction. Similarly, the double slit diffraction (Zeilinger *et al.*, 1982) and the action of Fresnel-zone plates (Klein *et al.*, 1981) have been observed. All these phenomena are, within their error bars, in agreement with the standard quantum mechanical predictions.

3. ELECTROMAGNETIC EFFECTS: SPINOR SYMMETRY AND SPIN SUPERPOSITION

The magnetic moment $\vec{\mu}$ of the neutron couples like a dipole to the magnetic field $\vec{B}$ with the Hamiltonian

$$H = -\vec{\mu}.\vec{B}. \tag{17}$$

The wave function therefore propagates as

$$\psi'(t) = e^{-\frac{iHt}{\hbar}}\psi(0) = e^{-\frac{i\vec{\mu}.\vec{B}t}{\hbar}}\psi(0) = e^{-\frac{i\vec{\alpha}.\vec{\alpha}}{2}}\psi(0) = \psi(\alpha), \tag{18}$$

where α is numerically equal to the Larmor precession angle

$$\alpha = \frac{2\mu}{\hbar} \int B\,dt \simeq \frac{2\mu}{\hbar v} \int B\,ds, \tag{19}$$

ψ now represents the two components of the spinor wave function, $\vec{\sigma}$ are the Pauli spin matrices, and $\int ds$ denotes the path integral along the neutron trajectory. It follows immediately that the wave function (Eq. (18)) has a characteristic 4π symmetry ($\psi(4\pi) = \psi(0)$), while for a 2π rotation a factor -1 appears ($\psi(2\pi) = -\psi(0)$) and all expection values show 2π symmetry ($\psi^2(2\pi) = \psi^2(0)$). Aharanov and Susskind (1967) and, independently, Bernstein (1967) predicted the observability of this 4π symmetry for interferometric experiments, while Eder and Zeilinger (1976) gave the theoretical framework for the neutron interferometric realization. On using Eq. (18) for the intensity calculation, Eq. (9) immediately yields

$$I_0 = |\psi_0(0) + \psi_0(\alpha)|^2 = 2|\psi_0^I|^2(1 + \cos\frac{\alpha}{2}), \tag{20}$$

which is valid equally for polarized and unpolarized neutrons and which indicates again the self-interference property in this type of experiment. The experimental verification of (20) succeeded in 1975 (Rauch et al., 1975; Werner et al., 1975). It has been also verified by a wave-front division interferometer (Klein and Opat, 1976), by a molecular beam system (Klempt, 1976), by an NMR system (Stoll et al., 1978), and repeated with the perfect crystal interferometer using a well defined magnetic field within Mu-metal sheets (Rauch et al., 1978). The most precise periodicity factor extracted from these measurements and corrected for up-to-date values of the physical constants is

$$\alpha_0 = (715.87 \pm 3.8)\mathrm{deg}, \tag{21}$$

which is in agreement with the predicted 4π symmetry because the error bars indicate the simple α limits only. At the entrance in the magnetic field a slight change of the kinetic energy occurs, due to the longitudinal Zeeman splitting (Eq. (3)), and, therefore, the velocity of the two subbeams ($\pm$) is slightly different (see Eq. (19)), which causes a very small correction to the formulas in the order of 10^{-5}, i.e., the ratio of the Zeeman energy to the kinetic energy of the neutrons (Bernstein, 1979; Berstein and Zeilinger, 1980). The interpretation of this type of experiment has been widely debated (Byrne, 1978; Mezei, 1979; Zeilinger, 1981; Berstein and Phillips, 1981). Recently, we repeated this experiment with polarized neutrons (Summhammer et al., 1983b), finding the same result and ascertaining again the measurability of the rotational periodicity of a spinor wave function.
Nuclear phase shift χ and magnetic spnor rotation α can be applied simultaneously to the coherent beams

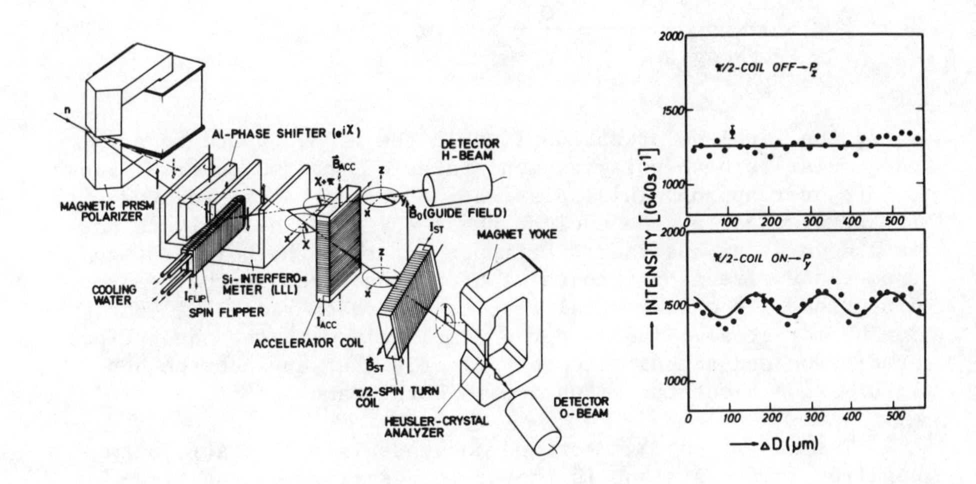

Fig. 5. Sketch of the experimental arrangement and characteristical results for the spin superposition experiment (Summhammer et al., 1983a)

$$\psi(\chi,\alpha) = e^{i\chi}e^{-i\vec{\alpha}.\vec{\alpha}/2}\psi(0,0), \tag{22}$$

which results for unpolarized neutrons in characteristic beat effects regarding the intensity and polarization modulation, also calculated by Eder and Zeilinger (1976) and experimentally verified by Badurek et al. (1976). Additional terms appears for polarized incident neutrons, where, for a special spinor rotation, the so-called Wigner (1963) phenomenon can be measured (Fig. 5). In this case polarized incident neutrons are produced by magnetic prism deflection (Badurek et al., 1979), and the polarization in one coherent beam bath is rotated by means of a static DC-flipper (Mezei, 1972). This flipper represents a region with an effective field in the x direction, produced by the magnetic guide field applied to the total beam paths and the field of the flipper coil, and permits a Larmor spin rotation around the x axis of 180° ($\vec{\alpha} = \pi\hat{x}$). Equation (22) therefore reduces to

$$\psi(\chi,\pi) = -i\sigma_x e^{i\chi}\psi(0,0) = U.\psi_0, \tag{23}$$

and the final polarization becomes ($\psi = \psi_0(1 + U)$)

$$\vec{P} = \frac{\psi_0^+(1+U^+)\,\vec{\sigma}\,(1+U)\psi_0}{I_0} = \hat{x}\sin\chi - \hat{y}\cos\chi, \tag{24}$$

i.e., the final polarization lies in the xy plane and is perpendicular to both polarization states of the two coherent beams before overlapping. This polarization vector can be rotated within the xy plane by the nuclear phase shift χ. The analysis has been done by an analyzer DC flipper rotating the polarization from the y direction into the z direction which is the analyzer direction. The experimental results (Summhammer et al., 1982, Summhammer et al., 1983a) agree completely with the theoretical prediction and demonstrate very clearly that any neutron has information about the situation in both beams.

The experiment described above deals with static beam handling components and is therefore described by the time-independent Schrödinger equation (Eq. (3)); it therefore influences the spatial and spinor part of the wave function only. An approved technique in polarized neutron physics consists in rotating the polarization by means of high frequency (HF) resonance fields (Alvarez and Bloch, 1940; Kendrick et al., 1970), which represents a time-dependent interaction (Eq. (2)); here we can again calculate the wave function (e.g., Krüger, 1980) and again have a coherent interaction. In this case one influences the time-dependent part of the wave function due to a change of the total energy of the neutron ($E \rightarrow E \pm \Delta E = E \pm 2\mu B = \hbar(\omega \pm \Delta\omega)$; Eq. (4), Alefeld et al., 1981). A straightforward calculation similar to that of Eq. (24) yields in this case

$$\vec{P}(t) = \hat{x}\cos(\Delta\omega t + \chi) + \hat{y}\sin(\Delta\omega t + \chi), \tag{25}$$

which shows a characteristic time dependence. Related measurements have been made recently using a time-resolved detector synchronized with the phase of the HF-field (Badurek et al., 1983), and the results are again in accordance with Eq. (25).
It shows that additional marks can be given to the wave function.

4. GRAVITATIONAL INTERACTION

The distances within the interferometer are very small compared to the radius of the earth and therefore the Hamiltonian due to the kinetic energy and gravitational interaction reads

$$H = \frac{\hbar^2 k^2}{2m} + m\vec{g}.\vec{r} - \vec{\omega}.\vec{L}, \tag{26}$$

where $\vec{g}$ is the gravitational acceleration, $\vec{r}$ the space vector from the center of the earth, $\vec{\omega}$ the angular rotation velocity of the earth, and $\vec{L}$ the angular momentum of the neutron relative to the center of the earth ($\vec{L} = \vec{r} \times \vec{p}$). From this Hamiltonian one obtains the equation of motion in the rotating frame

$$m\ddot{\vec{r}} = m\vec{g} - m\vec{\omega} \times (\vec{\omega} \times \vec{r}) - 2m\vec{\omega} \times \dot{\vec{r}} \qquad (27)$$

and therefore

$$\vec{r} \simeq \vec{r}_0 + \vec{v}_0 t + \frac{1}{2}\vec{g}\,t^2 + \frac{1}{3}t^3 \vec{\omega} \times \vec{g}', \qquad (28)$$

where

$$\vec{g}' = \vec{g} + \vec{\omega} \times (\vec{\omega} \times \vec{R}),$$

the effective acceleration at the position of the interferometer. The influence of the Coriolis term on the beam trajectory is very small compared to the usual gravitational term ($t \sim 30\ \mu s$, $\omega = 7.29 \times 10^{-5}\ s^{-1}$),

$$\frac{\frac{1}{2}|\vec{g}|t^2}{\frac{1}{3}t^3|\vec{\omega} \times \vec{g}'|} \sim 10^{-9}. \qquad (29)$$

The influence of both gravitational terms on the neutron beam trajectory is well-known (McReynolds, 1951; Dabbs, 1965), and it is the basic principle of the neutron-gravity refractometer (Koester, 1965; Koester, 1967; Koester and Nistler, 1975) and the gravity diffractometer for ultra cold neutrons (Scheckenhofer and Steyerl, 1981) These measurements have also been used to verify the equivalence of the gravitational and inert mass of the neutron (Koester, 1976; Sears, 1982).

The measurement of the gravitationally induced phase shift was first proposed by Overhauser and Colella (1974); the experiment was carried out, using a perfect-crystal neutron interferometer, by Colella et al. (1975) and repeated in more detail by Staudenmann et al. (1980). The gravitational phase shift β is obtained by the optical path difference (Eq. (8)), which can also be written in the form of a path integral along the neutron trajectory (Greenberger and Overhauser, 1979):

$$\beta = \oint \vec{k} \cdot d\vec{s}; \qquad (30)$$

here $\vec{k}$ is taken from Eq. (26) as the canonical momentum

$$\vec{k} = (m/\hbar)\,[\dot{\vec{r}} + (\vec{\omega} \times \vec{r})],$$

which gives two terms of gravitationally induced pahse shifts:

$$\beta = \beta_{grav} + \beta_{Sagnac} = (m/\hbar)[\oint \dot{\vec{r}} . d\vec{s} + \oint(\omega \times \vec{r}).d\vec{s}] =$$

$$= -\frac{2\pi m^2 g \lambda A \sin\phi}{h^2} + \frac{4\pi m}{h} \vec{\omega}.\vec{A}. \qquad (32)$$

In this equation ϕ denotes the angle between the horizontal plane and the surface A enclosed by the two coherent beams (see Fig. 1, Eq. A22, Shull et al., 1980):

$$A = (2\,T^2 + 2\,t\,T)\,tg\Theta_B. \qquad (33)$$

The Sagnac term due to the earth's rotation has also been derived by Page (1975), Anandan (1977), Stodolsky (1979), and Greenberger and Overhauser (1979). When the incident beam is vertical (Fig. 6a), the Sagnac rotational term can be written as

$$\beta_{Sagnac} = (4\,\pi\,m\,\omega\,A/h)\sin\phi_L \sin\varepsilon$$

where ϕ_L is the colatitude of the place where the experiment is performed and ε the rotation angle around the vertical direction ($\varepsilon = 0$ if $\vec{A}$ points to the west). Contrary to Eq. (29), the ratio of the rotation to the ordinary gravitation effect is much larger,

$$\frac{\beta_{Sagnac}}{\beta_{grav}} \simeq 0.025, \qquad (35)$$

and therefore more easily measureable. The related experiments were done at the University of Missouri Reactor in Columbia, and Fig. 6 shows the experimental arrangement and characteristic results for the measurement of the influence of the earth's rotation on the neutron phase (Werner et al., 1979; Staudemann et al., 1980) The experimental points were obtained by rotating the phase shifter at various orientations ε of the interferometer, which gives the beam modulation

$$I = I_0\{A + B\cos[\chi + \beta_{Sagnac}(\varepsilon)]\}. \qquad (36)$$

The ordinary gravitational effect β_{grav} was measured by rotating the interferometer crystal around a horizontal incident beam. In this case the bending of the perfect crystal, due to a change of its weight to the support had to be carefully corrected, which was done by observing X-ray interferences simultaneously (Colella et al., 1975; Staudenman et al., 1980).

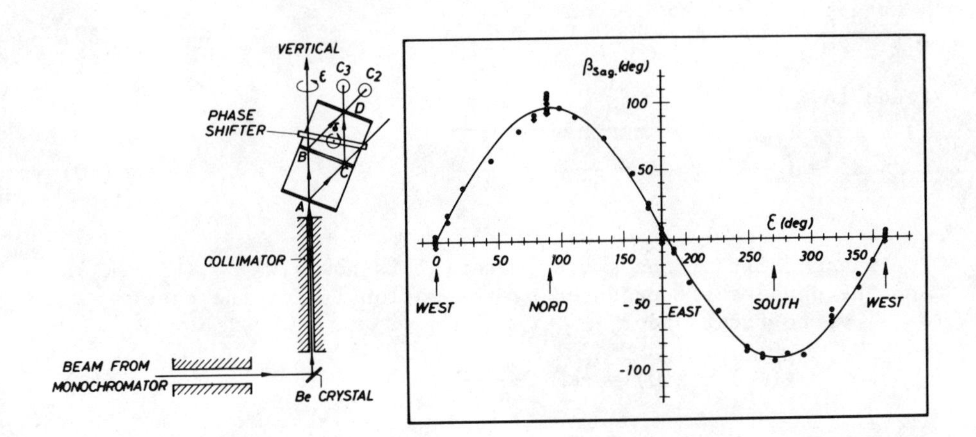

Fig. 6. Influence of the earth's rotation on the phase of the neutron wave: experimental arrangement and final results (Staudenmann et al., 1980).

5. THE NEUTRON FIZEAU EFFECT

The phase shifts that arise for a moving phase shifter depends only on the motion of its boundaries and not on that of its bulk or on its internal motion (e.g., Sears, 1982). The phase shift is described by an index of refraction n (Eq. (8)), which is determined by the mean interaction potential within the material (Eq. A8), which is independent of the wavelength when no resonance is nearby. A motion of the boundary, with a velocity component w_x in direction of the neutron velocity $\vec{v}$, results in a change of the wave vector $\vec{K}'$ and of the total energy $\hbar\omega$ of the neutron wave within the material of thickness D, because now one has a time-dependent interaction

$$V(x,y,t) = \bar{V}, \text{ for } w_x t < x < w_x t + D, \\ = 0, \text{ elsewhere.} \tag{37}$$

The boundary conditions, requiring that the wave functions (plane waves) agree at the moving boundaries, yield

$$(K'_x - k_x)w_x = \omega' - \omega, \tag{38}$$

which, together with

$$\frac{\hbar^2 k^2}{2m} = \hbar\omega \text{ and } \frac{\hbar^2 k^2}{2m} + \bar{V} = \hbar\omega', \tag{39}$$

leads to

$$K'_x = [\frac{mw_x}{\hbar k_x} + \sqrt{(1 - \frac{mw_x}{\hbar k_x}) - \frac{2m\,\bar{V}}{\hbar^2 k_x^2}}]\, k_x. \tag{40}$$

The phase shift of the moving material is now $\chi(w_x) = (K'_x - k_x)D$, and the phase shift produced by the motion is for the case $w_x \ll v$, to first order,

$$\Delta\chi = \chi(w_x) - \chi(0) = -k\,D\,\frac{\bar{V}}{2E}\,\frac{w_x}{v}\;. \tag{41}$$

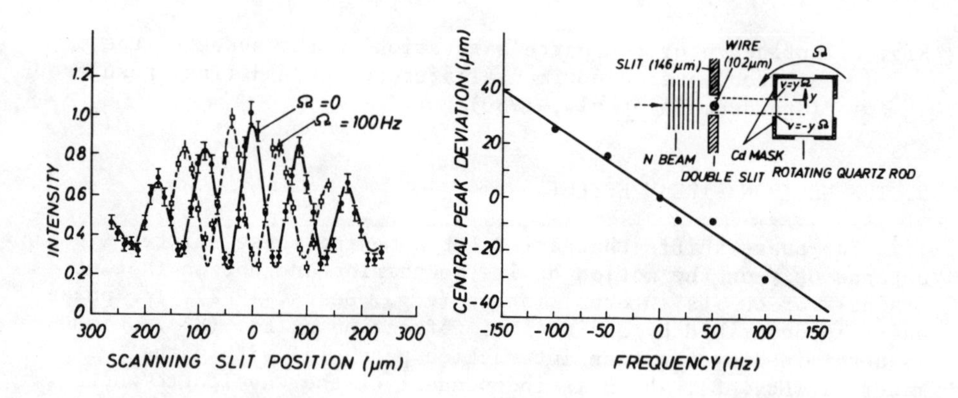

Fig. 7. Characteristic experimental result of the Fizeau experiment (left) and a comparison of calculated and measured Fizeau phase shifts, along with a sketch of the experimental arrangement (right) (Klein et al., 1981).

The relevant experiment has been done by Klein et al. (1981), using a two-slit interferometer based on wave-front division (Gähler et al., 1980) and rotating a small rectangular quartz rod within the coherent beams, which causes differently directed velocity components of the boundary for the two coherent beams. The experimental arrangement and the agreement between theory and experiment is visualized in Fig. 7. Various theoretical interpretations and the inclusion of relativistic effects were given recently by Horne et al., (1983).

APPENDIX: Perfect Crystal Neutron Optics

A1. Introduction

Most crystals and even so-called single crystals have a perfect crystal structure only within small mosaic blocks whose dimensions are in the order of 1 000 Å and whose mutual tilt angle is on the order of minutes of arc. In this case the phase relation of the beam is lost by the statistical nature of this mosaic structure, and the diffracted intensity of the individual blocks is additive and describable by the kinematic diffraction theory, commonly used by crystallographers and described in standard text books.

During the last decades large perfect crystals became available, mainly due to the success of semiconductor technology. In such crystals, mainly silicon and germanium, macroscopic regions of an undisturbed lattice exist without any dislocation lines, variation of the lattice constant, swirles, etc. These crystals represent a large coherent interaction volume, where the mutual interaction of the incident beam and the various diffracted beams have to be accounted for, including their phase relations. This can be described by the dynamical diffraction theory, developed first for X rays by Darwin (1914), Ewald (1916), and von Laue (1931), and summarized in the books of von Laue (1941), Zachariasen (1945), James (1950), and, recently, Pinsker (1978). The position of the diffraction maxima are given by the well-known Bragg relation $n\lambda = 2d_{hkl} \sin\Theta_B$, where λ is the neutron wavelength, n the order of diffraction, d_{hkl} the lattice space of the reflecting plane, and Θ_B the Bragg angle. These positions are rather well described by the kinematical and dynamical diffraction theory, but the intensity and the phases of the beams after diffraction at a perfect crystal are given by the dynamical diffraction theory only. The dynamical diffraction theory has been extended to electron diffraction by Bethe (1928) and von Laue (1948) and to neutron diffraction by Goldberger and Seitz (1947). Review articles have been written by Stern and Taub (1970), for electron diffraction, and by Rauch and Petrascheck (1978), Sears (1978), and Stassis and Oberteuffer (1974), for neutron diffraction. Due to the different kinds of interaction and the different wavelengths involved, the experimental results are quite different for these three types of radiation, which are briefly compared in the following table. Especially due to the weak absorption in the neutron case, perfect crystals are a unique tool for performing neutron optical experiments, using the coherent beams within or outside the perfect crystal; here crystals act as coherent beam splitters, mirrors, and analysers (see Bonse and Rauch, 1979; Klein and Werner, 1982).

	X-rays	electrons	neutrons
kind of radiation	electromagnetic	matter waves	
kind of interaction	electromagnetic		nuclear
interaction volume	atom		nucleus
absorption and inelastic interaction processes	medium	strong	weak
wave length	1 Å	$\frac{1}{100}$ Å	1 Å

A2. Basic Equations

We start with the Schrödinger equation

$$[-\frac{\hbar^2}{2m}\underline{\nabla}^2 + V(\vec{r})]\psi(\vec{r}) = E\,\psi(\vec{r}), \tag{A1}$$

using a strictly periodical potential $V(\vec{r}) = V(\vec{r}+\vec{R}_n)$ between the nuclei, at the positions $\vec{R}_n$, and a neutron of the energy E and mass m. The periodic structure suggests a Bloch "Ansatz"

$$\psi(\vec{r}) = u(\vec{r})\, e^{i\vec{k}.\vec{r}}, \tag{A2}$$

where one can expand $u(\vec{r})$ and $V(\vec{r})$ in a Fourier series, using the reciprocal lattice vector $\vec{G}$:

$$u(\vec{r}) = \sum_G u(\vec{G})\exp(i\vec{G}.\vec{r}), \quad V(\vec{r}) = \sum_G V(\vec{G})\exp(i\vec{G}.\vec{r}). \tag{A3}$$

Inserting these equations into the Schrödinger equation and considering the periodicity of the lattice, we get

$$[\frac{\hbar^2}{2m}(\vec{K}+\vec{G})^2 - E]u(\vec{G}) = -\sum_{\vec{G}'} V(\vec{G}-\vec{G}')u(G'), \tag{A4}$$

which is a homogeneous set of linear equations for $u(\vec{G})$. This set has nontrivial solutions only if the secular determinant is equal to zero, which requirement places restrictions on the $\vec{K}$ vector of the waves within the crystal.

The wavelength $\lambda(\lambda = 2\pi/k)$ of thermal neutrons is much smaller than the distance of the neutron-nucleus interaction, and therefore we can use the Fermi pseudopotential to describe this interaction:

$$V(\vec{r}) = \frac{2\pi\hbar^2}{m} \Sigma\, b_c \delta(\vec{r} - \vec{R}_r), \tag{A5}$$

with b_c denoting the bound coherent scattering length tabulated for most of the nuclei (e.g., Koester et al., 1981). The thermal motion of the nuclei can be treated by the Debye-Waller factor, which reduces the b_c value in Eq. (A5) slightly with increasing temperature. The absorption or inelastic scattering effects can be covered by an imaginary part of the scattering lengths. Both effects are commonly very small for neutrons and are not considered here. From equation (A3) and (A5) follows immediately that

$$V(\vec{G}) = \frac{1}{V} \int V(\vec{r}) \exp(-i\vec{G}.\vec{r}) d\vec{r} = \frac{2\pi\hbar^2 b_c F_g}{m v_c}, \tag{A6}$$

where v_c is the volume of the unit cell and F_g is the geometrical structure factor defined for any crystal structure (e.g., Kittel, 1966). For the most frequently used reflections on silicon and germanium crystals, we have the simple relation $|V(220)| = \sqrt{2}|V(111)| = V(000) = 2\pi\hbar^2 b_c N/m$, where we have used the Miller indices for the definition of the reciprocal lattice vectors and N for the number of atoms per unit volume. Notice that $|V(\vec{G})/E|$ is generally much smaller than 1, typically of the order 10^{-5}, which provides the basic for approximations enabling us to solve eq. (A4), because it states that the wave number K within the crystal differs only slightly from the vacuum wave number k. If the incident beam is far from the Bragg position, then $|(K + G^2) - k^2| \gg |2mV(0)/\hbar^2|$ and all $u(\vec{G} \neq 0)$ can be neglected compared to u(0); thus

$$\left[\frac{\hbar^2}{2m} K^2 - E\right] u(0) = -V(0) u(0), \tag{A7}$$

which is called the "one-beam approximation." Using the continuity condition for the tangential component of the waves at the boundary between crystal and vacuum, one gets the index of refraction

$$n = \frac{K}{k} \simeq 1 - \frac{V(0)}{2E} = 1 - \lambda^2 \frac{N b_c}{2\pi}. \tag{A8}$$

For all materials this value is very near to 1, but (A8) is nevertheless often used in neutron optics for describing prism deflection (Maier-Leibnitz and Springer, 1962; Schneider and Shull, 1971), total reflection (Fermi and Zinn, 1946; Maier-Leibnitz and Springer, 1963), the production of phase shifts between coherent beams (Rauch et al., 1974; Bonse and Rauch, 1979), and the realization of supermirrors (Schoenborn et. al.,

1974; Mezei and Daglesh, 1977). These experimental situations are schematically shown in Fig. A-1.

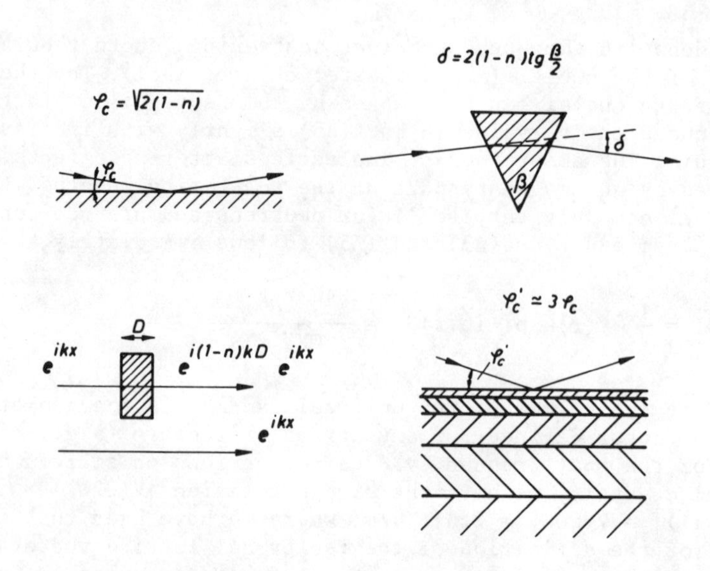

Fig. A.1. Neutron optical effects based on the index of refraction only (total reflection, prism deflection, phase shifter, and supermirror).

Here we are mainly interested in the two-beam case, where only one Bragg reflection (called G) is excited and, therefore, only u(0) and $u(\vec{G})$ are assumed to be different from zero. In this case Eq. (A4) reduces to

$$[\frac{\hbar^2}{2m} K^2 - E + V(0)]u(0) + V(-\vec{G})\ u(\vec{G}) = 0,$$

$$V(\vec{G})\ u(0) + [\frac{\hbar^2}{2m}\ (\vec{K}+\vec{G})^2 - E + V(0)]u(\vec{G}) = 0, \qquad \text{(A9)}$$

which are the fundamental equations of dynamical diffraction within the two beam approximation. Because K and k are expected to be only slightly different, we introduce the "Anregungsfehler" ε (von Laue, 1941)

$$K^2 \equiv k^2(1 + 2\varepsilon) \qquad \text{(A10)}$$

and obtained, from the boundary condition

$$\vec{K} = \vec{k} + \frac{\varepsilon\ k}{\cos\ \gamma}\ \vec{n},$$

that

$$(\vec{K} + \vec{G})^2 = k^2(1 + \frac{2\varepsilon}{b} + \alpha), \tag{A11}$$

where $\vec{n}$ is the normal on the surface while $b = \cos\gamma/\cos\gamma_G$ is the ratio of the cosines of the angles between the normal on the surface and the incident and reflected beams, respectively. The quantity α describes the deviation of the beam from the exact Bragg direction:

$$\alpha = (G^2 + 2\vec{k}\vec{G})/k^2 = 2(\Theta_B - \Theta)\sin 2\Theta_B. \tag{A12}$$

Using these quantities the secular determinant of Eq. (A9) reads

$$\begin{vmatrix} 2\varepsilon + \frac{V(0)}{E} & \frac{V(-\vec{G})}{E} \\ \frac{V(\vec{G})}{E} & \frac{2\varepsilon}{b} + \alpha + \frac{V(0)}{E} \end{vmatrix} = 0, \tag{A13}$$

which gives the solution for ε as

$$\varepsilon_{1,2} = \frac{1}{4}\{-\alpha b - (1-b)\frac{V(0)}{E} \pm \sqrt{[\alpha b - (1-b)\frac{V(0)}{E}]^2 + 4b\frac{V(\vec{G})V(-\vec{G})}{E^2}}\} \tag{A14}$$

Therefore, within the crystal four excited waves exists, with wave vectors K_1 and K_2, near the forward direction and vectors $(\vec{K}_1 + \vec{G})$ and $(\vec{K}_2 + \vec{G})$ near the Bragg direction (see Fig. A.2). By a variation of the direction of the incident beam (or of the wavelength), the allowed end points of these vectors lie on the hyperbolic dispersion surface marked in Fig. A.2 as S_1 and S_2, respectively. One notices that the exact solution not only shifts the end points from the Laue (La) to the Lorentz (Lo) points (index of refraction correction) but requires that these end points lie on the dispersion surfaces. From Eq. (A9) one obtains the amplitude relation

$$\frac{u_{1,2}(\vec{G})}{u_{1,2}(0)} = -\frac{2\varepsilon_{1,2} + \frac{V(0)}{E}}{\frac{V(-G)}{E}}, \tag{A15}$$

with the largest contributions appearing for the central region near the Lorentz point. The total wave function consists of two wave fields (1 and 2) such that both have components near the forward and near the Bragg direction:

$$\psi^{(1,2)} = \psi_0^{(1,2)} + \psi_G^{(1,2)} \tag{A16}$$

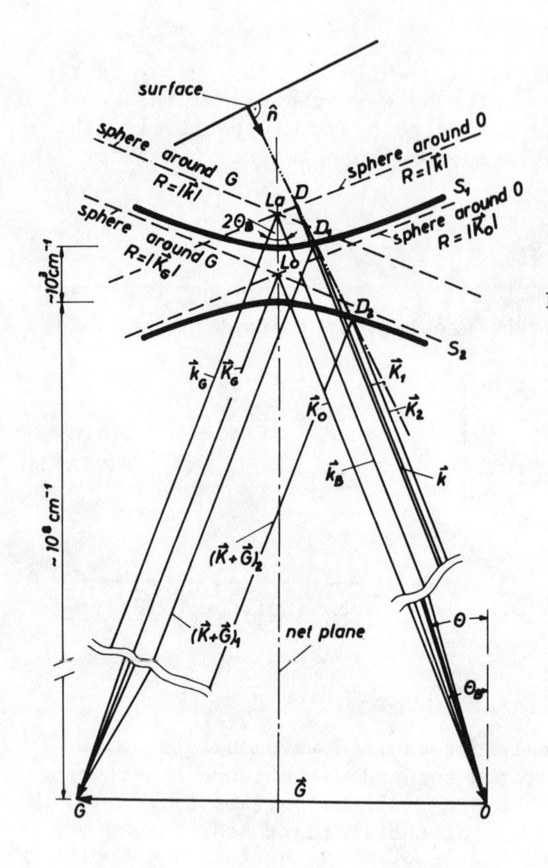

Fig. A.2. Indication of the dispersion surfaces and the internal neutron waves in the case of perfect-crystal neutron Bragg diffraction.

A3. Solution for a Plane Slab in Laue Position

In the Laue position the reflected beam leaves the crystal on the rear side of the crystal ($b > 0$). The boundary conditions at the entrance surface requires that there be only waves in the forward direction:

$$u_1(0) + u_2(0) = u_0,$$
$$u_1(\vec{G}) + u_2(\vec{G}) = 0, \tag{A17}$$

where u_0 is the amplitude of the incoming wave assumed to be a plane wave $\psi_e = u_0\, e^{ikr}$. In a similar sense we continue the internal waves at the rear surface with a plane wave in the forward direction ψ_0 and in the Bragg ψ_G direction ($\psi_0 = \psi_0^1 + \psi_0^2$,

$\psi_G = \psi_G^1 + \psi_G^2$). Using first Eqs. (A15) and (A17) and afterwards the boundary condition for the K-vectors, one gets the wave functions of the wave behind the crystal of thickness t:

$$\psi_0 = V_0\psi_e = [\cos(A\sqrt{1+y^2}) + \frac{y}{\sqrt{1+y^2}} \sin(A\sqrt{1+y^2})]e^{iPt}\psi_e,$$

$$\psi_H = V_H \exp(2\pi iyt/\Delta_0)\psi_e = [-i\frac{\sin(A\sqrt{1+y^2})}{\sqrt{1+y^2}}]. \tag{A18}$$

$$\cdot\, e^{it(P+2\pi iy/\Delta_0)}\psi_0.$$

Where we have used the generally accepted dimensionless parameters y and A instead of ε and t for the reduced angular deviation and the reduced crystal thickness, i.e.,

$$y = \frac{\alpha b - (1-b)V(0)/E}{2\sqrt{|b|}\sqrt{|V(\vec{G})V(-\vec{G})|/E}} = \frac{(\Theta_B - \Theta)k^2 \sin 2\Theta_B}{4\pi b_c N}$$

and

$$A = \pi t/\Delta_0 = \frac{2\pi t|b_c F_g|}{V_c k\sqrt{|\cos\gamma\cos\gamma_G|}} = \frac{2\pi b_c N}{k\cos\Theta_B}t, \tag{A19}$$

and

$$P = -\frac{\pi y}{\Delta_0} - \frac{k}{2\cos\gamma}\frac{V(0)}{E} = -\frac{\pi y}{\Delta_0} - \frac{Nb_c}{\cos\Theta_B},$$

where the second part of these equations correspond to a symmetrical Laue reflection at an even Si reflection (b = 1, V(G) = V(−G) = V(0)). From the wave functions, Eq. (A18), one easily finds the relative intensities ($P_0 = I_0/I_e$, $P_G = I_G/|b|I_e$) to be

$$P_G(y) = \frac{\sin^2[(\pi t/\Delta_0)\sqrt{1+y^2}]}{1+y^2}, \tag{A20}$$

$$P_0(y) = 1 - P_G(y).$$

The intensity variation in the center (y = 0) is determined by the "Pendellösung" length Δ_0 also defined in Eq. (A19). These

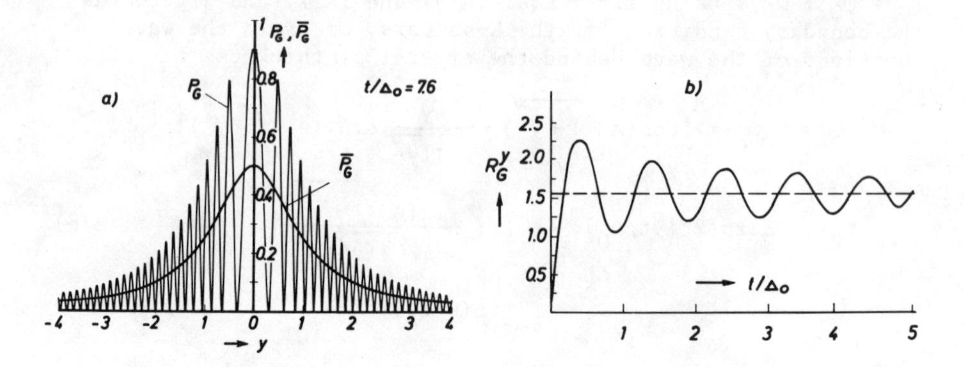

Fig. A.3. "Pendellösung" diffraction pattern for Laue diffraction and integrated intensity ($\bar{R} = \int P_G(y)dy$) as a function of the thickness of the crystal plates.

reflectivity curves show a very narrow fine structure depending on the crystal thickness (Fig. A.3), but most experiments are sensitive to only the averaged reflection curve,

$$\bar{P}_G(y) = \frac{1}{2}\frac{1}{1+y^2}, \tag{A21}$$

because often uncertainties in the crystal thickness or the unsharpness in wavelength or the angle cause this smearing effect. Its half-width is, on the y scale, $y_H = 2$, which corresponds for thermal neutrons to 1 - 5 seconds of an arc on the usual angular scale.

The direction of the neutron current is again calculated with the aid of Eq. (A18):

$$\vec{j}_{1,2} = \frac{\hbar}{2im}[\psi^{(1,2)*}\nabla\psi^{(1,2)} - \psi^{(1,2)}\nabla\psi^{(1,2)*}]$$

$$= \frac{\hbar k}{m}[\hat{k}_0|\psi_0^{(1,2)}|^2 + \hat{k}_G|\psi_G^{(1,2)}|^2], \tag{A22}$$

where interference terms, which oscillate with the lattice periodicity, dropped out after averaging over a unit cell. Because $\psi_0^{(1,2)}$ and $\psi_G^{(1,2)}$ strongly depend on y, the neutron current direction within the crystal changes drastically between the forward $\hat{k}_0$ direction and the Bragg direction $\hat{k}_G$ for rather small changes of y. The angle $\Omega_{1,2}$ between the neutron current and the reflecting net plane depends on y in accordance with

$$\Gamma_{1,2} = \frac{tg\Omega_{1,2}}{tg\Theta_B} = -\frac{1 - b|y \pm \sqrt{y^2 + b}|^2}{1 + b|y \pm \sqrt{y^2 + b}|^2}, \tag{A23}$$

which gives, for the symmetrical (220) Laue reflection on Si and 2 Å neutrons, an angle Ω of about 21° for a beam deflection $\Theta_B - \Theta$ of only 0.5 sec of arc; thus an angle magnification of about 10^5 exists. In order to observe the neutron currents one has to use a small entrance slit in front of the crystal; then the intensity within the crystal spreads about $-\Theta_B \leq \Omega \leq \Theta_B$ (Borrmann fan), and the intensity at the rear surface becomes

$$P_0(\Gamma) = \frac{(1-\Gamma)\cos^2(A\sqrt{1-\Gamma^2} + \frac{\pi}{4})}{(1-\Gamma)\sqrt{1-\Gamma^2}}$$

$$P_G(\Gamma) = \frac{\sin^2[A\sqrt{1-\Gamma^2} + \frac{\pi}{4}]}{\sqrt{1-\Gamma^2}} \tag{A24}$$

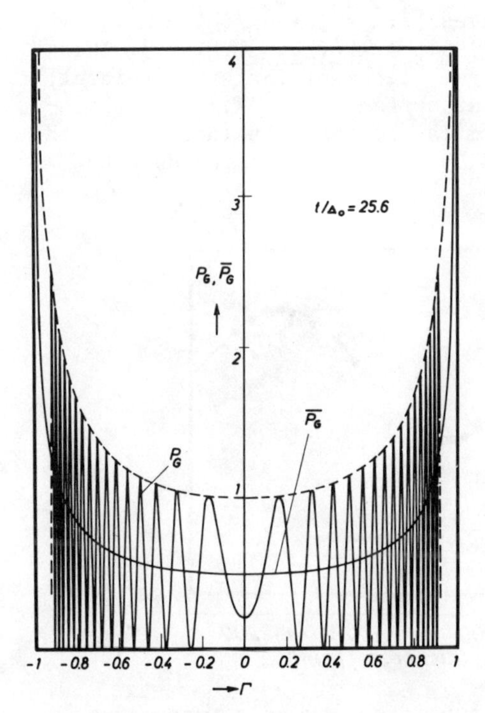

Fig. A.4. Intensity distribution within the Borrmann fan, showing again the "Pendellösung" interferences.

which is graphically shown in Fig. A.4. The intensities of Eq. (A24) are obtained from a coherent superposition of the wave functions of wave field 1 and 2 at the rear surface of the crystal and for +y and -y, respectively.

The oscillatory structure becomes narrower for thick crystals, which is analogous to the situation shown in Fig. A.3.

The pyhsical difference between the two wave fields consists in the fact that one wave field has its nodes at the position of the nuclei in the lattice and the other one has its nodes between these positions. This explains the slightly different wave vectors, because the mean interaction potential is different, and it explains that for strongly absorbing crystals one wave field is much more strongly attenuated than the other one, which also causes the "Pendellösung" fine structure to disappear.

The spatial limitation due to the slit causes the incident wave to be described by a superposition of plane waves,

$$\psi_e = \frac{1}{(2\pi)^3} \int d^3k' F(\vec{k},\vec{k}') \exp(i\vec{k}'\vec{r}), \tag{A25}$$

which can describe spherical waves $[F(\vec{k},\vec{k}') = u_0/(k'^2 - k^2)]$ or cylindrical waves (e.g., Kato, 1961; Azarov et al, 1974). This consideration is equivalent to the statement that the incident wave excites the whole dispersion surface (Fig. A.2.) simultaneously. The results are very similar to that obtained within the plane-wave theory (Eq.(A20) and Eq. (A24)) and coincide with them for thick crystals ($A \gg \pi$).

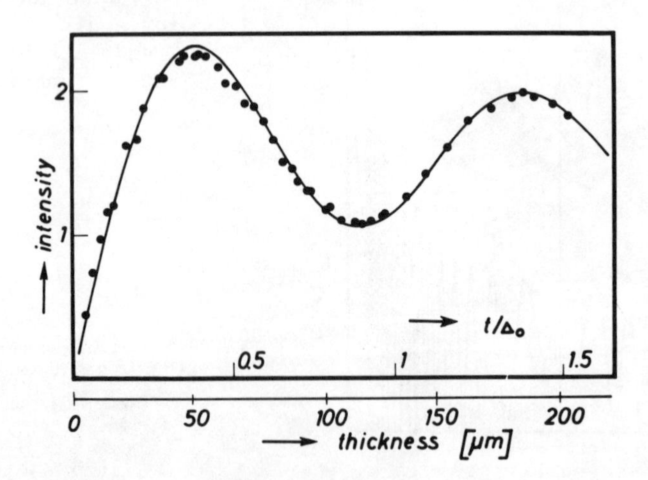

Fig. A.5. Observed integrated Laue intensity as a function of the crystal thickness (Sippel et al., 1965)

A4. Performed Experiments

The verification of the dynamical diffraction theory for neutrons started with the observation of the integrated diffracted intensity as a function of the crystal thickness (Sippel et al., 1965; Kikuta et al., 1971). This integrated intensity R_G is given by the Waller formula obtained from Eq. (A20) as $R_G = \int P_G(y)dy$ (Fig. A.5.). The influence of absorption on the reflection curve has been measured by Shil'shtein et al., (1971) and previously by the observation of the accompanying γ radiation by Knowles (1956).

The most detailed investigation of the beam profile within the Borrmann fan has been made by Shull (1968); he observed the

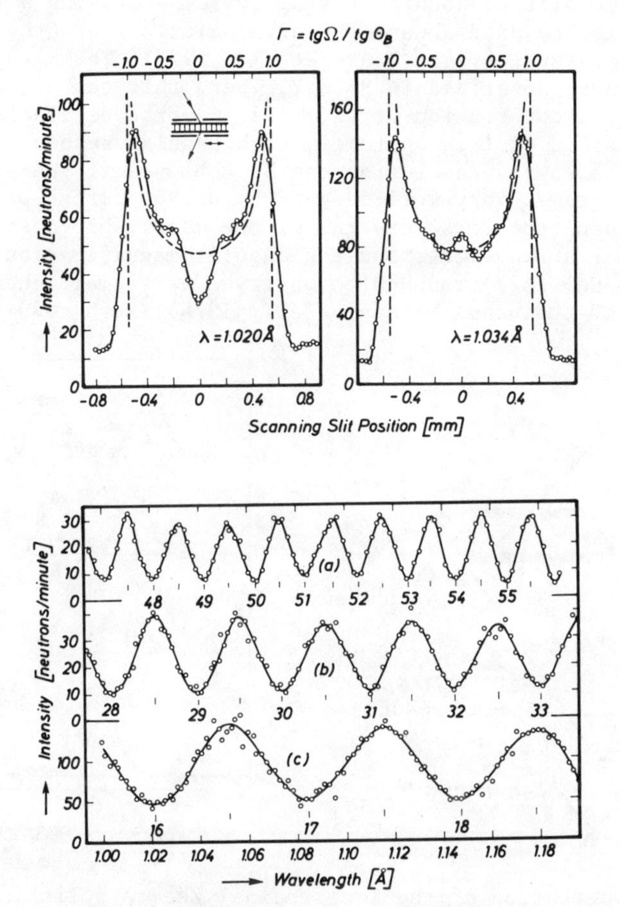

Fig. A.6. Observed Borrmann intensity profile and intensity variation of the central part as a function of λ (or A) (Shull, 1968).

distribution as a function of Γ (see Eq. (A24) and Fig. A.4.) for various wave lengths λ, which is equivalent to a variation of A (Eq. (A19)) and which is shwon in Fig. A.6.

The "Pendellösung" interference pattern becomes also visible on the angular scale if the rocking curves of two successive Laue reflections in a nondispersive arrangement are measured (Bonse et al., 1977). This rocking curve is given by the convolution of the reflection curves (Eq. (A20)),

$$P(y) = \int P_G(y')P_G(y-y')dy', \tag{A26}$$

and shows a marked needle structure near y = 0 with a half-width on the order of 0.001 sec of an arc. These measurements have been done with a monolithic double crystal system and using a wedge-shaped material rotated around the beam axis to reach the required angular sensitivity (Bonse et al., 1979). The results are similar to those described in Sec. 2, where this technique has been used for a observation of wide-slit neutron diffraction (Fig. 3). Instead of this extremely high angular resolution, an extremely high energy resolution can be achieved if the y value of the beam becomes very well-defined by narrow slits, placed in front and behind the first crystal of the monolithic-designed double crystal arrangement, and the angular magnification according to Eq. (A23) is used to observed very small changes of the k value of the beam (Kikuta et al., 1975). Such results,

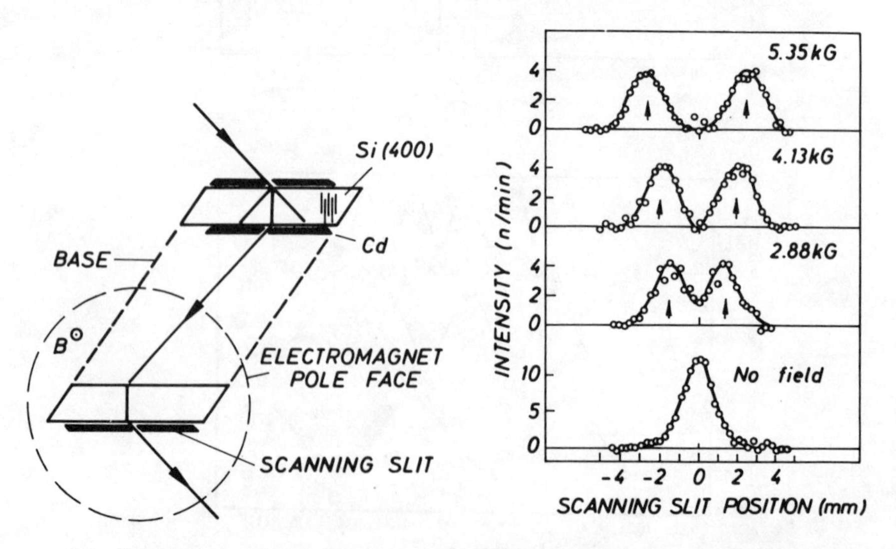

Fig. A.7. Observation of the longitudinal Zeeman splitting by observing the beam splitting within the Borrmann fan, using a monolithic two-plate crystal arrangement (Zeilinger and Shull, 1979).

obtained in connection with the observation of the longitudinal Stern-Gerlach effect of the neutron beam within a magnetic field, are shown in Fig. A.7. (Zeilinger and Shull, 1979).

REFERENCES

Y. Aharanov,and L. Susskind, Phys. Rev. 158 (1967) 1237.

B. Alefeld, G. Badurek, and H. Rauch, Z. Physik B41 (1981) 231.

L.W. Alvarez, and F. Bloch, Phys. Rev. 57 (1940) 111.

J. Anandan, Phys. Rev. D15 (1977) 1448.

L.V. Azarov, R. Kaplow, N. Kato, R.J. Weiss, A.J.C. Willson, and R.A. Young, X-Ray Diffraction (McGraw-Hill, London, 1974).

G. Badurek, H. Rauch, A. Zeilinger, W. Bauspiess, and U. Bonse, Phys. Rev. D14 (1976) 1177.

G. Badurek, H. Rauch, and J. Summhammer, Phys. Rev. Lett. 51 (1983), 1015.

W. Bauspiess, U. Bonse, and W. Graeff, J. Appl. Cryst. 9 (1976) 68.

W. Bauspiess, U. Bonse, and H. Rauch, Nucl. Instr. Meth, 157 (1978) 495.

H.J. Bernstein, Phys. Rev. Lett. 18 (1967) 1102.

H.J. Bernstein, in Neutron Interferometry, U. Bonse and H. Rauch, eds. (Clarendon Press, Oxford, 1979), p. 231.

H.J. Bernstein, and A.V. Phillips, Scientific American 245 (1981) 121.

H.J. Bernstein, and A. Zeilinger, Phys. Lett. 75A (1980) 169.

H. Bethe, Ann. Physik 87 (1928) 55.

U. Bonse, W. Graeff, and H. Rauch, Phys. Lett. 69A (1979) 420.

U. Bonse, W. Graeff, R. Teworte, and H. Rauch, Phys. Stat. Sol. (a)43 (1977) 487.

U. Bonse, and M. Hart, Appl. Phys. Lett. 6 (1965) 155.

U. Bonse, and U. Kischko, Z. Physik A305 (1982) 171.

U. Bonse, and H. Rauch, eds., Neutron Interferometry (Clarendon Press, Oxford, 1979).

L. de Broglie, Nature 112 (1923) 540.

J. Byrne, Nature 275 (1978) 188.

R. Colella, A.W. Overhauser, and S.A. Werner, Phys. Rev. Lett. 34 (1975) 1472.

J.W.T. Dabbs, J.A. Harvey, D. Paya, and H. Horstmann, Phys. Rev. B139 (1965) 765.

C.G. Darwin, Phil. Mag. 27 (1914) 315, 675.

G. Eder, and A. Zeilinger, Il Nuovo Cim. 34B (1976) 76.

P.P. Ewald, Ann. Phys. 49 (1916) 1, 117.

E. Fermi, and W.H. Zinn, Phys. Rev. 70 (1946) 103.

R. Gähler, J. Kalus, and W. Mampe, J. Phys. E13 (1980) 546.

R. Gähler, A.G. Klein, and A. Zeilinger, Phys. Rev. A23 (1981) 1611.

M.L. Goldberger, and F. Seitz, Phys. Rev. 71 (1947) 294.

W. Graeff, in Neutron Interferometry, U. Bonse and H. Rauch, eds. (Clarendon Press, Oxford, 1979), p. 34.
D.M. Greenberger, and A.W. Overhauser, Rev. Mod. Phys. 51 (1979) 43.
O. Hittmair, Lehrbuch der Quantentheorie, (Verlag K. Thiemig, München, 1972).
M.A. Horne, A. Zeilinger, A.G. Klein, and G.I. Opat, Phys. Rev. A28 (1983) 1.
R.W. James, The Optical Principles of the Diffraction of X-Rays, (Bell & Sons, London, 1950).
H. Kaiser, S.A. Werner, and E.A. George, Phys. Rev. Lett. 50 (1983) 560.
N. Kato, Acta Cryst. 14 (1961) 526.
H. Kendrick, J.S. King, S.A. Werner, and A. Arrott, Nucl. Instr. Meth. 79 (1970) 82.
S. Kikuta, K. Kohna, N. Minakawa, and K. Doi, J. Phys. Soc. Japan 31 (1974) 954.
S. Kikuta, I. Ishikawa, K. Kohna, and S. Hioshino, J. Phys. Soc. Japan 39 (1975) 471.
Ch. Kittel, Introduction to Solid State Physics, (John Wiley, New York, 1966).
A.G. Klein, P.D. Kearney, G.I. Opat, and R. Gähler, Phys. Lett. 83A (1981) 71.
A.G. Klein, and G.I. Opat, Phys. Rev. Lett. 37 (1976) 238.
A.G. Klein, G.I. Opat, A. Cimmino, A. Zeilinger, W. Treimer, and R. Gähler, Phys. Rev. Lett. 46 (1981) 1551.
A.G. Klein, and S.A. Werner, Rep. Progr. Physics 46 (1983) 259.
E. Klempt, Phys. Rev. D13 (1976) 3125.
J.W. Knowles, Acta Cryst. 9 (1956) 61.
L. Koester, Z. Physik 182 (1965) 328.
L. Koester, Z. Physik 198 (1967) 187.
L. Koester, Phys. Rev. D14 (1976) 907.
L. Koester, and W. Nistler, Z. Physik A272 (1975) 189.
L. Koester, H. Rauch, M. Herkens, and K. Schröder, Summery of Neutron Scattering Lengths, (JÜL-1755, KFA-Jülich, 1981).
E. Krüger, Nukleonika 25 (1980) 890.
M. von Laue, Ergeb. Exakt. Naturw. 10 (1931) 133.
M. von Laue, Materiewellen und Interferenzen (Akad. Verlagsges., Leipzig, 1948).
M. von Laue, Röntgenstrahlinterferenzen (Akad. Verlagsges., Leipzig, 1941), 2nd edn., Frankfurt/M., 1960.
H. Maier-Leibnitz, and T. Springer, Z. Physik 167 (1962) 386.
H. Maier-Leibnitz, and T. Springer, Reactor Science Techn. 17 (1963) 217.
A.W. McReynolds, Phys. Rev. 83 (1951) 172, 233.
A. Messiah, Quantum Mechanics (North-Holland, Amsterdam, 1965).
F. Mezei, Z. Physik 255 (1972) 146.
F. Mezei, in Neutron Interferometry (U. Bonse and H. Rauch, eds. Clarendon Press, Oxfors, 1979) p. 265.
F. Mezei, and P.A. Daglesh, Comm. Phys. 2 (1977) 41.

A.W. Overhauser, and R. Colella, Phys. Rev. Lett. 33 (1974) 1237.
L.A. Page, Phys. Rev. Lett. 35 (1975) 543.
D. Petrascheck, Acta Phys. Austr. 45 (1976) 217.
D. Petrascheck, and H. Rauch, Acta Cryst. 1984, in the press.
Z.G. Pinsker, Dynamical Scattering of X-Rays in Crystals (Springer Verlag, Berlin, 1978).
I. Pitowsky, Phys. Rev. Lett. 48 (1982) 1299.
N.F. Ramsey, Inst. Phys. Conf. Ser. 64 (1982) 5.
H. Rauch, in Neutron Interferometry, U. Bonse and H. Rauch, eds. Clarendon Press, Oxford, 1979) p. 161.
H. Rauch, U. Kischko, D. Petrascheck, and U. Bonse, Z. Physik B51 (1983) 11.
H. Rauch, and D. Petrascheck, in Neutron Diffraction, H. Dachs, ed., Top. Curr. Phys. 6 (1978) 303, (Springer-Verlag, New York).
H. Rauch, W. Treimer, and U. Bonse, Phys. Lett. A47 (1974) 369.
H. Rauch, A. Wilfing, W. Bauspiess, and U. Bonse, Z. Physik B29 (1978) 281.
H. Rauch, A. Zeilinger, G. Badurek, A. Wilfing, W. Bauspiess, and U. Bonse, Phys. Lett. 54A (1975) 425.
H. Scheckenhofer, and A. Steyerl, Nucl.Instr. Meth. 179 (1981) 393.
C.S. Schneider, and C.G. Shull, Phys. Rev. B3 (1971) 830.
B.P. Schoenborn, D.L.D. Caspar, and O.F. Kammerer, J. Appl. Cryst. 7 (1974) 508.
V.F. Sears, Can. J. Phys. 56 (1978) 1261.
V.F. Sears, Phys. Rev. 82 (1982a) 1.
V.F. Sears, Phys. Rev. D25 (1982b) 2023.
F. Selleri, Ann. Fond. Louis de Broglie 7 (1982) 45.
S.Sh. Shil'shtein, V.A. Somenkov, and V.P. Dokashenko, Zh. ETF Pis. Red. 13 (1971) 301.
C.G. Shull, Phys. Rev. Lett. 21 (1968) 1585.
C.G. Shull, Phys. Rev. 179 (1969) 752.
C.G. Shull, Inst. Phys. Conf. Ser. 64 (1982) 157.
C.G. Shull, A. Zeilinger, G.L. Squires, M.A. Horne, D.K. Atwood, and J. Arthur, Phys. Rev. Lett. 44 (1980) 1715.
D. Sippel, K. Kleinstück, and G.E.R. Schulze, Phys. Lett. 14 (1965) 174.
C. Stassis, and J.A. Oberteuffer, Phys. Rev. 10 (1974) 5192.
J.L Staudenman, S.A. Werner, R. Colella, and A.W. Overhauser, Phys. Rev. A21 (1980) 1419.
R.M. Stern, and H. Taub, CRC, Critical Rev. Solid State 221 (1970).
L. Stodolsky, in Neutron Interferometry, U. Bonse and H. Rauch, eds. (Clarendon Press, Oxford, 1979), p. 313.
M.E. Stoll, E.K. Wolff, and M. Mehring, Phys. Rev. A17 (1978) 1561.
J. Summhammer, G. Badurek, H. Rauch, and U. Kischko, Phys. Lett. 90A (1982) 110.
J. Summhammer, G. Badurek, H. Rauch, U. Kischko, and A. Zeilinger, Phys. Rev. A27 (1983a) 2523.

J. Summhammer, G. Badurek, H. Rauch, and U. Kischko, 1983b, in preparation.

S.A. Werner, R. Colella, A.W. Overhauser, and C.F. Eagen, Phys. Rev. Lett. 35 (1975) 1053.

S.A. Werner, J.L. Staudenmann, and R. Colella, Phys. Rev. Lett. 42 (1979) 1103.

E.P. Wigner, Am. J. Phys. 31 (1963) 6.

E.P. Wigner, Am. J. Phys. 38 (1970) 1005.

W.H. Zachariasen, Theory of X-Ray Diffraction in Crystals (Wiley, London, 1945).

A. Zeilinger, Nature 294 (1981) 544.

A. Zeilinger, R. Gähler, C.G. Shull, and W. Treimer, AIP Conf. Proc., Neutron Scattering, J. Faber ed. (Amer. Inst. Phys., 1982), p. 93.

A. Zeilinger, and C.G. Shull, Phys. Rev. B19 (1979) 3957.

A. Zeilinger, C.G. Shull, M.A. Horne, and J. Arthur, J. Phys. A (1983), in the press.

A. Zeilinger, C.G Shull, M.A. Horne, and G.L. Squires, in Neutron Interferometry, U. Bonse and H. Rauch eds. (Clarendon Press, Oxford, 1979), p. 48.

A UNIFIED EXPERIMENT FOR TESTING BOTH THE INTERPRETATION AND THE REDUCTION POSTULATE OF THE QUANTUM MECHANICAL WAVE FUNCTION

G. Tarozzi

Dipartimento di Fisica
Universita' di Bari
Bari, Italy

ABSTRACT

A new experiment on the wave-particle duality is discussed, unifying the two different kinds of experimental proposals recently put forward for revealing the physical properties of "quantum waves" of producing either interference or stimulated emission, but never the two at the same time, and which makes it possible in this way to carry out a single test both of the interpretation of the wave function and of the quantum mechanical postulate of the reduction of the wave packet, a well known basic consequence of the standard theory of measurement. We show moreover, how the hypothesis of quantum waves also enables us to solve Renninger's paradox of negative-result measurements, which has inspired the most explicitly spiritualistic and irrationalistic interpretations of quantum theory.

1. THE DUALISTIC NATURE OF QUANTUM THEORY

In its great conceptual effort to unify the two main theories of classical physics, Newton's mechanics and Maxwell's electromagnetism, and to provide a synthesis of their basic notions, matter and radiation, that is, particles and fields, quantum mechanics was born and still appears, more than half a century since its definitive formulation, as a dualistic theory, at least in a twofold sense.

The first kind of dualism is of a physical nature and concerns the question of the co-presence in quantum theory of two different physical entities, particles and waves, in terms radically changed with respect to the relationship between matter and

G. Tarozzi and A. van der Merwe (eds.), Open Questions in Quantum Physics, 377–390.

radiation peculiar to classical physics, as a result of the experimental confirmation of Einstein's theoretical predictions about the corpuscular properties of radiation and de Broglie's about the undulatory properties of matter.

The second kind of dualism, on the other hand, is mathematical and originates in the coexistence, within the logical structure of quantum mechanics, between a deterministic equation of motion -- directly derived from d'Alembert's classical equation for waves, through the application of the correspondence principle -- regulating the time-evolution of unobserved physical systems and a probabilistic description of the results of the measuring operations performed to determine the observable properties of the same physical systems, without being given another unambiguous specification of precise mutually exclusive conditions for these two different types of transition (*1*).

These two kinds of (physical and mathematical) dualism are at the origin of the most controversial epistemological problems of microphysics: the interpretation of the wave function ψ , which, as everbody knows, is the basic theoretical entity of quantal formalism and the theory of measurement.

We should not, however, forget to mention the existence of a third kind of dualism, which is characteristic of the quantum-mechanical treatment of two separate systems, which can be described both through nonfactorable second-type and factorable first-type state vectors, and lies at the logical roots of the well known Einstein-Bell contradiction, i.e., of the Einstein-Podolsky-Rosen argument and of the Bell theorem, and the resulting question of the compatibility between the relativistic description, where Einstein's locality principle, forbidding the propagation of instantaneous actions at a distance between space-time separate systems, can never be violated, and the quantum-mechanical description, where such a principle does not seem always to be satisfied (*2*).

The discussion of this third kind of dualism represents, nevertheless, the object of the first part of this volume, whereas the aim of the present paper is to show how a solution proposed with respect to the physical duality between waves and particles, through the hypothesis recently put forward, according to which "quantum waves" correspond to physical objects devoid of the fundamental properties of all other physical systems, like energy and momentum, and characterizable only by means of relational properties with the particles, has not only the advantage of providing a non-contradictory interpretation of the wave function, as already stressed, but also contains implications of major importance for the problem of measurement, i.e., even for the "mathematical" dualism of quantum theory.

In the first place it is stressed how, *independently of the result* that could be obtained from the experiments on the interpretation of the wave function, testing Selleri's hypothesis that quantum waves (QW) can produce stimulated emission, one is *always* able to test *also* the validity of one of the fundamental axioms of quantum theory, the so-called reduction of the wave packet (or wave function, or state vector), via the discussion of a new experiment, unifying all the preceding proposals for detecting physical properties of QW.

Secondly, as we shall see in the *Appendix*, it will be shown that the *hypothesis of quantum waves* (QWH) enables us to solve one of the -- in my opinion -- most serious paradoxes of the theory of measurement, such as to inspire the most explicitly spiritualistic and irrationalistic interpretations of microphysics, i.e., the paradox of negative-result measurements, due originally to Renninger (*3*) and subsequently taken up by Wigner (*4*) as the conclusive objection to those theories of measurement which attempted to explain this process as physical interaction between the (microscopic) measured object and the (macroscopic) measuring apparatus (*5*), and as a decisive confirmation of von Neumann's theory, on the basis of which the reduction of the wave function resulting from the measuring process can only be justified through recourse to the intervention of an extraphysical entity, such as the mind or conscioussness of a human observer (*6*).

The relevance therefore emerges of an interpretative approach to the foundations of physics, whose main purpose is to carry on a critical analysis aimed at a conceptual clarification of its fundamental principles and theoretical terms and which appears, even with respect to a problem, such as measurement, always considered a purely syntactical one, methodologically and epistemologically more fertile than the usual formal approaches, according to which the only way of eliminating the conceptual problems of quantum mechanics consists either in modifying the logical structure of the theory or in restricting the validity of its domain of application.

2. TWO CLASSES OF EXPERIMENTS ON THE WAVE-PARTICLE DUALITY

The first proposal for testing physical properties of QW was advanced in 1969 by F. Selleri (*7*) and marked a crucial development with respect to the much debated and still open question of the wave-particle duality. As a consequence of this proposal, the interpretation of the fundamental theoretical concept of quantal formalism was shifted from a mere matter of metaphysical choice, giving rise to an endless dispute in the course of which all the points of view logically conceivable have been maintained, to a genuine empirical problem which can be solved through a direct appeal to experimental evidence.

The basic idea behind this experiment was the acceptance of the physical dualism of waves and particles, but not of its symmetric nature. As in the case of Karl Popper's propensity field notion (*8*), the QWH implies some kind of "ontological priority" of particles with respect to waves, in the sense that waves without particles cannot be characterized by means of the basic properties possessed by all other physical objects, like energy, momentum, charge, and mass, but only through *relational properties with the particles* : the observable properties of *producing interference* and *stimulating emission*. This means that QW would have to belong to a *weaker level* of physical reality, containing objects which are sensible carriers of exclusively relational predicates in the language of quantum mechanical events (*9*).

Such a point of view is not in contrast to, but perfectly consistent even with the opinion expressed by the very proponent of the standard statistical interpretation of the function, that "... both particles and waves have some sort of reality, but it must be admitted that the waves are not carriers of energy and momentum" (*10*).

Nevertheless, the QWH differs conceptually in a significant way both from Born's purely statistical interpretation, since it denies the wave-particle dilemma of the Copenhagen School and also Heisenberg's "withdrawal into mathematical formalism" by trying to explain the physical nature of the dualism, and from de Broglie's concept of the pilot wave possessing a (very) small amount of energy-momentum, whereas QW are a zero-energy undulatory phenomenon.

On the basis of Selleri's original proposal, two different classes of experiments for detecting physical properties of QW have developed, whose logical and epistemological implications go beyond the problem of the interpretation of the wave function and involve the other very controversial and still unsolved problem of measurement, and more precisely the famous quantal postulate of the reduction of the wave function, which -- despite repeated attempts by authoritative authours to eliminate it, on account of its completely *ad hoc* formal structure, not founded on any reasonable physical explanation -- still remains one of the fundamental axioms of standard quantum mechanics.

The experiments of the first type set out to demonstrate that QW, according to the original Selleri hypothesis, have the property of producing stimulated emission in systems of excited atoms, whose excitation energy is the same as the one possessed by the particles (*11,12,13*).

We were concerned, as already stressed (*13*), with a merely interpretative hypothesis, which in no way contrasted with the logical structure of quantum theory, but only with the standard interpretation of its fundamental theoretical entity. This aspect of the conceptual clarification of the QWH was moreover pointed out by the explanation which, in the light of such a hypothesis, one is able to provide with respect to the paradox of negative-result measurements (*14*). Such a paradox, which will be discussed in detail in the Appendix to the present paper, originates from the existence of physical situations in which the reduction of the state vector is not produced by any physical process of particle detection, but seems to occur only as a consequence of the change of our knowledge, that is of the knowledge of the observer about the behaviour of the particle, legitimizing in this way an interactionistic and mentalistic metaphysics, with respect to the mind-body problem, which Newton's classical mechanics, asserting itself against Cartesianism, and Berkeley's subjectivistic idealism had already appeared to have exorcised, once and for all, from the domain of natural science.

In a similar way the explanatory role of the QWH was stressed, in providing a reasonable interpretation of apparently contradictory physical situations, like the two-slits experiment, without any need to weaken the laws of classical logic (*15*), thus preserving the requirement of their analyticity, as one of the basic principles of scientific philosophy.

The experiments of the second type, on the other hand, aim to reveal the interference property of QW, by trying to prove experimentally the persistence of the interference pattern, even in the physical situations where one is able to establish the path followed in an interferometer by the wave without the particle and, consequently, of the particle itself (*16*,*17*).

In this way, however, we are not only testing an interpretative hypothesis with respect to a theoretical entity of quantum mechanics, but also the validity of a basic postulate contained in the formalism of the theory, i.e., the above mentioned axiom of the reduction of the wave packet, one of whose main physical consequence is the wiping out of the interference, when we know which is the path followed by each particle in the interferometer.

The most advanced experiment of this type, which has developed out of other earlier proposals, eliminating the difficulties and contradictions the met with, is due to Garuccio, Rapisarda, and Vigier (GRV) (*17*).

The peculiar feature of this proposed experiment, as already stressed by Selleri and as we shall see in the discussion which follows, is that of being realizable solely and exclusively in the case that the failure of the first-type experiments has shown that quantum waves do not possess the property of producing stimulated emission.

Such a conceptual state of affairs leads to unsatisfactory implications from the epistemological point of view, because in this way an alternative interpretation of the wave function like Selleri's QWH, would allow the orthodox formulation of the theory to be upheld against the possibility of experimental refutation of one of its fundamental postulates, whereas this is not so for the standard interpretation, which would be therefore more scientific on the grounds of Popper's well known principle of falsifiability. In addition, the fact that the two different properties of QW, of producing interference or stimulated emission can be revealed in two mutually exclusive physical situations might justify something like a modified notion of Bohr's complementarity principle, no longer between the undulatory and corpuscular properties of physical systems, but between the two previous undulatory properties and would lead therefore to a kind of "meta complementarity".

We shall show nevertheless the possibility of going beyond such a theoretical impasse, through the proposal of a unified test of the properties of QW, which demonstrate that the two previous classes of experiments, though conceptually very different, as has already been pointed out, are not mutually exclusive, but can be combined in a single experiment, which contains, as special cases, Selleri's experiment, Garuccio, Rapisarda, and Vigier's and moreover, a new variant allows us to test the validity of the quantum mechanical postulate of the reduction of the wave function, even in the case that the first class of experiments, leads to a positive result.

3. A PROPOSAL FOR A NEW UNIFIED EXPERIMENT

In this regard let us consider the experimental device shown in Fig. 1, where we have a very low intensity incoherent source of monochromatic photons, which are made to cross a semireflecting mirror M_1. Both in the transmitted and in the reflected path, there are other two semireflecting mirrors M_2 and M_3, splitting the beam again in two parts: The transmitted ones propagate towards the photomultipliers P_2 and P_1, respectively whereas the reflected ones are concentrated in the interference detector IR.

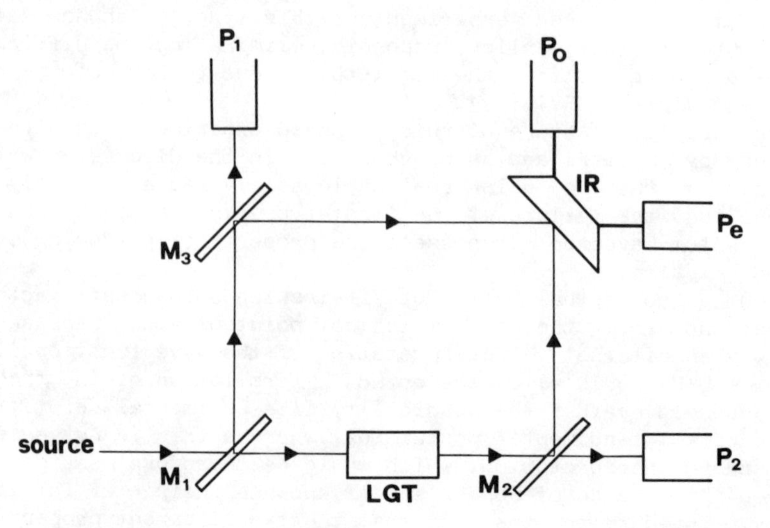

Fig. 1: Experimental set-up for performing three different kinds of experiments on the physical properties of quantum waves.

For our purposes we can use, as already proposed by GRV (*17*), an interference detecting screen devised by Pfleegor and Mandel, which is an instrument built with a stack of thin glass plates, each of which has a thickness corresponding to a half fringe and whose plates are cut and arranged so that any photon falling on the odd plates is fed to one photomultiplier P_o, while photons falling on the even plates are fed to the other P_e. In this way the interference effect correspond to the differences in the detection of P_o and P_e, that is, to $N(P_o) \neq N(P_e)$. Another device for revealing interference has been proposed by A. Gozzini (*18*), who has suggested the use, instead of the Pfleegor and Mandel detector, of a beamsplitter with two photomultipliers P_c and P_d placed in the constructive and destructive interference regions, respectively.

Finally, on the path of the transmitted photon, between M_1 and M_2, we insert a laser gain tube LGT, for the sake of amplification of the particle or of the wave without the particle.

By means of this experimental device we can perform the three following kinds of measurements, corresponding to the three previously mentioned experiments.

1. Measurement of the coincidences $P_1 \wedge P_2$ and $P_1 \wedge \sim (P_o \leftrightarrow P_e)$: Selleri's experiment; by removing M_2 and M_3 we have exactly the same physical situations of the original Selleri's proposal.

If one finds coincidences between the detection of a photon in P_1 and in any other photomultiplier, this implies that a wave without the particle (detected by P_2) has propagated in the transmitted beam and, according to Selleri's assumption, has produced in LGT the stimuled emission of another particle.

At this point, depending on the result obtained, we can perform one or the other one of the following experiments.

2a. Measurement of the differences between the number of coincidences $P_2 \wedge P_o$ and $P_2 \wedge P_e$, that is of $N(P_2 \wedge P_o)$ vs. $N(P_2 \wedge P_e)$: the GRV experiment.

If the first experiment fails, i.e., if QW are not able to produce stimulated emission, then we can try to demonstrate that they can, at least, produce interference, isolating on the reflected beam the wave without the particle and showing that this wave cooperates in the creation of the interference pattern in IR.

In this case, LGT crossed by a photon will duplicate it, according to GRV, in two identical photons, which will be detected by P_2 and IR with probability 1/3 (of course, they could both be detected by either P_2 or IR with equal probability, but these cases are devoid of interest for this experiment).

Now the thesis of the GRV experiment is that one will find interference in IR, that is:

$$N(P_2 \wedge P_o) \neq N(P_2 \wedge P_o),$$

in contrast with the quantum mechanical predictions, according to which

$$N\ (P_2 \wedge P_o) = N\ (P_2 \wedge P_e)$$

as a consequence of the reduction of the wave function produced by our knowledge of the path followed by the particle, obtained from the measurement with P_2.

It must, however, be stressed that the GRV experiment introduces, with respect to the first-type experiments, the additional assumption that every single act of amplification does not, in general, give rise to a random change of the phase of the wave packet crossing LGT. But such a condition, as already pointed out (*12*), is known to be satisfied in every case in which the amplifying gas volume has dimensions small compared with a radiation wavelength.

2b. Measurement of the differences in the number of the coincidences between $P_1 \wedge P_o$ and $P_1 \wedge P_e$, that is:
$N\ (P_1 \wedge P_o)$ vs. $N\ (P_1 \wedge P_e)$.

This third experiment is based on the idea, opposite to the one of GRV, that the Selleri experiment has shown the QW property of producing stimulated emission. In such a case, the result: $N\ (P_1 \wedge P_o) \neq N\ (P_1 \wedge P_e)$ would no longer contradict the quantum mechanical postulate of the reduction of the wave function, since in this physical situation we would not be able to establish whether the photon falling on IR has been transmitted or reflected by M_1, because we do not know whether the photon detected by P_2 is a consequence of the duplication of the particle transmitted or of the stimulated emission produced by the wave without the (reflected) particle.

Nevertheless, by detecting the particle in P_1, we are sure that the QW is propagating from M_3 to IR, whereas the particle is arriving from M_2. So we can use the property of stimulated emission for testing both the interpretation of the state vector and the postulate of the reduction of the wave packet.

We are, of course, perfectly aware of the fact that modern formulations of quantum mechanics try to eliminate this postulate because of its meagrely satisfactory physical nature, but, as already stressed, we cannot forget that it still remains a basic axiom in the standard formulation of the theory.

I conclude with a brief remark concerning the problem of the change in the phase of QW, which could be produced by the amplification in a LGT. Now, if in the GRV experiment we duplicate the incoming photon and we are faced therefore with a measuring operation, even if of a special kind (Herbert has suggested for these cases the term "third-type measurements", and this terminology has also been adopted by A. Garuccio (*19*) in his lecture) due to the interaction and energy exchange between the photon and

LGT, implying an unavoidable disturbance in the considered physical situation, in our new proposal no energy-momentum exchange, which is one of the basic features of every measuring process, and in particular of the quantum mechanical, can occur between the quantum wave and LGT, when the former crosses the latter, since, according to our fundamental assumption, this wave represents a zero-energy undulatory phenomenon.

Thus no wave phase change can in this new experimental proposal be related to a perturbation produced in our LGT, by any kind of measuring process.

4. APPENDIX: THE RESOLUTION OF THE PARADOX OF NEGATIVE-RESULT MEASUREMENTS

The most troubling difficulty of the quantum mechanical theory of measurement is certainly the one so clearly outlined by Professor Popper in his opening lecture and can be condensed in the following question: "Can the state of a physical system be modified by the knowledge we have of it?"

To pose the question in a more precise way, we ask: "Is the fact that, after every measuring operation, a physical system occupies a well-defined state a consequence of the process of interaction between measuring instrument and measured system or rather of the observer's becoming conscious of the result obtained?".

At first sight this would seem more a question of metaphysical disputes on the mind-body problem than one of the basic epistemological questions of quantum theory. Nevertheless, the serious mathematical contradictions arising in the various theories of measurement, and the several physical paradoxes deriving from them, would appear to indicate the inevitability of such an intrusion of the mind or consciousness of the human observer into the logical context of contemporary physics, as in the case of Schrödinger's half alive and half dead cat, which seems to have to die or live once and for all only after an act of observation performed by a human mind, or of Wigner's friend, whose brain would be in a state of superposition of different mental states, until he communicates to another "privileged" observer the result of his own observation (*20*).

There is however a paradox in the theory of measurement that more than any other provides a basic support for these metaphysical speculations: the already mentioned paradox of negative-result measurements, whose first formulation is due to K. Renninger, but which was subsequently discussed in detail both by Wigner (*4*) and by de Broglie (*21*), from opposite points of view. As a matter of fact, the former appealed to the paradox for defending the psychophysical character of the von Neumann theory against the alternative theories of measurement elaborated in the spirit of Bohr's philosophy, whereas the latter analyzed it in order to show the necessity of going beyond these contradictory features of orthodox quantum mechanics.

The paradox consists in the proposal of a *gedanken* experiment in which we consider a source P emitting isotropically photons in all directions, partially surrounded by a hemispheric screen E_1 of center P and radius R_1 subtending a solid angle Ω around P and completely surrounded by a second and, in this case, spherical, screen E_2 of radius $R_2 > R_1$ both covered by a photon-sensitive substance. In this way the photon emitted by P can be absorbed by E_1 and E_2, with probabilities respectively given by:

(1) $\omega_1 = \Omega / 4\pi$ and (2) $\omega_2 = (4\pi - \Omega)/4\pi$

We can therefore write, for the initial state, at time t_0, the following wave function:

$$(3) \qquad |\psi_{t_0}\rangle = \sqrt{\omega_1}\,|\psi_1\rangle + \sqrt{\omega_2}\,|\psi_2\rangle$$

Now, at the subsequent time $t_1 = R_1/c$, two different events, whose respective probability is given by ω_1 and ω_2, may occur.

The first possibility is the detection of the photon on E_1, implying that the probability ω_1 is nullified, whereas ω_2 becomes equal to unity: According to the postulate of the reduction of the wave packet, the new physical situation will be described by:

$$(4) \qquad |\psi_{t_1}\rangle = |\psi_1\rangle$$

In this case no particular contrast seems to arise between mathematical description and (its) physical interpretation: The reduction of the initial state (3) of superposition to the well defined final state (4) can, in some way, be explained as the result of the physical process of interaction between the revealed photon and the detecting screen E_1. One would, of course, rightly object that if the detecting apparatus E_1, like any other physical system, must be described by quantal formalism, one has really no process of reduction and is simply led to the new state of superposition:

$$(5) \qquad |\psi_{t_1}\rangle = \sqrt{\omega_1}\,|\psi_1\rangle\,|\phi_1\rangle + \sqrt{\omega_2}\,|\psi_2\rangle\,|\phi_2\rangle$$

where $|\phi_1\rangle$ and $|\phi_2\rangle$ are the states for measuring device E_1, and if one wants to avoid that E_1 registers simultaneously two different results, one would have to resort, as in von Neumann's theory of measurement, to the intervention of some extraphysical entity. But this is the standard problem of the reduction of the wave packet, with respect to which some ways out have been conceived as, for example, the theories of measurement of Jauch, or of Daneri, Loinger, and Prosperi, according to which the process of reduction is the result of the physical interaction between the measured object and the *macroscopic* measuring apparatus, respectively described, in these theories either as a *classical* system, whose observable must all be represented by commuting operators, or as a *thermodynamic* one, subjected to an irreversible evolution.

The second possibility is that, at the time t_1 one has no detecting in E_1, implying $\omega_1 = 0$ and $\omega_2 = 1$. For the postulate of the reduction of the wave packet, the physical situation will be then described by

(6) $$|\psi_{t_1}\rangle = |\psi_2\rangle$$

Nevertheless, in this case, the change in the mathematical description can in no way be related to any observable physical process of detection of some physical event, but merely appears as a consequence of the knowledge we, as human observers, have obtained about a certain physical situation, by *not observing* the occurrence of a given phenomenon. We cannot, therefore, make appeal to any theory of measurement, like the ones previously mentioned, to explain such a change of description.

It is just this negative character, i.e., the fact that it is not a measurement, but the absence of any measurement, at the origin of the reduction process, which makes Renninger's *gedanken* experiment one of the most serious paradoxes of the whole quantum physics: The screen E_1 is a measuring device not interacting with the photon, which will, in fact, be revealed by E_2 and acts instead only as a source of knowledge allowing us to learn something about the behaviour of this physical object. But such a possibility would imply as a direct consequence that our mental acts can modify real physical situations or, alternatively, that the wave function does not describe the physical world, but the evolution of our mental states, legitimizing the previously mentioned interactionistic and mentalistic metaphysics.

It can be easily shown that the QWH allows to replace this metaphysical and, according to the approaches of some authors, parapsychological interpretation of Renninger's *gedanken* experiment with a consistent physical explanation.

As a matter of fact, let us consider a trivial variant to the first experiment discussed in the preceding section, that can be performed with the device of Fig.1, by removing the semi-reflecting mirrors M_2 and M_3 and increasing the distance between the photomultiplier P_1 and the remaining mirror M_1. We have thus a physical situation conceptually equivalent to Renninger's experiment, whose wave function will be similarly given by (3), with the only difference that ω_1 and ω_2 now correspond to the probabilities, both equal to 1/2, of the detection of the photon emitted by the source P, by P_1 and P_2.

According to the QWH, the laser gain tube LGT on the path of the transmitted beam will behave either as an ideal duplicator for the incoming particle, or as a source of a single photon, when crossed by the wave without the particle, for the process of stimulated emission. Moreover, it is possible, at least in principle, to distinguish the detection of the twin photons from the revelation of the single one, on the grounds of the *pulse height* of the photomultiplier P . In any case, even if such a discrimination were not possible, one could use a system of semireflecting mirrors splitting the beam coming out of LGT, in several secondary beams, each of which directed towards a different photomultiplier, in such a way as to separate the twin photons, when they are present in the primary beam.

In this way, the reduction of the state of superposition (3) to one of the well defined states (4) or (5) will take place *in both cases* as a consequence of the physical process of interaction, of the amplified particle or wave without particle, with the measuring instrument P_2 . We are therefore no longer concerned, as in the case of Renninger's paradox, with two different kinds of reduction, the first corresponding to a physical process, the second to a psychophysical or merely psychological one that follows from the change of our knowledge of the probabilities for the occurrence of certain physical events.

This means that, also in the original Renninger experiment, the reduction of the wave packet did not follow from a change of our knowledge, but from a real change in the physical situation.

The second and third experiments discussed in section 3 will probably allow us to go beyond the very solution of the paradox of negative-result measurements, since they try to show that, when the change in the mathematical description is not the result, as in the case of Renninger's experiment, of some physical interaction, but only of the modification of our knowledge, such a change has to correspond to a *wrong prediction* of quantum theory, as it would be demonstrated by the persistence of the interference pattern, even when we know which is the path followed by each photon in our experimental device.

A positive result, even from only one of the experiments (2a) and (2b) would therefore prove in a conclusive way that the reduction of the wave function is neither a psychophysical or psychological event, as in von Neumann's and Wigner's theory of measurement, nor a purely "mathematical process" connected to a "sudden change of our knowledge", as maintained by Heisenberg (*22*) in his proposed solution to Renninger's paradox, but a concrete empirical fact occurring in the physical world.

REFERENCES

1. G. Tarozzi, *Chance and Order's Coexistence as an Insuperable Dualism in the Logical Structure of Quantum Physics*, paper presented at the International School of Logic and Scientific Methodology on "Chance and Order in Natural Science," Erice, 1981.
2. F. Selleri and G. Tarozzi, "Quantum Mechanics, Reality and

Separability," *Riv.Nuovo Cimento* 4(2), (1981)
3. M. Renninger, "Messungen ohne Störung des Meßobjekts," *Z. Phys.* 158, 417 (1960).
4. E.P. Wigner, "The Problem of Measurement," *Am.J.Phys.* 31, 6 (1963).
5. A. Daneri, A. Loinger and G.M. Prosperi, "Quantum Theory of Measurement and Ergodicity Conditions," *Nucl. Phys.* 33, 297 (1962); "Further Remarks on the Relations between Statistical Mechanics and Quantum Theory of Measurement," *Nuovo Cimento* 44 B, 119 (1966).
6. J. von Neumann, *Mathematische Grundlagen der Quantenmechanik* (Springer, Berlin,1932); F. London and E. Bauer,*La Théorie de l'Observation en Mécanique Quantique* (Hermann, Paris, 1939).
7. F. Selleri, "On the Wave Function of Quantum Mechanics," *Lett. Nuovo Cimento* 1, 908 (1969); "Realism and the Wave Function of Quantum Mechanics," in *Foundations of Quantum Mechanics*, ed. B. d'Espagnat, (Academic Press, New York, 1971).
8. K.R. Popper, *Quantum Theory and the Schism in Physics* (Hutchinson, London, 1983).
9. G. Tarozzi, "From Ghost to Real Waves: a Proposed Solution to the Wave-Particle Dilemma," in *The Wave-Particle Dualism*, eds. S. Diner, D. Fargue, G. Lochack and F. Selleri (Reidel, Dordrecht, 1984).
10. Letter from M. Born to M. Renninger, dated May 23, 1955, in M. Jammer, *The Philosophy of Quantum Mechanics* (Wiley, New York, 1973).
11. F. Selleri, "Can an Actual Existence Be Granted to Quantum Waves?," *Ann. Fond. L. de Broglie* 7, 45 (1982).
12. F. Selleri, "Gespensterfelder," in *The Wave-Particle Dualism*, S. Diner et al. (Reidel, Dordrecht, 1984).
13. G. Tarozzi, "Two Proposals for Testing Physical Properties of Quantum Waves," *Lett. Nuovo Cimento* 35, 53 (1982).
14. G. Tarozzi, "Réalisme d'Einstein et Mécanique Quantique," *Rev. Synthèse* 101-102, 125 (1981).
15. F. Selleri and G. Tarozzi, "Is Nondistributivity for Micro-systems Empirically Founded?," *Nuovo Cimento* 43 B, 31 (1978).
16. K.R. Popper, A. Garuccio, and J.-P. Vigier, "An Experiment to Interpret EPR Action-at-a-Distance: the Possible Detection of Real de Broglie's Waves," *Epist.Lett.* 30, 21 (1981); A. Garuccio, K.R. Popper, and J.-P. Vigier, "Possible Direct Physical Detection of de Broglie Waves," *Phys. Lett.* 86A, 397 (1981); A. Garuccio and J.-P. Vigier, "Possible Experimental Test of the Causal Stochastic Interpretation of Quantum Mechanics: Physical Reality of de Broglie Waves," *Found. Phys.* 10, 797 (1980).
17. A. Garuccio, V. Rapisarda, and J.-P. Vigier, "New Experimental Set-Up for the Detection of de Broglie Waves," *Phys. Lett.* 90A, 17 (1982).

18. A. Gozzini, "On the Possibility of Realising a Low Intensity Interference Experiment with a Determination of the Particle Trajectory," in *The Wave-Particle Dualism* S.Diner et al. (Reidel Dordrecht, 1984).
19. A. Garuccio, *Third Kind Measurements and Wave-Particle Dualism* published in the present volume.
20. E. Schrödinger, "Die gegenwärtige Situation in der Quantenmechanik," *Naturwiss.* 23, 807 (1935); E.P. Wigner, "Remarks on the Mind-body Problem," in *Symmetries and Reflections* (Indiana University Press, Bloomington, 1967).
21. L. de Broglie, "De la mécanique ondulatoire à la mécanique quantique: l'aller et le retour," in *Sa Conception du Monde Physique* (Gauthier-Villars, Paris, 1973).
22. Letter from H. Heisenberg to M. Renninger dated February 2, 1960, in M. Jammer, *The Philosophy of Quantum Mechanics* (Wiley, New York, 1973).

EPILOGUE

At the end of the conference a public lecture was given by Professor Popper in the Teatro Piccinni of Bari. The text of this lecture has been reproduced in the following pages not only because it represents the most recent, systematic exposition of the philosophy of one of the greatest human thinkers of our time, but also because it provides a general philosophical perspective from which all the contributions to the present volume can be viewed.

As a matter of fact, it is in the light of the dynamic conception of science perceived, according to Popper's evolutionary epistemology, as an endless process of revision and criticism of the basic concepts and principles of scientific theories, that the existence of still-open questions in quantum mechanics -- more than half a century after what most physicists persist in considering as its final formulation -- can be recognized. Such an acknowledgement of unsolved conceptual problems in the foundations of microphysics is by contrast inadmissable within the purview of the stagnant philosophy of the Copenhagen interpretation, which culminated in the absurd myth of the completeness of the quantum formalism (based on the dogmatic acceptance of the von Neumann proof against hidden variables, a proof that implied the "mathematical impossibility" of any future more satisfying generalization of the existing theory) and in Bohr's notion of complementarity (conceived as a dogma that epistemologically renounces any possibility to understand the dualistic nature of quantum "phenomena" and to overcome the related dilemmas between waves and particles, macroscopic and microscopic entities, observer or measuring instrument and observed or measured system, a causal and space-time description of microscopic processes).

It therefore appears evident that a radical emancipation from the negative philosophy of the Copenhagen school and from its irrationalistic aspects is a necessary precondition if one is to look for a real solution, i.e., for something completely different from their legitimation, of the main quantum paradoxes which, according to the suggestive thesis advanced by Professor Popper in the opening lecture of this conference, can be overcome -- contrary to a merely formal point of view -- neither by introducing more or less relevant changes in the mathematical struc-

G. Tarozzi and A. van der Merwe (eds.), Open Questions in Quantum Physics, 391–393.

ture of the present theory, nor by limiting its domain of application to the physical world, but rather through a critical process of reinterpretation, in the light of a realist and objectivist philosophy, of the basic principles and concepts of quantum mechanics, like Heisenberg's indeterminacy relations, the wave function ψ, and the very notion of probability. In this way, on the basis of a new, different philosophical interpretation of quantum formalism, it would seem possible to eliminate the most serious contradictions of standard quantum mechanics: the mysterious dilemma of the wave-particle duality; the paradoxes of the theory of measurement and the inadmissible interference of the human observer with the laws of physics; the Einstein-Bell contradiction and the possibility of the existence of superluminal signals propagating with an efficiency completely independent of the distance and therefore contradicting not only special relativity but even a nonlocal theory like classical mechanics, where the efficiency of the instantaneous actions decreases with increasing distance.

Starting from his thesis on the interpretative nature of the foundational problems of quantum mechanics, Professor Popper has developed the main lines of a program of research into the philosophy of empirical science, whose relevance can be compared with that of the investigations conducted by analytic philosophers into the foundations of mathematics, which did lead to a "philosophical" resolution of the paradoxes of set theory and other conceptual problems in classical mathematics, but which, when applied to physical science, proved inadequate for solving physical paradoxes such as linguistic contradictions. As shown by Professor Popper, the solution of the physical paradoxes can only be empirical: The contrast between the two opposite interpretations of quantum theory, that is, the Einstein-de Broglie versus the Bohr-Heisenberg approach, giving rise to that "schism of physics" which is the subject of the third volume of Popper's *Postscripts* to his *Logic of Scientific Discovery*, is a matter neither of conventional linguistic nor of arbitrary metaphysical choice, but one that can be decided through an appeal to experience.

The major relevance of this program of research into the philosophy of physics can nevertheless not signify, we think, a complete reduction of the present open questions in quantum physics to the problem of a correct philosophical interpretation of the existing formalism and a resultant generalized philosophical resolvability of quantum paradoxes.

The first reason for this lies, in our opinion, in the existence of at least one foundational problem of quantum mechanics, that is not a consequence of the subjectivist philosophy of the Copenhagen interpretation, but originates rather in the logical structure of the generally accepted formulation of the theory and can therefore be solved by altering this mathematical formulation. We are referring to the earlier mentioned Einstein-Bell contradiction, which we believe is not merely due to the anti-realist perspective of the orthodox interpretation -- according to which the predictability, with certainty and without disturbance, of a property of a physical system does not represent a sufficient guarantee for attributing that property to the system -- but arises instead, as has been pointed out in some papers contained in the first part of this volume, from our inability to provide a noncontradictory description of two separate physical systems that have interacted in the past on the basis of the formalism of quantum theory of spin (which as has been demonstrated, is the only so far known mathematical description of the physical world whose predictions violate Bell-type inequalities). It seems more likely to us that the subjectivist conception of the Copenhagen interpretation is invoked to the extent that it proposes a philosophically unconvincing but logically consistent way out of the previous formal difficulty, just as the von Neumann-Wigner idealistic interpretation of the theory of measurement appeals to the observer's consciousness in order to justify the inability of the quantum formalism to describe the wave-packet collapse as a physical process of interaction between measuring instrument and measured system.

The second reason is our belief that an uncritical acceptance of Popper's thesis, implying a generalized reducibility of the foundational problems of quantum mechanics in terms of a correct re-interpretation of its basic theoretical concepts and principles, would involve the risk of contradicting the very dynamic view of scientific knowledge that lies at the basis of his evolutionary epistemology and of supporting, on the other hand, a legitimative and conservative conception of science bent on preserving the present formal structure of the existing empirical theories at the mere cost of modifying in a suitable way their philosophical interpretation.

This conception is of course not intrinsic to Popper's approach, but could well be the result, as has already happened in some misinterpretations of Popper's epistemology, of an improper use of his fundamental thesis, whose great methodological fertility, when properly applied, very clearly emerges both in his opening lecture of the conference (which presents a variant of the EPR argument discriminating experimentally between the Einstein and Copenhagen interpretations) and in other contributions reported in the second and third parts of this volume which show how a new realist interpretation of the quantum mechanical wave function opens up novel prospects for the resolution of the wave-particle dilemma and even of some very serious paradoxes in the theory of measurement.

EVOLUTIONARY EPISTEMOLOGY*

Karl Popper

London School of Economics & Political Science
Houghton Street
London WC2 2AE

1. INTRODUCTION

Epistemology is the English term for the theory of knowledge, especially of scientific knowledge. It is a theory that tries to explain the status of science and the growth of science. Donald Campbell called my epistemology 'evolutionary' because I look upon human language, upon human knowledge, and upon human science, as a product of biological evolution, especially of Darwinian evolution through natural selection.

I look upon the following as the main problems of evolutionary epistemology: the evolution of human language and the part it has played (and still plays) in the growth of human knowledge; the ideas of truth and falsity; the description of states of affairs; and the way states of affairs are picked out by language from the complexes of facts that constitute the world; that is, 'reality'.

To put this at once briefly and simply in the form of two theses:

First thesis: The specifically human ability to know, and also the ability to produce scientific knowledge, are the results of natural selection. They are closely connected with the evolution of a specifically human language. This first thesis is almost trivial. My second thesis is perhaps slightly less trivial.

Second thesis: The evolution of scientific knowledge is, in the main, the evolution of better and better theories. This is,

*This chapter is based on a public lecture given after the conference on *Open Questions on Quantum Physics*, in Bari, Italy, on May 7, 1983.

G. Tarozzi and A. van der Merwe (eds.), Open Questions in Quantum Physics, 395–413.

again, a Darwinian process. The theories become better adapted through natural selection: they give us better and better information about reality. (They get nearer and nearer to the truth.) All organisms are problem solvers: problems arise together with life.

We are always faced with practical problems; and out of these grow sometimes theoretical problems; for we try to solve some of our problems by proposing theories. In science these theories are highly competitive. We discuss them critically; we test them and we *eliminate* those theories which we judge to be less good in solving the problems which we wish to solve: so only the best theories, those which are most fit, survive in the struggle. This is the way science grows.

However, even the best theories are always our own inventions. They are beset by errors. What we do when we test our theories is this: we try to detect the errors which may be hidden in our theories. That is to say: we try to find the weak spots of our theories, the place where they break down. This is the critical method.

There is often much ingenuity needed in this critical testing process.

We can sum up the evolution of theories by the following diagram:

$$P_1 \rightarrow TT \rightarrow EE \rightarrow P_2$$

A problem (P_1) gives rise to attempts to solve it by tentative theories (TT). These are submitted to a critical process of error elimination (EE). The errors which we detect give rise to new problems (P_2). The distance between the old and the new problem is often very great: it indicates the progress made.

It is clear that this view of the progress of science is very similar to Darwin's view of natural selection by way of the elimination of the unfit: of the errors in the evolution of life, the errors in the attempts at *adaptation*, which is a trial and error process. Analogously, science works by trial (theory making) and by the elimination of the errors.

One may say: from the amoeba to Einstein there is only one step. Both work with the method of tentative trials (TT) and of error elimination (EE). *Where is the difference?*

The main difference between the amoeba and Einstein is not in the power of producing TT, tentative theories, but in EE, in the way of error elimination.

The amoeba is not aware of the process of EE. The main errors of the amoeba are eliminated by eliminating the amoeba: this is just natural selection.

As opposed to the amoeba, Einstein was aware of the need for EE: he criticized and tested his theories severely. (Einstein said that he produced and eliminated a theory every few minutes.)

What was it that enabled Einstein to go beyond the amoeba?

The main thesis of this chapter will be the answer to this question.

Third thesis: What enables a human scientist like Einstein to go beyond the amoeba is the possession of what I call the *specifically human language*.

While the theories of the amoeba are part of the organism of the amoeba, Einstein could formulate his theories in language; if needed, in written language. He could in this way put his theories outside his organism. This enabled him to look upon a theory *as an object*; to look at it *critically*; to ask himself whether it can solve his problem, and whether it could possibly be true; and to eliminate it if he found that it could not stand up to this criticism.

It is *only* the specifically human language which can be used for this kind of purpose.

My three theses together give an outline of my evolutionary epistemology.

2. THE TRADITIONAL THEORY OF KNOWLEDGE

What is the usual approach to the theory of knowledge, to epistemology? It is totally different from my evolutionary approach here sketched in Section 1. The usual approach demands *justification* by *observation*. I reject both parts of this position.

It usually starts from a question like: 'how do we know?', and it usually takes this question to mean the same as: 'what kind of perception or observation is the justification of our assertions?'. In other words, it is concerned with the justification of our assertions (in my way of putting it, of our theories); and it looks for this justification to our perceptions or to our observations. This epistemological approach may be called *observationism*.

Observationism assumes that out senses or our sense organs are the sources of our knowledge; that we are'given' some so-called 'sense data' (a sense datum is something that is given to us by our senses) or some perceptions and that our 'knowledge' is the result of, or the digest of, these sense data, or of our perceptions, or of the information received. The place where the sense data are digested is, of course, the head, shown below.

That is why I like to call observationism 'the bucket theory of the mind'.

The theory can also be put as follows. The sense data enter the bucket through the well-known seven holes, two eyes, two ears, one nose with two holes, the mouth; and also through our skin, the sense of touch. In the bucket, they are digested; or more especially, they are connected, associated, and classified. And then, from those which recur repeatedly, we obtain - by repetition, association, and by generalization and induction - our scientific theories.

The bucket theory, or observationism, is the standard theory of knowledge from Aristotle to some of my own contemporaries, for example, Bertrand Russell; or the great evolutionist J. B. S. Haldane or Rudolf Carnap.

But it is also the theory held by the man in the street.

The man in the street can put it very briefly: 'how do I know? Because I have had my eyes open, I have seen, and I have heard'. Carnap also identifies the question 'how do I know?' with 'what perceptions or observations are the sources of my knowledge?'.

These straightforward questions and answers by the man in the street give, of course, a reasonably true picture of the situation as he sees it. But it is not a view that may be elevated and transformed into a theory of knowledge that can be taken seriously.

Before proceeding to criticize the bucket theory of the human mind, I wish to mention that the opposition to it goes back to Greek times (Heraclitus, Xenophanes, Parmenides). Kant saw the problem fairly clearly; he stressed the difference between non-observational or *a priori* knowledge and observational or *a posteriori* knowledge. The idea that we may have *a priori* knowledge shocked many people; and the great ethologist and evolutionary epistemologist Konrad Lorenz proposed that the Kantian knowledge *a priori* may be knowledge that was, some thousands or millions of years ago, first acquired *a posteriori* (Lorenz (1941)), and that it became later genetically fixed by natural selection. However, in a book written between 1930 and 1932, and still available only in German (Popper (1979); it is the book to which Donald Campbell referred when he described my theory of knowledge as 'evolutionary') I proposed that the *a priori* knowledge had never been *a posteriori*, and that, from the historical and genetical point of view, *all* our knowledge was the *invention* of animals and therefore *a priori* in its conception (although not, of course, *a priori* valid, in the Kantian sense). It becomes *adapted* to the environment by natural selection: the apparent *a posteriori* knowledge is always the result of the *elimination* of badly fitting *a priori* invented hypotheses, or adaptations. In other words, all knowledge is the result of *trial* (invention) and of the *elimination of error*, of badly fitting *a priori* inventions.

The trial and error method is the one by which we are

actively seeking information from our environment.

3. CRITICISM

My *fourth thesis* (which I have been teaching and preaching for more than 60 years) is:

Every single aspect of the justificationist and observationist philosophy of knowledge is mistaken:

1) Sense data or similar experiences do not exist.
2) Associations do not exist.
3) Induction by repetition or generalization does not exist.
4) Our perceptions may mislead us.
5) Observationism or the bucket theory is a theory which says that knowledge can stream from outside into the bucket through our sense organs. In fact, we, the organisms, are most active in our acquisition of knowledge - perhaps more active than in our acquisition of food. Information does not stream into us from the environment. Rather, it is we who explore the environment and suck information from it actively, like food. And humans are not only active but sometimes critical.

A famous experiment which refutes the bucket theory and especially the sense datum theory is due to Held and Hein (1963); it is reported in a book which Sir John Eccles and I wrote together (Popper and Eccles (1977)). It is the experiment of the active and the passive kitten. These two kittens are so linked that the active kitten causes the passive kitten to be moved in a kind of perambulator, passively, through exactly the same environment in which the normal kitten moves actively. So the passive kitten gets with great approximation the same perceptions as the active kitten. But tests of the kittens show that, while the active kitten learns a lot, the kitten kept passive has learned nothing.

The defenders of the observationist theory of knowledge could argue against this criticism that there is also a kinaesthetic sense, the sense of our movements, and that the absence of the kinaesthetic sense data in the sensory input of the passive kitten can explain, within the observationist theory, why the passive kitten learns nothing. What our experiment shows, the observationist might say, is nothing more than this: unless the kinaesthetic sense data are associated with the optical and acoustic sense data, the latter will not be useful.

In order to make my rejection of observationism or the bucket theory or sense datum theory independent of any such defence, I shall now introduce an argument which I regard as decisive. It is an argument that is typical for my evolutionary theory of knowledge.

The argument may be put as follows. The idea that theories

are the digest of sense data or perceptions or observations *cannot be true* for the following reasons.

Theories are (and in fact all knowledge is) from the evolutionary point of view, part of our tentative *adaptations* to our environment. They are like expectations or anticipations. This is indeed their function: the biological function of all knowledge is to try to anticipate what will happen in our environment. Now our sense *organs*, for example our eyes, are also such adaptations. They are, seen in this light, theories: animal organisms have invented eyes, and evolved them in great detail, like an expectation or theory that light in the visible wavelength will be useful for extracting information from the environment; for sucking from the environment information that can be interpreted as indicating the *state* of the environment, both the long-term state and the short-term state.

Now our sense *organs*, obviously, are logically prior to our alleged sense *data* (even though there would be a feedback if sense data did exist; just as there may be a feedback from our perceptions upon the sense organs).

It is therefore impossible that all theories or theory-like entities are reached by induction or generalization from the alleged 'data', the allegedly 'given' influx of information, from perception or observation. For the sense organs that suck information from the environment are genetically, as well as logically, prior to this information. The camera, and its structure, comes before the photograph, and the organism, and its structure, comes before any information.

I think that this argument is decisive and that it leads to a new view of life.

4. LIFE AND THE ACQUISITION OF KNOWLEDGE

Life is usually characterized by the following powers or functions which largely depend on one another:

1) Procreation and heredity.
2) Growth.
3) Absorption and assimilation of food.
4) Sensitivity to stimuli.
   I suggest that this might be replaced by:
   a. Problem solving (problems which may emerge from the environmental or the internal situation of the organism). *All organisms are problem solvers.*
   b. Active exploration of the environment, often helped by random trial movements. (Even plants explore their environment.)
5) Construction of theories about the environment, in the form of physical organs or other changes of the anatomy or in the form of new behavioural repertoires, or changes to the exist-

ing behavioural repertoire.

All these functions originate in the organism. This is very important. They are all actions. They are not reactions to the environment.

This can also be put as follows. It is the organism, and the state in which it happens to be, that determines or chooses or selects what kind of changes in the environment may be significant for it, so that it may 'react' to them, as 'stimuli'.

One usually speaks of the stimulus which releases a reaction; and one usually implies that the environmental stimulus comes first and that it causes the release of the reaction of the organism. This leads to the (mistaken) interpretation that the stimulus is a bit of information, flowing into the organism from the outside, and, altogether, that the stimulus comes first; it is the cause, which comes before the response, the effect.

I think this is fundamentally mistaken.

It is a mistake due to a traditional model of physical causation that does not work with organisms, and that does not even work with motor cars or wireless sets; indeed, not with things that have some access to a source of energy which they can expend in *different* ways, and in different amounts.

Even a motor car or a wireless set *selects* according to its internal state the stimuli to which it responds. The motor car may not react properly to pressure on the accelerator if it is held still by its handbrake. And the wireless set will not be tempted by the most beautiful symphony unless it is tuned to the proper frequency.

The same holds for organisms; even more strongly, insofar as they have to be tuned in or programmed by themselves. They are tuned in, for example, by their gene structure, or by some hormone, or by the lack of food, or by curiosity, or by the hope of learning something interesting.

All this speaks very strongly against the bucket theory of the mind, which was often formulated by the following phrase: 'Nothing is in the intellect that was not before in the senses' - or in Latin: *'Nihil est in intellectu quid non antea fuerat in sensu'*. This is the slogan of observationism, of the bucket theory. Few people know of its prehistory. It stems from a contemptuous remark of the anti-observationist Parmenides who said something like this: 'There is just nothing at all in the erring intellects (*plakton noon* should be *plankton noon* in Parmenides, Diels and Kranz (1960)) of these people except what was previously in their much-erring (*polyplanktos*) sense organs'. (See my *Conjectures and Refutations*, pp. 410-413 of the Routledge & Kegan Paul editions from 1968 on.) I suggest that it may have been Protagoras who countered Parmenides by turning this into a proud slogan of observationism.

5. LANGUAGE

From these considerations we see the significance of the active and explorative behaviour of animals and men. And this insight in its turn is of great importance for the theory of evolution in general, not only for evolutionary epistemology. I must, however, come now to a central point of evolutionary epistemology - the evolutionary theory of the human language.

The most important contribution to the evolutionary theory of language known to me lies buried in a little paper published in 1918 by my former teacher Karl Bühler (1918). In this paper, too little regarded by modern students of linguistics, Bühler distinguishes three stages in the evolution of language (Table 1, to which I have added a fourth function). In each of these stages, language has a task, a certain biological function.

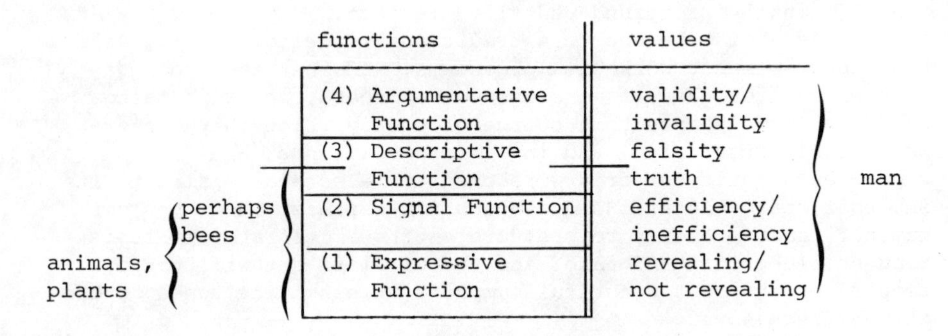

	functions	values	
	(4) Argumentative Function	validity/ invalidity	man
	(3) Descriptive Function	falsity truth	man
perhaps bees	(2) Signal Function	efficiency/ inefficiency	man
animals, plants	(1) Expressive Function	revealing/ not revealing	man

Table 1

The lowest stage is the one where the only biological function of language is to *express* outwardly the inward state of an organism; perhaps by certain noises, or by certain gestures.

It is probably only for a comparatively short period that this is the *only* function of language. Very soon some other animals (either of the same species or of other species) will take notice of the *expressions* and *adapt* themselves to them: they discover how to suck information from them, how to include them into the stimuli of their environment to which they may respond with advantage. More especially, they may use the expression as a warning of some impending danger. For example, the roar of a lion which is a *self-expression* of the lion's inner state may be used, in this way, as a warning by the lion's possible prey. Or a cry of a goose expressing fear may be interpreted by other geese as a warning against a hawk; and a different cry as a warning against a fox. So the *expressions* may *release* in the receiving or responding animal a typical pre-formed reaction. The responding animal takes the *expression* as a *signal*, as a *sign* that releases a definite response. It thus comes in *communication* with the expressing animal.

At this stage, the original expressive function has changed. And what was originally an outward sign or symptom, although the expression of an inner state, has acquired a signal function or release function. It may now even be used by the expressor as a signal and may thus change its biological function from expressing to signalling, even to conscious signalling.

So we have now two evolutionary levels: *first*, the *pure expression*, and *second*, the expression which tends to become a *signal* since there will be receiving animals that will respond: or react to it as a *signal*: so we have *communication*.

Bühler's *third* evolutionary level is that of the human language. According to Bühler, human language and human language alone, introduces a revolutionary novelty: it can *describe*; that is, describe a state of affairs, a situation. It may be a state of affairs that is present at the instant of time at which the state of affairs is being described, such as: 'our friends are coming now'; or a state of affairs that has no present relevance, such as: 'my brother-in-law died 13 years ago', or a state of affairs that may have never taken place and may never take place, such as: 'far behind this mountain there is another mountain that is made of solid gold'.

Bühler calls the power of human language to describe possible or real states of affairs the '*descriptive function*' of human language. And he rightly stresses its immense importance.

Bühler shows that language never loses its expressive function. Even a highly unemotional description will retain some of it. Nor does language ever lose its signalling or communicative function. Even an uninteresting (and false) mathematical equation, such as '$10^5 = 1{,}000{,}000$' may perhaps provoke a correction and thus a reaction, even an angry emotional reaction, from a mathematician.

But neither the expressiveness nor the sign-character - the signal that produces the reaction - is characteristic of the human language; nor is it characteristic of it that it serves a family of organisms to communicate. The *descriptive character* is characteristic of the human species. And it is something new, and something truly revolutionary: *human language can convey information about a state of affairs, about a situation that may or may not be present or biologically relevant. It may not even exist.*

Bühler's simple and most important contribution has been neglected by almost all linguists. They still talk as if the essence of human language would be self-expression; or as if words like 'communication' or 'sign language', or 'symbol language' would sufficiently characterize the human language. (But signs or symbols are also used by other animals.)

Bühler of course never asserted that there are no further functions of human language: language may be used to beg, or to implore, or to admonish. It may be used for commanding, or for advising. It may be used to offend people or to hurt people, or to frighten them. And it may be used to comfort people or to make people feel loved and at ease. But on the human level, the descriptive language can be basic to all these uses.

6. HOW DID THE DESCRIPTIVE FUNCTION EVOLVE?

It is easily seen how the signalling function of language evolved once the expressive function had appeared. But it is very difficult to see how the descriptive function could have developed out of the signalling function. It must be admitted that the signalling function can come close to the descriptive function. A characteristic warning cry of a goose may mean 'hawk!' and another one may mean 'fox!'. And this is in many ways very close to a descriptive statement. 'A hawk is coming! hide!' or 'take to the air! a fox is coming!'. But there are great differences between such descriptive warning cries and human descriptive languages. These differences make it difficult to believe that the human descriptive languages evolved from warning cries, and other signals, such as war cries.

Admittedly, the dancing language of bees comes close to the human descriptive use of language. Bees can convey, by their dance, information about the direction and distance from the beehive concerning a place where food can be found and concerning the character of the food.

But there is a most important difference between the biological situations of the bee language and the human language: the descriptive information conveyed by the dancing bee is part of a signal for the other bees; and it is its main function to incite these other bees to an action that is useful here and now; the information conveyed is closely related to the present biological situation.

As opposed to this, the information conveyed by human language may not be immediately useful. It may not be useful at all, or useful only years later and in a totally different situation.

There is also a possible *playfulness* in the use of human descriptive language which makes it so different from warning cries or mating cries or from the bee language, which all serve in very serious biological situations. Natural selection might explain that the system of warning cries becomes richer, more differentiated. But one should expect that if this happens, it should also become more rigid. Human language, however, must have evolved by a process that combined a great increase in differentiation with a still greater increase in the *degrees of freedom* of its use. (The term 'degrees of freedom' may be taken either in the ordinary sense or in its mathematical or in its physical sense.)

All this becomes clear if we look at one of the oldest ways of using human language: their use in story telling and in the invention of religious myths. Both these uses no doubt have functions that are biologically serious. But these functions are fairly remote from the situational urgency and the rigidity of warning cries.

It is, more especially, the rigidity of these biological signals (as we may call them) which creates the difficulty: it is difficult to conceive that the evolution of the biological signals may lead to the human language with its chatter, the variety of its uses, and its playfulness on the one hand, and, on the other hand, its most serious biological functions, such as *its function in the acquisition of new knowledge*, such as the discovery of the use of fire.

However, there may be some ways out of this impasse, even though these ways out are purely speculative hypotheses. What I have to say are conjectures; but they may indicate how things may have happened in the evolution of language.

The *playfulness of young animals*, especially mammals, to which I wish to draw special attention, and its biological functions, raises tremendous problems, and some excellent books deal with this great subject. (See, for example, Baldwin (1895), Eigen and Winkler (1975), Groos (1896), Hochkeppel (1973), Lorenz (1973, 1977), and Morgan (1908).) The subject is too vast and important to go into here in any detail. But I conjecture that it may be the key to the problem of the evolution of freedom and of human language. I shall only refer to some recent discoveries which show the creativeness of the playfulness of young animals and its significance for new inventions. We can read about Japanese monkeys in Menzel (1965): for example, where he writes:

> '[It] is usually juveniles rather than adults who are the originators of group adaptation processes and "pro-cultural" changes in relatively complex behaviours such as coming into a newly established feeding area, acquiring new food habits, or adopting new methods of collecting foods...'.

(See also, Frisch (1959), Itani (1958), Kawamura (1959), and Miyadi (1964).)

I suggest that the main phonetic apparatus of human language does not arise from the closed system of warning calls or war cries and similar signals (systems that need to be rigid and may be genetically fixed), but rather from playful babbling and chatter of mothers with babies, and of gangs of children; and that the descriptive function of the human language - its use for describing states of affairs in the environment - may arise from make-believe plays, so-called 'representation plays' or 'imitation plays'; and especially from the play-acting of children trying to imitate playfully the behaviour of adults.

These imitation plays are widely established among many mammals: there are mock fights, mock war cries, also mock cries for help: and mock commands, impersonating certain adults. (This may lead to giving them names, possibly names intended to be descriptive.)

Play-acting may go with babble and chatter; and it may create the *need* for something like a descriptive or explanatory commentary. In this way a *need* for storytelling may develop, together with a situation in which the descriptive character of the story is clear from the beginning. Thus the human language, the descriptive language, may have been first invented by children playing or play-acting, perhaps as a secret gang language (they still invent such things sometimes). It may then have been taken over by the mothers (like the inventions of Japanese monkey children, see above), and only later, with modifications, by the adult males. (There are still some languages that contain grammatical forms indicating the sex of the speaker.) And out of the storytelling - or as part of it - and out of the description of a state of affairs may have developed *the explanatory story; the myth; and later the linguistically formulated explanatory theory.*

The *need* for the descriptive story, perhaps also for the prophecy, with its immense biological significance, may in time become genetically fixed. The tremendous superiority, especially in warfare, given by the possession of a descriptive language, creates a new selection pressure; and this, perhaps, explains the astonishingly rapid growth of the human brain.

It is unfortunate that it is hardly possible that a speculative conjecture like the foregoing could ever become *testable.* (It would not even be a test if we were to succeed in inducing the Japanese monkey children to do all I suggested above.) Yet even so, it has the advantage of telling us an explanatory story of how things *may* have happened - how a flexible and descriptive human language *may* have arisen: a descriptive language which from the start would have been open, capable of almost infinite development, stimulating the imagination, and leading to fairy tales, to myths, and to explanatory theories: to 'culture'.

I feel that I should draw attention here to the story of Helen Keller (see Popper and Eccles (1977)): it is one of the most interesting cases to show the inborn need of a child for actively acquiring the human language, and for its humanizing influence. We may conjecture that the need is encoded on the DNA, together with various other dispositions.

7. FROM THE AMOEBA TO EINSTEIN

Animals, and even plants, acquire new knowledge by the method of trial and error; or, more precisely, by the method of trying out certain active movements, certain *a priori* inventions, and the elimination of those which do not 'fit', which are not well-adapted. This holds for the amoeba (see H. S. Jennings (1906)); and it also holds for Einstein. Wherein lies the main difference between these two?

I think the elimination of errors works in different ways. In the case of the amoeba, any gross error may be eliminated by eliminating the amoeba. Clearly this does not hold for Einstein: he knows he will make mistakes and he is actively looking out for them. But it is not surprising to find that most men have inherited from the amoeba a strong aversion to making mistakes, and also to admitting that they have made them! However, there are exceptions: some people do not mind making mistakes as long as there is a chance of discovering them, and trying again if one is discovered. Einstein was one of these; and so are most creative scientists: in contradistinction to other organisms, human beings use the method of trial and error *consciously* (unless it has become 'second nature' to them). We have, it seems, two types of people: those who are under the spell of an inherited aversion to mistakes, and who therefore fear them and fear to admit them; and those who also wish to avoid mistakes, but know that we make mistakes more often than not, who have learned (by trial and error) that they may counter this by *actively searching for their own mistakes*. The first type of people are *thinking dogmatically;* the second are those *who have learned to think critically*. (By saying 'learned' I wish to convey my conjecture that the difference between the two types is not based on inheritance but on education.) I now come to my

Fifth thesis: in the evolution of man, the descriptive function of the human language has been the prerequisite for critical thinking: it is the descriptive function that makes critical thinking possible.

This important thesis can be established in various ways. Only a descriptive language of the kind described in the foregoing Section 6 raises *the problem of truth and falsity*: the problem whether or not some description corresponds to the facts. It is clear that the problem of truth precedes the evolution of critical thinking. Another argument is this. Prior to human descriptive language, all theories could be said to be part of the structure of organisms which were their carriers. They were either inherited organs, or inherited or acquired dispositions to behave, or inherited or acquired unconscious expectations. So they were part and parcel of their carrier.

But in order to make it possible to criticize a theory, the organism must be able *to regard the theory as an object*. The only way known to us to achieve this is to formulate the theory in a descriptive language and, preferably, in a written language.

In this way, our theories, our conjectures, the trials of our trial-and-error attempts, can become objects; objects like dead or living physical structures. They may become objects of critical investigation. And they may be killed by us without killing their carriers. (Strangely enough, even critical thinkers frequently develop hostile feelings towards the carriers of theories they criticize.)

I may perhaps insert here a brief note on what I regard as a very minor problem: whether or not the two types of people - the dogmatic thinkers and the critical thinkers - are hereditary types. As hinted above, I conjecture that they are not. My reason is, simply, that these 'types' are an invention. One may perhaps classify actual people according to this invented classification. But there is no reason to think that the classification is based on DNA - no more than there is reason to believe that liking or disliking golf is so based. (Or that what is called IQ measures 'intelligence': as pointed out by Peter Medawar, no competent agriculturist would dream of measuring the fertility of the soil by a measure depending on *one variable*. But some psychologists seem to believe that 'intelligence', which involves creativity, could be so measured.)

8. THREE WORLDS

Human language is, I suggest, the product of human inventiveness. It is a product of the human mind, of our mental experiences and dispositions. And the human mind, in its turn, is the product of its products: its dispositions are due to a feedback effect. One particularly important feedback effect, mentioned above, would be the disposition to invent arguments, to *give reasons* for accepting a story as true, or for rejecting it as false. Another very important feedback effect is the invention of the sequence of natural numbers.

First comes the dual and the plural: one, two, many. Then come the numbers up to 5; then come the numbers up to 10 and to 20. And then comes the invention of the principle that we can extend any series of numbers by adding one: the successor principle, the principle of forming a successor to every numeral.

Each such step is a *linguistic* innovation, an invention. The innovation is linguistic, and it is totally different from counting (for example, a shepherd carving one mark on a stick for each sheep that goes past). Each such step changes our mind - our mental picture of the world, our consciousness.

Thus there is feedback, or interaction, between our language and our mind. And with the growth of our language and of our mind, we can see more of our world. Language works like a searchlight: just as a searchlight picks up a plane, language may bring into focus certain aspects, certain states of affairs which it describes from the continuum of facts. Thus language not only interacts with our mind, it helps us to see things and possibilities, which we would never have seen without it. I suggest that early inventions such as the igniting and controlling of fire and, very much later, the invention of the wheel (unknown to certain highly cultured peoples) were made with the help of language: they were made possible (in the case of fire) by identifying very dissimilar situations. Without a language, only biological situations to which we *react* in the same way (food, dangers, etc.) can be identified.

There is at least one good argument in favour of the conjecture that descriptive language is much older than the control of fire: if deprived of language, children are scarcely human. Deprivation of language has even a physical effect on them, possibly worse than the deprivation of some vitamin, to say nothing of the devastating mental effect. Children deprived of language are mentally abnormal. But in a mild climate, nobody is dehumanized by being deprived of fire.

In fact, learning a language and learning to walk upright seem to be the only skills whose acquisitions are vital to us and, no doubt, are genetically based; and both of these are eagerly acquired by small children, largely on their own initiative, in almost any social setting. Learning a language is also a tremendous intellectual achievement. And it is one that all normal children master; probably because it is deeply needed by them (a fact that may be used as an argument against the doctrine that there are physically normal children of very low innate intelligence).

About twenty years ago I introduced a theory that divides the universe into three sub-universes which I called World 1, World 2, and World 3.

World 1 is the world of all physical bodies and forces, and fields of forces; also of organisms, of our own bodies and their parts, our brains, and all physical, chemical, and biological processes within living bodies.

World 2 I call the world of our mind: of conscious experiences of our thoughts, of our feelings of elation or depression, of our aims, of our plans of action.

World 3 I call the world of the products of the human mind, and especially the world of our human language; of our stories, our myths, our explanatory theories; the world of our mathematical and physical theories, and of our technologies and of our biological and medical theories. But beyond this, also the world of human creation in art, in architecture and in music - the world of all those products of our minds, which, I suggest, could never have arisen without human language.

World 3 may be called the world of culture. But my theory, which is highly conjectural, stresses the central role played by descriptive language in human culture. World 3 comprises all books, all libraries, all theories, including, of course, false theories, and even inconsistent theories. And it attributes a central role to the ideas of truth and of falsity.

As indicated before, the human mind lives and grows in interaction with its products. It is greatly influenced by the feedback from the objects or inmates of World 3. And World 3, in its turn, consists largely of physical objects, such as books and buildings and sculptures.

Books, buildings, and sculptures, which are products of the human mind, are, of course, not only inmates of World 3 but also inmates of World 1. But in World 3 there are also symphonies, and mathematical proofs, and theories. And symphonies, proofs, and theories are strangely abstract objects: Beethoven's Ninth is not identical with his manuscript (which may get burned without the Ninth getting burned) or with any or all of the printed copies, or records, or with any or all of its performances. Nor is it identical with human experiences or memories. The situation is analogous to Euclid's proof of the prime number theorem and to Newton's theory of gravitation.

The objects that constitute World 3 are highly diverse. There are marble sculptures like those of Michelangelo. These are not only material, physical bodies, but unique physical bodies. The status of painting, of architectural works of art, and of manuscripts of music is somewhat similar; and so is even the status of rare copies of printed books. But as a rule, the status of a book as a World 3 object is utterly different. If I ask a physics student whether he knows Newton's theory of gravity, then I do not refer to a material book and certainly not to a unique physical body, but to the objective *content* of Newton's thought or, rather, to the objective content of his writings. And I do not refer to Newton's actual thought processes which of course belong to World 2, but to something far more abstract; to something that belongs to World 3, and which was developed by Newton by a critical process of improving upon it, again and again, at different periods of his life.

It is difficult to make this quite clear, but it is very important. The main problem is the status of a statement: and the logical relations between statements, or more precisely between the logical *contents* of statements.

Now all the purely logical relations between statements, such as contradictoriness, compatibility, deducibility (the relation of logical consequence) are World 3 relations. They are relations holding *only* between World 3 objects. They are, decidedly, *not* psychological World 2 relations: they hold independently of whether anybody has ever thought about them, or has ever believed that they hold. On the other hand, they can very easily be 'grasped': they can be easily understood; we can, mentally, think it all out, in World 2, and we may experience that a deducibility relation holds and that it is trivially convincing; and this is a World 2 experience. Of course, with difficult theories, such as mathematical or physical theories, it may even happen that we grasp them, understand them, without at the same time being convinced that they are true.

Our minds, belonging to World 2, are thus capable of standing in close contact with World 3 objects. Yet World 2 objects - our subjective experiences - should be clearly distinguished from the objective World 3 statements, theories, conjectures, and also open problems.

I have spoken before about the interaction between World 2 and World 3, and I will illustrate this by another arithmetical example. The sequence of natural numbers, 1, 2, 3... is a human invention. As I emphasized before it is a *linguistic invention*, as opposed to the invention of counting. Spoken and perhaps written languages co-operated in inventing and perfecting the system of natural numbers. But we do not invent the distinction between odd numbers and even numbers: we *discover* this within the World 3 object - the sequence of natural numbers - which we have produced, or invented. Similarly we discover that there are divisible numbers and prime numbers. And we discover that the prime numbers are at first very frequent (up to 7, even the majority): 2, 3, 5, 7, 11, 13...; but after 13 they slowly become rarer and rarer. These are facts which we have not made, but which are unintended and unforeseeable and inescapable consequences of the invention of the sequence of natural numbers. They are objective facts of World 3. That they are unforeseeable may become clear when I point out that there are open problems here. For example, we have found that primes sometimes come in pairs, such as 11 and 13, or 17 and 19, or 29 and 31. These are called twin primes, and they become very rare when we proceed to large numbers.

But in spite of much research, we do not know yet whether twin primes fizzle out completely or whether they turn up again and again; or in other words, we do not know yet whether or not there exists a greatest pair of twin primes. (The so-called twin prime conjecture proposes that no such greatest pair exists; or in other words, that the sequence of twin primes is infinite.)

Thus there are open problems in World 3; and we work on discovering these open problems, and on attempts to solve them. This shows very clearly the objectivity of World 3, and the way World 2 and World 3 interact: not only can World 2 work on discovering and solving World 3 problems, but World 3 can act upon World 2 (and through World 2 on World 1).

We can distinguish between the almost always conjectural knowledge in the World 3 sense - knowledge in the objective sense - and the World 2 knowledge, that is, information carried by us in our heads: knowledge in the subjective sense. The distinction between knowledge in the subjective sense (or World 2 sense) and knowledge in the objective sense (World 3 knowledge: formulated knowledge, for example, in books, or stored in computers, and possibly unknown to any person) is of the greatest importance. What we call 'science', and what we aim for, is, in the first instance, true *knowledge in the objective sense* (although we must never forget that we do not have a criterion of truth and that our objective knowledge remains conjectural knowledge). But it is most important, of course, that knowledge in the subjective sense should also spread - together with the knowledge how little we know.

The incredible thing about the human mind, about life, evolution, and mental growth is the interaction, the feedback, the give and take, between World 2 and World 3; between our mental growth and the growth of the objective World 3 which is the result of our endeavours, and our talents, our gifts, and which helps us to transcend ourselves.

It is this self-transcendence which appears to me to be the most important fact of all life and of all evolution: in our interaction with World 3 we can learn, and thanks to our invention of language, our fallible human minds can grow into lights that illuminate the universe.

9. REFERENCES

Baldwin, J. M. (1895) *Mental Development in the Child and in the Race*, MacMillan and Co., New York.

Bühler, K. (1918) 'Kritische Musterung der neueren Theorien des Satzes', *Indogermanisches Jahrbuch*, 6, 1-20.

Diels, H., and Kranz, W. (1964) *Fragmente der Vorsokratiker*, Weidmann, Dublin and Zürich.

Eigen, M., and Winkler, R. (1975) *Das Spiel*, R. Piper and Co. Verlag, München.

Frisch, J. E. (1959) 'Research on Primate Behavior in Japan', *American Anthropologist*, 61, 584-596.

Groos, K. (1896) *Die Spiele der Tiere*, Jaence.

Held, R., and Hein, A. (1963) 'Movement produced stimulation in the development of visually guided behaviour', *Journal of Comparative Physiological Psychology*, 56, 872-876.

Hochkeppel, W. (1973) *Denken als Spiel*, Deutscher Taschenbuch Verlag, München.

Itani, J. (1958) 'On the acquisition and propagation of a new food habit in the natural group of the Japanese monkey at Takasakiyama', *Primates*, 1, 84-98.

Jennings, H. E. (1906) *The Behaviour of the Lower Organisms*, Columbia University Press, New York.

Kawamura, S. (1959) 'The process of sub-culture propagation among Japanese macaques', *Primates*, 2, 43-60.

Lorenz, K. Z. (1941) 'Kants Lehre vom Apriorischen im Lichte gegenwärtiger Biologie', *Blätter f. dt. Philos.*, 15, 1941. New impression in *Das Wirkungsgefüge und das Schicksal des Menschen*, Serie Piper 309 (1983).

Lorenz, K. Z. (1973) *Die Rückseite des Spiegels*, Piper, München.

Lorenz, K. Z. (1977) *Behind the Mirror*, Methuen, London.

Lorenz, K. Z. (1978) *Vergleichende Verhaltungsforschung, Grundlagen der Ethologie*, Springer Verlag, Wien/New York.

Menzel, E. W. (1965) 'Responsiveness to objects in free-ranging Japanese monkeys', *Behaviour*, 26, 149.

Miyadi, D. (1964) 'Social life of Japanese monkeys', *Science*, 143, 783-786.

Morgan, C. (1908) *Animal Behaviour*, Edward Arnold, London.

Popper, K. R. (1934) *Logik der Forschung*, Julius Springer, Vienna; 8th edition, 1984, J. C. B. Mohr (Paul Siebeck), Tübingen; also (1983) *The Logic of Scientific Discovery*, 11th impression, Hutchinson, London.

Popper, K. R. (1963, 1981) *Conjectures and Refutations*, Routledge and Kegan Paul, London.

Popper, K. R. (1979, written 1930-32) *Die beiden Grundprobleme der Erkenntnistheorie*, J. C. B. Mohr (Paul Siebeck), Tübingen.

Popper, K. R., and Eccles, J. C. (1977) *The Self and Its Brain*, Springer International, Berlin, Heidelberg, London, New York, 404-405. Now also as a paperback at Routledge and Kegan Paul, London (1984).

SUBJECT INDEX